ENCYCLOPEDIA OF MATHEMATICS AND ITS APPLICATIONS

EDITED BY G.-C. ROTA

Interval methods for systems of equations

ENCYCLOPEDIA OF MATHEMATICS AND ITS APPLICATIONS

1 Luis A. Santalo *Integral geometry and geometric probability*
2 George E. Andrews *The theory of partitions*
3 Robert J. McEliece *The theory of information and coding: a mathematical framework for communication*
4 Willard Miller, Jr *Symmetry and separation of variables*
5 David Ruelle *Thermodynamic formalism: the mathematical structures of classical equilibrium statistical mechanics*
6 Henryk Minc *Permanents*
7 Fred S. Roberts *Measurement theory with applications to decisionmaking, utility, and the social sciences*
8 L. C. Biedenham and J. D. Louck *Angular momentum in quantum physics: theory and application*
9 L. C. Biedenham and J. D. Louck *The Racah–Wigner algebra in quantum theory*
10 W. Dollard and Charles N. Friedman *Product integration with application to differential equations*
11 William B. Jones and W. J. Thron *Continued fractions: analytic theory and applications*
12 Nathaniel F. G. Martin and James W. England *Mathematical theory of entropy*
13 George A. Baker, Jr and Peter R. Graves-Morris *Padé approximants. Part I: Basic theory*
14 George A. Baker, Jr and Peter R. Graves-Morris *Padé approximants, Part II: Extensions and applications*
15 E. C. Beltrametti and G. Cassinelli *The logic of quantum mechanics*
16 G. D. James and A. Kerber *The representation theory of the symmetric group*
17 M. Lothaire *Combinatorics on words*
18 H. O. Fattorini *The Cauchy problem*
19 G. G. Lorentz, K. Jetter, and S. D. Riemenschneider *Birkhoff interpolation*
20 Rudolf Lidl and Harald Niederreiter *Finite fields*
21 William T. Tutte *Graph theory*
22 Julio R. Bastida *Field extensions and Galois theory*
23 John R. Cannon *The one-dimensional heat equation*
24 Stan Wagon *The Banach–Tarski paradox*
25 Arto Salomaa *Computation and automata*
26 Neil White (ed) *Theory of matroids*
27 N. H. Bingham, C. M. Goldie and J. L. Teugels *Regular variation*
28 P. P. Petrushev and V. A. Popov *Rational approximation of real functions*
29 Neil White (ed) *Combinatorial geometries*
30 M. Pohst and H. Zassenhaus *Algorithmic algebraic number theory*
31 J. Aczel and J. Dhombres *Functional equations containing several variables*
32 Marek Kuczma, Bogden Chozewski and Roman Ger *Iterative functional equations*
33 R. V. Ambartzumian *Factorization calculus and geometric probability*
34 G. Gripenberg, S.-O. Londen and O. Staffans *Volterra integral and functional equations*
35 George Gasper and Mizan Rahman *Basic hypergeometric series*
36 Erik Torgersen *Comparison of statistical experiments*
37 Arnold Neumaier *Interval methods for systems of equations*

ENCYCLOPEDIA OF MATHEMATICS AND ITS APPLICATIONS

Interval methods for systems of equations

ARNOLD NEUMAIER

Institute of Applied Mathematics
University of Freiburg

The right of the University of Cambridge to print and sell all manner of books was granted by Henry VIII in 1534. The University has printed and published continuously since 1584.

CAMBRIDGE UNIVERSITY PRESS

Cambridge

New York Port Chester

Melbourne Sydney

CAMBRIDGE UNIVERSITY PRESS
Cambridge, New York, Melbourne, Madrid, Cape Town, Singapore, São Paulo, Delhi

Cambridge University Press
The Edinburgh Building, Cambridge CB2 8RU, UK

Published in the United States of America by Cambridge University Press, New York

www.cambridge.org
Information on this title: www.cambridge.org/9780521331968

First published 1990
This digitally printed version 2008

A catalogue record for this publication is available from the British Library

Library of Congress Cataloguing in Publication data
Neumaier, A.
Interval methods for systems of equations / by Arnold Neumaier.
p. cm. – (Encyclopedia of mathematics and its applications)
Includes bibliographical references.
ISBN 0-521-33196-X
1. Interval analysis (Mathematics) 2. Equations – Numerical
solutions. I. Title. II. Series.
QA297.75.N49 1990
519.4 – dc20 89-70812 CIP

ISBN 978-0-521-33196-8 hardback
ISBN 978-0-521-10214-8 paperback

CONTENTS

PREFACE

The present book gives a presentation of the interval arithmetical tools and methods for the solution of linear and nonlinear systems of equations in the presence of uncertainties in parameters (data errors) and computer arithmetic (rounding errors). It is based on lectures which I gave repeatedly at Freiburg University. A standard background in linear algebra, analysis and numerical analysis is required.

Since there are now over 2000 publications on interval arithmetic (Garloff, 1985, 1987) I have been rather selective in the choice of the material. The major restrictions are:

(1) Only finite-dimensional problems are treated; some reasons for this limitation are discussed below. Thus we also do not touch recent applications of interval arithmetic to computer-assisted proofs in analysis (cf. Eckmann & Wittwer, 1985; Lanford, 1986; Matsumoto, Chua & Ayaki, 1988).
(2) Eigenvalue problems are not discussed; this omission is mainly due to a lack of time on my part. (For some references see Remarks to Chapter 5.)
(3) Range enclosure problems are treated only to the extent needed for an understanding of the basic principles, and to allow applications to implicit functions; the subject essentially belongs to the field of global optimization and a systematic presentation of the state of the art might well fill another book of this size. (For some references see Remarks to Chapter 2.)

In writing this book I tried to achieve several objectives. My first goal was to develop the tools which are necessary to solve the basic problems of finite-dimensional numerical analysis by interval methods. Here 'solving a problem' means obtaining a rigorous verification of existence (and uniqueness), and finding guaranteed enclosures with narrow bounds (i.e. with little overestimation).

My second goal was to present the solution methods themselves. Here I discussed mainly those methods which I consider of greatest value for practical work. Since there are hardly any numerical comparisons in the literature, this evaluation was sometimes difficult, but I hope that my intuitive choice turns out to be a good one.

My third goal was to assist in assessing the quality of the results produced by various methods. This involves considerations about convergence, speed, overestimation, and the influence of finite precision arithmetic.

My final goal was to exhibit the beauty of interval analysis by giving an exposition of the material in a uniform and coherent manner. I tried to write the book in such a way as to please both the pure mathematician by being elegant and the applied mathematician by being constructive and useful. This resulted in the deletion of material which did not fit this pattern.

My presentation of the subject differs from that of the books by Moore (1966, 1979) and Alefeld & Herzberger (1983) in that linear systems are treated much more extensively. This is justified by the fact that it allows me to deduce much of the known nonlinear theory as simple corollaries of the linear theory.

Moreover, to make the book accessible to a wider audience I have provided a self-contained exposition of the relevant background material on nonnegative matrices (§3.2), M-matrices (§3.6), fixed point theorems (§3.3 and §5.3); those already familiar with the results might still enjoy the unconventional proofs given here.

As regards references, I have given only those which are required by the text (Alefeld & Herzberger, 1983, give fairly complete references for the time before 1982). So I refer to proofs of assertions which are stated but not proved in the book, to background and supplementary material, to further details on practical implementations of a method, and to original sources for important theorems. However, no references are given to the many variations of a method, to methods of proofs, or to ideas that influenced the book in an indirect way by helping to shape my thoughts on the subject. Let me just mention a few books which I used during the preparation: Alefeld & Herzberger (1983), Aubin & Ekeland (1984), Berman & Plemmons (1979), Bohl (1974), Clarke (1983), Collatz (1964), Kuratowski (1966), Moore (1966, 1979), Ortega & Rheinboldt (1970), Schwetlick (1979), Stoer & Bulirsch (1980), Varga (1962), Wilkinson (1965), Zangwill & Garcia (1981).

Theory and algorithms for a rigorous error control are fairly well developed for finite-dimensional systems. This is reflected by the existence of reliable methods which possess excellent overestimation properties when the input intervals are sufficiently narrow. In the nonlinear case, some of the methods even have a global convergence behavior which surpasses that of classical methods of numerical analysis. Moreover, it is remarkable that so-called nonsmooth problems, where the derivative does not exist everywhere, pose no extra difficulty to interval algorithms. This contrasts with the special modifications required in classical methods (cf. Aubin & Ekeland, 1984; Clarke, 1983).

At present the only limitations that arise in the linear case concern the need for an explicit inverse when the matrix is not diagonally dominant or an M-matrix (but see Alvarado, 1990a,b, for new developments). This restricts applications to systems of moderate dimensions (problems with $n = 100$, say, can still be solved in a few seconds). In the nonlinear case, the main limitation consists of the restriction to problems given by arithmetical expressions involving only the rational operations, absolute values, and 'elementary' functions for which the range and the range of the first derivative can be computed exactly (or, in finite-precision arithmetic, with directed – outward – rounding).

The situation is rather different for systems of equations in infinite-dimensional spaces. Since my present knowledge about the infinite-dimensional case is still rather limited, I have not attempted to treat this more general situation. Practical methods for the enclosure of infinite-dimensional systems seem to be available at present only when the corresponding linear operators are inverse monotone; then differential inequality techniques apply. My impression is that much more research is needed before wide-ranging practical applications are possible.

For the interested reader we give here a (necessarily incomplete) list of references to some work on interval methods for problems in function spaces.

For *numerical quadrature* see Caprani, Madsen & Rall (1981), Corliss (1987a,b), Corliss & Rall (1987), Eiermann (1986, 1989a), Krenz (1987), Nickel (1968), and Petković & Petković (1985, 1988/9). Eiermann (1989b) gives enclosures for integrals over unbounded integrals.

For *ordinary differential equations* see Adams (1980b), Ams (1987), Cordes (1987), Cordes & Adams (1987), Dobronets (1988), Eijgenraam (1981), Gambill & Skeel (1988), Kaucher & Miranker (1984, 1988), Krückeberg & Leisen (1985), Lohner (1987, 1988), Nickel (1979), Schröder (1987, 1990), Spreuer (1981), the bibliographies Nickel (1986a) and Corliss, Krenz & Davis (1988). Eigenvalue problems are considered in Behnke (1988), Lohner (1988), and Ohsmann (1988). A nonsmooth boundary value problem with endpoint singularities is treated in Fefferman & de la Llave (1986).

For *partial differential equations* see Adams & Ames (1982), Adams & Spreuer (1975), Appelt (1973, 1974), Bauch (1977, 1980a,b), Dobner

(1987), Dobner & Kaucher (1987), Faass (1975), Kaucher (1984, 1987), Kaucher & Miranker (1984), Lohner, Adams & Ames (1985), Nakao (1988, 1989), Scharf (1968), Scheu (1975), Scheu & Adams (1975), and Tost (1970).

For *integral equations* see Cryer (1969), Dobner (1989), Kenney, Linz & Wang (1989), Klein (1973), Moore (1984), Nazarenko & Marchenko (1983), Neuland (1969), and Rall (1984b, 1985). For some other *functional equations* see Eckmann & Wittwer (1985) and Lanford (1986).

At present, publications in interval analysis use a multitude of different notations, due to the need to distinguish at many places real and interval quantities. There was a reasonably well-established tradition of writing real quantities as lower case variables and interval quantities as upper case variables; but this conflicts with the more standard convention of using lower case letters for scalars and vectors, and capitals for matrices. So I kept the latter convention and gave up the typographical distinction between reals and intervals. Together with the identification of reals with thin (degenerate) intervals, this is quite natural. However, for real quantities which vary generically in an interval quantity x or A, the notations $\tilde{x}$ and $\tilde{A}$, respectively, are used; so we shall talk, e.g., of an arbitrary vector $\tilde{x} \in x$, etc.

Several people helped me during the preparation of this book. I want to thank Karl Nickel for introducing me to the subject and for the support he gave me during the years that I worked with him at Freiburg University; Jürgen Garloff for his continuous effort in updating and completing the 'Interval Library' (a collection of all published and many unpublished papers about the subject) at Freiburg (now at Düsseldorf), which I used extensively; Shen Zuhe for giving me the opportunity to give a series of lectures on selected parts of the book at Nanjing University; Eckart Baumann, Martin Eiermann, Günter Mayer, Friedhelm Heizmann and Michael Wohlfahrt for reading large portions of the manuscript carefully and for pointing out a number of inaccuracies; Doris Norbert for writing the program for drawing the cover illustration, and Helga Sturm for the remaining drawings; Lilly Wüst-Huber for typing the manuscript on our sometimes all too obstinate word processor; and my students, who were patient enough to listen to my lectures on interval analysis on which this book is based.

Finally, I want to express my gratitude to the Creator of the world who provides mankind with the means to understand and control His creation, to my master Jesus Christ who fills my life with peace and satisfaction and who gave me the patience and strength needed to write this book, and to the Holy Spirit who guided me on the sometimes difficult road to arrive at the insight into this beautiful piece of mathematics.

The Lord says: *I have given you the choice between a blessing and a curse, between good and evil, between life and death.* (Deuteronomy 30). May the methods communicated, to cope rigorously with uncertainties and errors in mathematical models, not be used for the destruction of mankind but contribute to make the world one day a safer place in which to live.

A. Neumaier

SYMBOL INDEX

1

Basic properties of interval arithmetic

1.1 Motivation

Interval arithmetic is an elegant tool for practical work with inequalities, approximate numbers, error bounds, and more generally with certain convex and bounded sets. In this section we give a number of simple examples showing where intervals and ranges of functions over intervals arise naturally.

(i) *Physical constants or measurements.* These are often known only approximately. As an example, the acceleration of fall, whose normed value is

$$g = 9.80665 \text{ m/s}^2,$$

depends on height, and ranges in the southern Black Forest between $g_{\text{Feldberg}} = 9.8045 \text{ m/s}^2$ and $g_{\text{Freiburg}} = 9.8082 \text{ m/s}^2$. Therefore, a realistic description of g in this area is

$$g \in [9.8045, 9.8082] \text{ m/s}^2$$

(ii) *Representation of numbers.* In hand or machine calculation all numbers must be written with a finite number of digits, e.g.

$$\tfrac{1}{3} \approx 0.33333, \quad \sqrt{2} \approx 1.4142, \quad \pi \approx 3.1416,$$

which usually implies a small rounding error. A more precise representation would specify the error, e.g. by saying that the displayed digits are correct (i.e. the error is at most half a unit in the last place). A still sharper representation is specified by the tightest representable lower and upper bounds. Thus, with five significant digits,

$$\tfrac{1}{3} \in [0.33333, 0.33334],$$
$$\sqrt{2} \in [1.4142, 1.4143],$$

$$\pi \in [3.1415, 3.1416].$$

(iii) *Conversion error*. Many decimal numbers, like 0.1, for example, do not have an exact representation in a computer if the arithmetic is based on binary or hexadecimal numbers; e.g. in five-digit hexadecimal arithmetic we have

$$0.1 \text{ (decimal)} \approx 0.1999\text{A (hexadecimal)}.$$

To specify the error we can write

$$0.1 \text{ (decimal)} \in [0.19999, 0.1999\text{A}] \text{ (hexadecimal)}.$$

(iv) *Rounding errors*. If we perform arithmetic operations with decimal numbers it is likely that the result is not representable with the same number of digits and has to be rounded. We have

$$1/0.12345 \approx 8.1004,$$

but the inclusion

$$1/0.12345 \in [8.1004, 8.1005]$$

conveys more information.

(v) *Mean value theorems*. For a real continuously differentiable function f, the mean value theorems of differential and integral calculus state that for $x \neq y$,

$$\frac{f(x) - f(y)}{x - y} = f'(\xi)$$

for some ξ between x and y, and

$$\int_a^b f(x)\,\mathrm{d}x = (b - a)f(\xi)$$

for some ξ between a and b. To make more than only theoretical or qualitative use of these formulae one would like to be able to compute or at least bound the range of f' (or f) over the interval defined by x and y (or a and b).

(vi) *Lipschitz constants*. In the context of initial value problems for ordinary differential equations (and, of course, at many other places), real functions are often assumed to be Lipschitz continuous in some interval $[a, b]$; i.e. there is a constant L such that

$$|f(x) - f(y)| \leq L|x - y| \quad \text{for all } x, y \in [a, b].$$

Sometimes, explicit values for L are required (e.g. to evaluate error bounds). If f is continuously differentiable we use the mean value theorem to get

$$L = \max_{\xi \in [a,b]} |f'(\xi)|.$$

Again a bound for the range of f' in $[a, b]$ is required.
(vii) *Verifying monotony and convexity*. A twice continuously differentiable real function f is monotone increasing if

$$f'(x) \geq 0 \quad \text{for all } x \in [a, b],$$

and convex if

$$f''(x) \geq 0 \quad \text{for all } x \in [a, b].$$

This can be checked if the range of f' and f'' in $[a, b]$ is known, respectively.
(viii) *Error terms*. Formulae for the approximation of a function f by a Taylor series, of a derivative of f by a difference quotient, or of an integral of f by a weighted sum of function values contain an error term usually expressible with a suitable derivative of f at an unknown intermediate point. For example, if the derivatives involved exist and are continuous, we have

$$f(x) = \sum_{i=0}^{n} \frac{f^{(i)}(x_0)}{i!}(x - x_0)^i + \frac{f^{(n+1)}(\xi)}{(n+1)!}(x - x_0)^{n+1}$$

with ξ between x_0 and x;

$$f'(x) = \frac{f(x+h) - f(x-h)}{2h} + \frac{h^2}{6} f'''(\xi),$$

with ξ between $x - h$ and $x + h$; and

$$\int_{x-h}^{x+h} f(t)\,\mathrm{d}t = \frac{h}{3}(f(x-h) + 4f(x) + f(x+h)) - \frac{h^5}{90} f^{(4)}(\xi),$$

with ξ between $x - h$ and $x + h$. To determine the error quantitatively, an enclosure of $f^{(i)}(\xi)$ is needed, where ξ ranges over the relevant interval.
(ix) *A posteriori error bounds*. The mean value theorem implies the following *a posteriori* bound for the error of an approximation $\tilde{x}$ to a zero x^* of a real, continuously differentiable function f in an interval $[a, b]$ in which f is strictly monotone:

$$|\tilde{x} - x^*| = \left|\frac{f(\tilde{x})}{f'(\xi)}\right| \leq \frac{|f(\tilde{x})|}{\inf_{\xi \in [a,b]} |f'(\xi)|}.$$

(x) *Verifying the hypothesis of Brouwer's fixed point theorem*. This theorem states that if K is a convex, compact subset of $\mathbb{R}^n$ and $F: K \rightarrow K$ is continuous then F has a fixed point x^* in K, i.e. there is a vector $x^* \in K$ with $F(x^*) = x^*$ (Theorem 5.3.13). To apply this theorem in specific cases, one has to verify that the image $F(x)$ of each $x \in K$ indeed belongs to K. This again is a range determination problem – now in several dimensions – and can be reduced to interval arithmetic if K is a rectangular box, i.e. an interval $[\underline{x}, \overline{x}] = \{\tilde{x} \in \mathbb{R}^n \mid \underline{x} \leq \tilde{x} \leq \overline{x}\}$ in the componentwise partial order of $\mathbb{R}^n$.

(xi) *Sensitivity analysis*. If x is a vector of approximate data and Δx is a vector containing the bounds for the error in the components of x, one is often interested in the influence of these errors on the result $F(x)$ of a computational process; i.e. one is interested to find a vector ΔF such that, for given Δx,

$$|F(\tilde{x}) - F(x)| \leq \Delta F \quad \text{for } |\tilde{x} - x| \leq \Delta x.$$

Sometimes the dependence of ΔF on Δx is also sought. Stated in the form

$$F(\tilde{x}) \in [F(x) - \Delta F, \quad F(x) + \Delta F] \quad \text{for } \tilde{x} \in [x - \Delta x, x + \Delta x],$$

we again have a range problem.

(xii) *Tolerance problem*. As converse counterpart to (xi) one is often interested in the (maximal) allowable tolerance Δx which guarantees a certain accuracy ΔF of the result $F(x)$; i.e. one is interested to find a (maximal) vector Δx such that, for given ΔF,

$$|F(\tilde{x}) - F(x)| \leq \Delta F \quad \text{for } |\tilde{x} - x| \leq \Delta x.$$

Problems like those mentioned above make it desirable to find a method of calculating with intervals similar to that of calculating with real numbers. This would allow us to evaluate a function like

$$f(x) = e^x - x - 1$$

at an interval $x = [-1, 1]$ by simple substitution, resulting in

$$f([-1, 1]) = e^{[-1,1]} - [-1, 1] - 1.$$

It is a vain hope to expect to obtain the range of f over an interval in this way. However, as we shall see in the following, operations and elementary functions for intervals can be defined in such a way that the evaluation of an expression at an interval gives at least an enclosure of the range. With suitable refined techniques this enclosure can be made arbitrarily close to the range. This encourages the treatment of the fundamental problems of numerical analysis with interval methods. We shall see that we obtain rigorous and realistic enclosures for the solution of numerical problems in the presence of rounding errors and of inaccurate data lying in specified intervals.

1.2 Intervals

In this section we define real intervals and operations with intervals. $\mathbb{R}$ denotes the set of real numbers. A (real) *interval* is a set of the form

$$x \equiv [\underline{x}, \overline{x}] := \{\tilde{x} \in \mathbb{R} \mid \underline{x} \le \tilde{x} \le \overline{x}\},$$

where $\underline{x}, \overline{x}$ are elements of $\mathbb{R}$ with $\underline{x} \le \overline{x}$. In particular, intervals are closed and bounded subsets of $\mathbb{R}$, and we can use standard set theoretic notation. The (possibly empty) *open interval* $]\underline{x}, \overline{x}[$, the *interior* of x, is denoted by

$$]\underline{x}, \overline{x}[\equiv \operatorname{int}(x) := \{\tilde{x} \in \mathbb{R} \mid \underline{x} < \tilde{x} < \overline{x}\}.$$

The set of all real intervals is denoted by $\mathbb{IR}$. We shall use x as a generic symbol for an interval; its lower bound $\underline{x}$ is denoted by a bar below x, and its upper bound $\overline{x}$ by a bar above x. If x is a more complex expression, we also write

$$\underline{x} \equiv \inf(x), \quad \overline{x} \equiv \sup(x).$$

Note that we also use int, inf, sup for sets that are more general than intervals.

The symbol $\tilde{}$ is often used to denote a generic element $\tilde{x} \in x$. An interval is called *thin* if $\underline{x} = \overline{x}$, and *thick* if $\underline{x} < \overline{x}$. Thin intervals contain only one real number, and we shall tacitly identify a thin interval with the unique real number contained in it; thus $x = [x, x]$ for thin intervals. In particular, real numbers and intervals need not be distinguished notationally, and we use the distinction of real numbers with $\tilde{}$ only when appropriate.

As explained in the previous section, a thick interval x is often interpreted as the formalization of the intuitive notion of an unknown number $\tilde{x}$ known to lie in x. Motivated by the representation of an unknown number as an approximation plus/minus an error bound, we introduce the *midpoint* $\check{x}$ of an interval x,

$$\check{x} \equiv \operatorname{mid}(x) := (\overline{x} + \underline{x})/2,$$

and the *radius* $\operatorname{rad}(x)$ of x,

$$\operatorname{rad}(x) := (\overline{x} - \underline{x})/2.$$

Again we use the operator mid only when a longer expression follows. The membership relation can now be expressed as

$$\tilde{x} \in x \Leftrightarrow |\tilde{x} - \check{x}| \le \operatorname{rad}(x).$$

Thus, the radius of an interval x is a measure for the absolute accuracy of the midpoint $\check{x}$ considered as an approximation of an unknown number $\tilde{x}$ contained in x.

The absolute value of a real number has several extensions to intervals. The most useful extension is the *magnitude* of x, defined as

$$|x| \equiv \text{mag}(x) := \max\{|\tilde{x}| \mid \tilde{x} \in x\}.$$

However, another relevant extension is the *mignitude* of x, defined as

$$\langle x\rangle \equiv \text{mig}(x) := \min\{|\tilde{x}| \mid \tilde{x} \in x\}.$$

We shall use the terms magnitude and *absolute value* synonymously. We note that both magnitude and mignitude can be described in terms of the endpoints:

$$|x| = \max\{|\underline{x}|, |\overline{x}|\},$$
$$\langle x\rangle = \min\{|\underline{x}|, |\overline{x}|\} \quad \text{if } 0 \notin x, \langle x\rangle = 0 \text{ otherwise.}$$

If S is a nonempty bounded subset of $\mathbb{R}$ we denote by

$$\square S := [\inf(S), \sup(S)]$$

the *hull* of S, i.e. the tightest interval enclosing S. For example, $\square\{a, b\}$ is the interval with endpoints a and b, i.e.

$$\square\{a, b\} = [a, b] \text{ if } a \le b, \quad \square\{a, b\} = [b, a] \text{ if } a > b.$$

We extend the *order relations* $\omega \in \{<, \le, >, \ge\}$ to interval arguments by defining

$$x \,\omega\, y :\Leftrightarrow \tilde{x} \,\omega\, \tilde{y} \quad \text{for all } \tilde{x} \in x, \tilde{y} \in y.$$

It is easy to see that

$$x < y \Leftrightarrow \overline{x} < \underline{y}, \quad x > y \Leftrightarrow \underline{x} > \overline{y},$$
$$x \le y \Leftrightarrow \overline{x} \le \underline{y}, \quad x \ge y \Leftrightarrow \underline{x} \ge \overline{y}.$$

The order relations are antisymmetric and transitive, and $\le$ and $\ge$ are reflexive. But two intervals need not be comparable since, e.g., $[1, 3] \nleq [2, 4] \nleq [1, 3]$.

Elementary operations $\circ \in \Omega := \{+, -, *, /, **\}$ are defined on the set of intervals by putting

$$x \circ y := \square\{\tilde{x} \circ \tilde{y} \mid \tilde{x} \in x, \tilde{y} \in y\} = \{\tilde{x} \circ \tilde{y} \mid \tilde{x} \in x, \tilde{y} \in y\}$$

for all $x, y \in \mathbb{IR}$ such that $\tilde{x} \circ \tilde{y}$ is defined for all $\tilde{x} \in x$, $\tilde{y} \in y$. This restricts the definition of the division x/y to intervals y with $0 \notin y$. Similarly, the exponentiation $x ** y$ is restricted to one of the cases (i) $\underline{x} > 0$, (ii) $\underline{x} \ge 0, \underline{y} > 0$, (iii) $0 \notin x$, y an integer ≤ 0, or (iv) y a positive integer.

Elementary functions are the members of a predefined set Φ of real functions, continuous on every closed interval on which they are defined.

We extend any $\varphi \in \Phi$ to interval arguments by putting

$$\varphi(x) := \square\{\varphi(\tilde{x}) \mid \tilde{x} \in x\} = \{\varphi(\tilde{x}) \mid \tilde{x} \in x\}$$

for all $x \in \mathbb{IR}$ such that $\varphi(\tilde{x})$ is defined for all $\tilde{x} \in x$. In this book, the set Φ shall consist of the following functions:

> abs (absolute value), sqr (square) sqrt (square root), exp (exponential), ln (natural logarithm), sin (sine), cos (cosine), arctan (arc tangent).

However, other real functions may be included into Φ provided that they are continuous on each closed interval on which they are defined. Note that the operators mid, rad are *not* elementary functions. In order to preserve standard notation we shall often write

$$\begin{aligned} -x &\text{ for } 0 - x, & x^{-1} &\text{ for } 1/x, \\ xy &\text{ for } x * y, & \sqrt{x} &\text{ for sqrt}(x), \\ x^n &\text{ for } x ** n, & e^x &\text{ for exp}(x). \end{aligned}$$

However, we distinguish between $|x|$ (the real-valued absolute value of an interval) and abs(x) (the interval extension of the real absolute value). Moreover we shall delete brackets if this does not lead to confusion.

By noting monotonicity properties of operations and elementary functions we find that – where defined –

$$\begin{aligned} x \circ y &= \square\{\underline{x} \circ \underline{y}, \underline{x} \circ \overline{y}, \overline{x} \circ \underline{y}, \overline{x} \circ \overline{y}\} \quad \text{for } \circ \in \{+, -, *, /\}, \\ \text{abs}(x) &= [\text{mig}(x), \text{mag}(x)], \\ \varphi(x) &= [\varphi(\underline{x}), \varphi(\overline{x})] \quad \text{for } \varphi \in \{\text{sqrt}, \exp, \ln\}, \\ \text{sqr}(x) &= [0, |x|^2] \quad \text{or} \quad \square\{\underline{x}^2, \overline{x}^2\} \quad \text{depending on whether } 0 \in x \text{ or not.} \end{aligned}$$

For the power $**$ and the other elementary functions, $\varphi(x)$ can be expressed in terms of the endpoint values $\varphi(\underline{x})$, $\varphi(\overline{x})$ and the local maxima and minima of φ in a similar way as for $\varphi = $ sqr. The details are not difficult and are left to the reader.

In most cases it is possible to decide in advance which of the four possible endpoints $\underline{x} \circ \underline{y}, \underline{x} \circ \overline{y}, \overline{x} \circ \underline{y}, \overline{x} \circ \overline{y}$ define the lower and upper bounds of $x \circ y$, respectively. For addition and subtraction one has

$$\begin{aligned} x + y &= [\underline{x} + \underline{y}, \overline{x} + \overline{y}], \\ x - y &= [\underline{x} - \overline{y}, \overline{x} - \underline{y}]. \end{aligned}$$

For multiplication and division, the result depends on the signs of x and y as displayed in Tables 1.1*a* and *b*.

Table 1.1*a*. *Multiplication xy*.

	$y \geq 0$	$y \ni 0$	$y \leq 0$
$x \geq 0$	$[\underline{x}\underline{y}, \overline{x}\overline{y}]$	$[\overline{x}\underline{y}, \overline{x}\overline{y}]$	$[\overline{x}\underline{y}, \underline{x}\overline{y}]$
$x \ni 0$	$[\underline{x}\overline{y}, \overline{x}\overline{y}]$	$[\min(\underline{x}\overline{y}, \overline{x}\underline{y}), \max(\underline{x}\underline{y}, \overline{x}\overline{y})]$	$[\overline{x}\underline{y}, \underline{x}\underline{y}]$
$x \leq 0$	$[\underline{x}\overline{y}, \overline{x}\underline{y}]$	$[\underline{x}\overline{y}, \underline{x}\underline{y}]$	$[\overline{x}\overline{y}, \underline{x}\underline{y}]$

Table 1.1*b*. *Division x/y*.

	$y > 0$	$y < 0$
$x \geq 0$	$[\underline{x}/\overline{y}, \overline{x}/\underline{y}]$	$[\overline{x}/\overline{y}, \underline{x}/\underline{y}]$
$x \ni 0$	$[\underline{x}/\underline{y}, \overline{x}/\underline{y}]$	$[\overline{x}/\overline{y}, \underline{x}/\overline{y}]$
$x \leq 0$	$[\underline{x}/\underline{y}, \overline{x}/\overline{y}]$	$[\overline{x}/\underline{y}, \underline{x}/\overline{y}]$

As special cases we mention the formulae

$$\operatorname{rad}(x) = 0 \Rightarrow xy = \begin{cases} [x\underline{y}, x\overline{y}] & \text{if } x \geq 0, \\ [x\overline{y}, x\underline{y}] & \text{if } x \leq 0, \end{cases}$$

$$0 \notin x \Rightarrow x^{-1} = [\overline{x}^{-1}, \underline{x}^{-1}].$$

The reader not familiar with interval arithmetic should check the following examples:

$$\begin{aligned} [1,2] + [3,4] &= [4,6], & [1,2]/[-1,2] &\ \text{undefined} \\ 1 + [-3,2] &= [-2,3], & 2/[1,2] &= [1,2], \\ [2,5] - [0,2] &= [0,5] & [-2,4]/[1,2] &= [-2,4], \\ [1,2] * [3,4] &= [3,8], & [2,4]/[-2,-1] &= [-4,-1], \\ [-2,-1] * [0,2] &= [-4,0], & [2,3] ** 2 &= [4,9], \\ [-1,2] * [-1,2] &= [-2,4], & [-1,2] ** 2 &= [0,4], \\ [1,2] * [-1,1] &= [-2,2], & \operatorname{sqrt}([4,9]) &= [2,3], \\ -3 * [1,2] &= [-6,-3], & \sin([0,10]) &= [-1,1]. \end{aligned}$$

1.3 Rounded interval arithmetic

When realizing interval arithmetic on a computer, care has to be taken to guarantee a proper implementation. This is due to the fact that there is only a finite set $\mathbb{M} \subseteq \mathbb{R}$ of *machine-representable numbers*, so that in general intervals have to be rounded. Since we want to preserve the intuition that an interval x stands for an unknown number $\tilde{x} \in x$ we require that x is contained in the interval x' obtained by rounding x (outward rounding). And since we want to lose as little information as possible we want x' to be the tightest such interval. With the notation

$$\nabla\tilde{x} := \sup\{x_{\mathrm{m}} \in \mathbb{M} \mid x_{\mathrm{m}} \le \tilde{x}\},$$
$$\Delta\tilde{x} := \inf\{x_{\mathrm{m}} \in \mathbb{M} \mid \tilde{x} \le x_{\mathrm{m}}\}$$

for the *optimal directed roundings*, the tightest interval with endpoints in $\mathbb{M}$ containing $x \in \mathbb{IR}$ is given by the *optimal outward rounding*

$$\Diamond x := [\nabla\underline{x}, \Delta\overline{x}].$$

Thus, the optimal outward rounding $\Diamond$ can be realized if the optimal directed roundings ∇ and Δ are available. Clearly, the optimal roundings satisfy (for $\tilde{x}, \tilde{y} \in \mathbb{R}$, $x, y \in \mathbb{IR}$):

$$\nabla\tilde{x} \le \tilde{x}, \text{ with equality iff } \tilde{x} \in \mathbb{M}, \tag{1a}$$
$$\tilde{x} \le \tilde{y} \Rightarrow \nabla\tilde{x} \le \nabla\tilde{y}, \tag{1b}$$
$$\Delta\tilde{x} \ge \tilde{x}, \text{ with equality iff } \tilde{x} \in \mathbb{M}, \tag{2a}$$
$$\tilde{x} \le \tilde{y} \Rightarrow \Delta\tilde{x} \le \Delta\tilde{y}, \tag{2b}$$
$$x \subseteq \Diamond x, \text{ with equality iff } \underline{x}, \overline{x} \in \mathbb{M}, \tag{3a}$$
$$x \subseteq y \Rightarrow \Diamond x \subseteq \Diamond y. \tag{3b}$$

It is not difficult to see that, conversely, (1), (2) and (3) characterize the operators Δ, ∇ and $\Diamond$; we also note that if, as usual $x_{\mathrm{m}} \in \mathbb{M}$ implies $-x_{\mathrm{m}} \in \mathbb{M}$ then the relation $\Delta\tilde{x} = -\nabla(-\tilde{x})$ holds for $\tilde{x} \in \mathbb{R}$.

With an optimal outward rounding we may now define operations $\diamond\!\!\!\circ$ and elementary functions $\varphi^{\Diamond}$ on a computer by

$$x \mathbin{\diamond\!\!\!\circ} y := \Diamond(x \circ y) \quad \text{for } \circ \in \Omega,$$
$$\varphi^{\Diamond}(x) := \Diamond\varphi(x) \quad \text{for } \varphi \in \Phi;$$

e.g. if $\mathbb{M}$ is the set of decimal floating point numbers with five significant digits then

$$[1,2] \mathbin{\diamond\!\!\!/} [2,3] = \Diamond[1/3, 1] = [0.33333, 1],$$
$$\exp^{\Diamond}([1,2]) = \Diamond[\mathrm{e}, \mathrm{e}^2] = [2.7182, 7.3891].$$

The basic inclusions rules

$$\tilde{x} \circ \tilde{y} \in x \mathbin{\diamond\!\!\!\circ} y \quad \text{for all } \tilde{x} \in x, \tilde{y} \in y,$$
$$\varphi(\tilde{x}) \in \varphi^{\Diamond}(x) \quad \text{for all } \tilde{x} \in x, \varphi \in \Phi$$

remain valid. However, since $\mathbb{M}$ is a finite set, it is easy to see that there are intervals z for which $\Diamond z$ does not exist. Therefore it may happen that an operation or elementary function leads to *overflow*, i.e. that $x\diamond\!\!\!\circ\, y$ or $\varphi^{\diamond}(x)$ does not exist even if x and y have machine-representable endpoints and $x \circ y$ or $\varphi(x)$ exists.

In order to be able to bound the amount of overestimation in outward rounding we suppose that $0 \in \mathbb{M}$ and any two consecutive machine numbers $\tilde{x}_1, \tilde{x}_2 \in \mathbb{M}$ satisfy the inequality

$$|\tilde{x}_1 - \tilde{x}_2| \le \varepsilon_{\mathbb{M}}|\tilde{x}_1| + \eta_{\mathbb{M}} \tag{4}$$

for suitable numbers $\eta_{\mathbb{M}}$ and $\varepsilon_{\mathbb{M}}$ with

$$0 \le \eta_{\mathbb{M}} \ll \varepsilon_{\mathbb{M}} \ll 1.$$

This is satisfied for the commonly used sets of floating-point numbers with basis B, mantissa length L, and exponent range $[-E, F]$. If $\mathbb{M}$ consists of zero and normalized numbers only, i.e.

$$\mathbb{M} = \{0\} \cup \{m \cdot B^{\mathrm{e}} \mid -E \le \mathrm{e} \le F, \quad B^{-1} \le |m| < 1, \quad m \cdot B^L \in \mathbb{Z}\},$$

then (4) holds with

$$\eta_{\mathbb{M}} = B^{-1-E}, \quad \varepsilon_{\mathbb{M}} = B^{1-L},$$

and if $\mathbb{M}$ contains normalized and denormalized numbers, i.e.

$$\mathbb{M} = \{m \cdot B^{\mathrm{e}} \mid -E \le \mathrm{e} \le F, \quad |m| < 1, \quad m \cdot B^L \in \mathbb{Z}\},$$

then (4) holds with

$$\eta_{\mathbb{M}} = B^{-L-E}, \quad \varepsilon_{\mathbb{M}} = B^{1-L}.$$

1.3.1 Proposition For all intervals $x \in \mathbb{IR}$ for which $\diamond x$ exists we have

$$x \subseteq \diamond x \subseteq x * [1 - \varepsilon_{\mathbb{M}}, 1 + \varepsilon_{\mathbb{M}}] + [-\eta_{\mathbb{M}}, \eta_{\mathbb{M}}].$$

Proof We begin by showing

$$\diamond \tilde{x} \subseteq x * [1 - \varepsilon_{\mathbb{M}}, 1 + \varepsilon_{\mathbb{M}}] + [-\eta_{\mathbb{M}}, \eta_{\mathbb{M}}] \quad \text{for } \tilde{x} \in x. \tag{5}$$

Since (5) holds trivially if $\tilde{x} \in \mathbb{M}$ we suppose that $\tilde{x} \notin \mathbb{M}$. If $\tilde{x} \ge 0$ then $0 \le \nabla\tilde{x} < \tilde{x} < \Delta\tilde{x}$, and $\nabla\tilde{x}$ and $\Delta\tilde{x}$ are consecutive machine numbers. Hence $\Delta\tilde{x} - \nabla\tilde{x} = |\nabla\tilde{x} - \Delta\tilde{x}| \le \varepsilon_{\mathbb{M}}\nabla\tilde{x} + \eta_{\mathbb{M}} \le \varepsilon_{\mathbb{M}}\tilde{x} + \eta_{\mathbb{M}}$. Therefore

$$\nabla\tilde{x} \ge \tilde{x} - (\Delta\tilde{x} - \nabla\tilde{x}) \ge \tilde{x}(1 - \varepsilon_{\mathbb{M}}) - \eta_{\mathbb{M}},$$

$$\Delta\tilde{x} \le \tilde{x} + (\Delta\tilde{x} - \nabla\tilde{x}) \le \tilde{x}(1 + \varepsilon_{\mathbb{M}}) + \eta_{\mathbb{M}},$$

so that

$$\begin{aligned}\diamond\tilde{x} = [\nabla\tilde{x}, \Delta\tilde{x}] &\subseteq [\tilde{x}(1 - \varepsilon_{\mathbb{M}}) - \eta_{\mathbb{M}}, \tilde{x}(1 + \varepsilon_{\mathbb{M}}) + \eta_{\mathbb{M}}] \\ &= \tilde{x} * [1 - \varepsilon_{\mathbb{M}}, 1 + \varepsilon_{\mathbb{M}}] + [-\eta_{\mathbb{M}}, \eta_{\mathbb{M}}].\end{aligned}$$

This implies (5) for $\tilde{x} \geq 0$. The case $\tilde{x} < 0$ is treated similarly. Finally, the proposition follows from (5) by observing that

$$\Diamond x = [\nabla \underline{x}, \Delta \overline{x}] = \square\{\Diamond \underline{x}, \Diamond \overline{x}\} \subseteq x * [1 - \varepsilon_{\mathbb{M}}, 1 + \varepsilon_{\mathbb{M}}] + [-\eta_{\mathbb{M}}, \eta_{\mathbb{M}}]. \quad \square$$

If optimal directed roundings are not available on the computer, a suboptimal artificial outward rounding can be defined using the auxiliary function

$$\mathrm{rd}(x) := \begin{cases} [\underline{x}\underline{q} - \eta, \overline{x}\overline{q} + \eta] & \text{if } \underline{x} \geq 0, \\ [\underline{x}\overline{q} - \eta, \overline{x}\underline{q} + \eta] & \text{if } \overline{x} \leq 0, \\ [\underline{x}\overline{q} - \eta, \overline{x}\overline{q} + \eta] & \text{otherwise,} \end{cases}$$

where $q = [1 - \varepsilon_{\mathbb{M}}, 1 + \varepsilon_{\mathbb{M}}]$, $\eta = \eta_{\mathbb{M}}$. Then

$$\Diamond x \subseteq \mathrm{rd}(x), \quad x \subseteq \mathrm{rd}([\Delta \underline{x}, \nabla \overline{x}]),$$

and on most computers we have

$$\begin{aligned} x \Diamond\!\!\!\circ\, y &\subseteq \mathrm{rd}(\mathrm{rd}(x) \circ \mathrm{rd}(y)) && \text{for } \circ \in \Omega, \\ \varphi^{\Diamond}(x) &\subseteq \mathrm{rd}(\varphi(\mathrm{rd}(x))) && \text{for } \varphi \in \Phi, \end{aligned}$$

even when the operations involved in the formation of the right-hand sides are replaced by the corresponding machine operations. However, if the specific rounding characteristics of a computer are known, it is usually possible to devise formulae providing a sharper enclosure. The execution time for and the accuracy of interval arithmetic depends on the implementation of interval arithmetic and on the accuracy of the outward rounding; good modern implementations require between one and five times the execution time for ordinary floating-point arithmetic.

1.4 Interval vectors and arithmetical expressions

In this section we define arithmetical expressions and their interval evaluation, and describe some of the characteristic phenomena of interval arithmetic. Since we want to cover expressions in several variables we begin by introducing some notation for interval vectors.

$\mathbb{R}^n$ denotes the vector space of (column) vectors with n real components. $\mathbb{IR}^n$ denotes the set of *interval vectors* $x = (x_1, \ldots, x_n)^T$ with n components $x_1, \ldots, x_n \in \mathbb{IR}$. We interpret $x \in \mathbb{IR}^n$ as the set of all vectors $\tilde{x} \in \mathbb{R}^n$ with $\tilde{x}_i \in x_i$ for $i = 1, \ldots, n$. For $n = 2$, this set is a rectangle in the plane (see Fig. 1.1).

In general, we refer to an interval vector as a *rectangular box* or simply a *box* if we want to emphasize the geometric interpretation as a region in n-space. The *vertices* of a box $x \in \mathbb{IR}^n$ are the vectors of the set

$$\mathrm{vert}(x) := \{\tilde{x} \in \mathbb{R}^n \mid \tilde{x}_i \in \{\underline{x}_i, \overline{x}_i\} \quad \text{for } i = 1, \ldots, n\}.$$

We use the *comparison* and *inclusion relations* ω in the set $\{\leq, <, \geq, >, \subseteq, \supseteq\}$ componentwise, i.e. for $x, y \in \mathbb{IR}^n$,

$$x \,\omega\, y :\Leftrightarrow x_i \,\omega\, y_i \quad \text{for } i = 1, \ldots, n;$$

similarly, infimum, supremum, midpoint, radius, absolute value, and outward rounding are defined componentwise:

$$\begin{aligned}
\underline{x} \equiv \inf(x) &:= (\underline{x}_1, \ldots, \underline{x}_n)^{\mathrm{T}}, \\
\overline{x} \equiv \sup(x) &:= (\overline{x}_1, \ldots, \overline{x}_n)^{\mathrm{T}}, \\
\check{x} \equiv \mathrm{mid}(x) &:= (\check{x}_1, \ldots, \check{x}_n)^{\mathrm{T}}, \\
\mathrm{rad}(x) &:= (\mathrm{rad}(x_1), \ldots, \mathrm{rad}(x_n))^{\mathrm{T}}, \\
|x| &:= (|x_1|, \ldots, |x_n|)^{\mathrm{T}}, \\
\Diamond x &:= (\Diamond x_1, \ldots, \Diamond x_n)^{\mathrm{T}}.
\end{aligned}$$

In particular, we have

$$\begin{aligned}
x &= \{\tilde{x} \in x \mid \underline{x} \leq \tilde{x} \leq \overline{x}\}, \\
\tilde{x} \in x &\Leftrightarrow |\tilde{x} - \check{x}| \leq \mathrm{rad}(x).
\end{aligned}$$

Since we want to prove statements about the evaluation of arithmetical expressions like $\sin(\xi^2 - 2) + \xi + 2$ we give rigorous definitions of the intuitive notions of arithmetical expressions and their evaluation. For a simple way of working with partially defined functions we introduce the symbol NaN ('not a number', which may be thought of as 'undefined' or 'unspecified value'). We put $\mathbb{R}^* := \mathbb{R} \cup \{\mathrm{NaN}\}$, $\mathbb{IR}^* := \mathbb{IR} \cup \{\mathrm{NaN}\}$, and we define the elementary operations and elementary functions of $\mathbb{IR}^*$ by

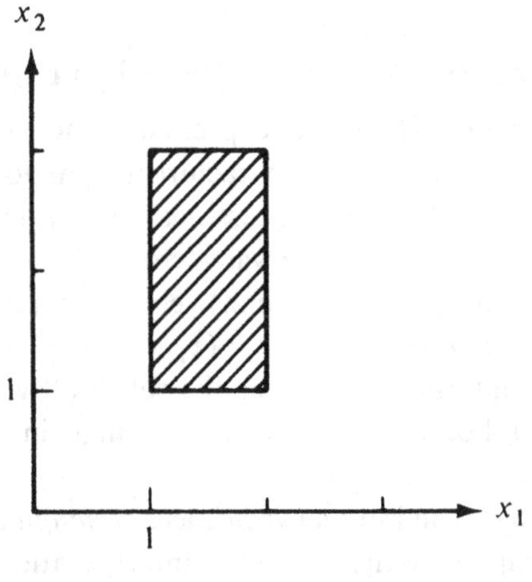

Fig. 1.1. The interval vector $x = ([1, 2], [1, 3])^{\mathrm{T}}$.

declaring NaN as the value of $x \circ y$ ($\circ \in \Omega$) or $\varphi(x)$ ($\varphi \in \Phi$) whenever these expressions are not defined in $\mathbb{IR}$. In particular,

$$\text{NaN} \circ x = x \circ \text{NaN} = \varphi(\text{NaN}) = \text{NaN}$$

for all $x \in \mathbb{IR}^*$, $\circ \in \Omega$, $\varphi \in \Phi$. We also extend the inclusion relations $\in$, $\subseteq$ to $\mathbb{IR}^*$ by the rules

$$\begin{aligned} \tilde{a} &\in \text{NaN for all } \tilde{a} \in \mathbb{R}, \\ a &\subseteq \text{NaN for all } a \in \mathbb{IR}^*, \\ \text{NaN} &\subseteq a \text{ only for } a = \text{NaN}. \end{aligned}$$

Clearly, all rules of the preceding section remain valid for arguments in $\mathbb{IR}^*$. We also put

$$\text{rad}(\text{NaN}) := \text{NaN}, \quad |\text{NaN}| := \text{NaN},$$

where NaN is taken to be greater than any real number.

If D is a subset of $\mathbb{R}^n$ we write $\mathbb{I}D$ for the set of $x \in \mathbb{IR}^n$ with $x \subseteq D$. An interval function, f: $\mathbb{IR}^n \to \mathbb{IR}^*$ is called *inclusion isotone* if, for $x, y \in \mathbb{IR}^n$,

$$x \subseteq y \Rightarrow f(x) \subseteq f(y).$$

We say that an interval function f: $\mathbb{IR}^n \to \mathbb{IR}^*$ is an *interval extension* of the real function f_0: $D \subseteq \mathbb{R}^n \to \mathbb{R}$ if

$$f(\tilde{x}) = f_0(\tilde{x}) \quad \text{for } \tilde{x} \in D, \tag{1}$$

$$f_0(\tilde{x}) \in f(x) \quad \text{for all } \tilde{x} \in x \in \mathbb{I}D, \tag{2}$$

$$f(x) = \text{NaN} \quad \text{for } x \notin \mathbb{I}D. \tag{3}$$

f is called a *weak interval extension* or an *interval enclosure* if only (2) and (3) hold. Weak interval extensions provide the enclosure

$$\{f_0(\tilde{x}) \mid \tilde{x} \in x\} \subseteq f(x)$$

for the range of f_0 over rectangular boxes $x \in \mathbb{IR}^n$. Clearly, an inclusion isotone interval function f which has real values for real arguments in $D \subseteq \mathbb{R}^n$ is an interval extension of the real function $f_0 = f|_D$.

A large class of inclusion isotone interval extensions of real functions can be defined through arithmetical expressions.

An *arithmetical expression* in the (formal) variables $\xi_1, \ldots, \xi_n$ is a member of the set $\mathfrak{A} = \mathfrak{A}(\xi_1, \ldots, \xi_n)$ defined by

(i) $\mathbb{R} \subseteq \mathfrak{A}$,
(ii) $\xi_l \in \mathfrak{A}$ for $l = 1, \ldots, n$,
(iii) $g, h \in \mathfrak{A} \Rightarrow (g \circ h) \in \mathfrak{A}$ for all $\circ \in \Omega$,
(iv) $g \in \mathfrak{A} \Rightarrow \varphi(g) \in \mathfrak{A}$ for all $\varphi \in \Phi$,

(v) among the sets $\mathfrak{A}$ satisfying (i)–(iv), $\mathfrak{A}(\xi_1, \ldots, \xi_n)$ is minimal with respect to inclusion.

A *rational expression* is an arithmetical expression in which no power ** and no elementary function occur.

Subexpressions of an expression f are those expressions which occur within the recursive definition of f (including f itself); in formal terms, an expression k is a *subexpression* of f if either $f = k$, or $f = g \circ h$ and k is a subexpression of g or h, or $f = \varphi(g)$ and k is a subexpression of g.

We use the symbol $f\{\xi_1, \ldots, \xi_n)$, in short $f(\xi)$ or just f, to denote an arithmetical expression in $\xi_1, \ldots, \xi_n$. When writing down arithmetical expressions we shall make use of standard mathematical conventions regarding the deletion of brackets, writing $-\xi$ for $(0 - \xi)$, $\xi_1\xi_2$ for $(\xi_1 * \xi_2)$, etc. Thus we shall use, e.g., the familiar notation $\sin(\xi^2 - 2) + \xi + 1$ for the formally precise expression $(\sin((\xi ** 2) - 2) + (\xi + 1))$ in the variable ξ.

Note that, while the absolute value function may be part of an arithmetical expression, expressions involving magnitude, mignitude, midpoint or radius are *not* arithmetical expressions.

For an arithmetical expression $f = f(\xi)$ in n variables the *value* $f(x)$ of f at $x \in \mathbb{IR}^n$ is obtained by substituting the intervals x_l for the corresponding formal parameters ξ_l (*interval evaluation*). More precisely, $f(x)$ is the element of $\mathbb{IR}^*$ defined recursively as follows:

$$\begin{aligned}
f(x) &= c && \text{if } f = c \text{ is a real constant,}\\
f(x) &= x_l && \text{if } f = \xi_l \text{ is a variable,}\\
f(x) &= g(x) \circ h(x) && \text{if } f = (g \circ h), \circ \in \Omega,\\
f(x) &= \varphi(g(x)) && \text{if } f = \varphi(g), \varphi \in \Phi.
\end{aligned}$$

In the above examples, $f(\xi) = \sin(\xi^2 - 2) + \xi + 1$ has the subexpressions

$$f_1 = \xi^2, \quad f_2 = f_1 - 2, \quad f_3 = \sin f_2, \quad f_4 = \xi + 1, \quad f = f_3 + f_4;$$

thus for $x = [0, 10]$ we get

$$f_1(x) = [0, 100], \quad f_2(x) = [-2, 98], \quad f_3(x) = [-1, 1], \quad f_4(x) = [1, 11],$$

and therefore $f([0, 10]) = [0, 12]$. For the expression $f(\xi) = 1/(\xi - 2)$, on the other hand, the value $f([0, 10]) = 1/[-2, 8] = \text{NaN}$ is 'undefined'.

On a computer with a finite set $\mathbb{M}$ of machine-representable numbers the interval evaluation is simulated by a rounded evaluation. The *outward rounded value* $f^{\Diamond}(x)$ of an arithmetical expression f in n variables at $x \in \mathbb{IR}^n$ is defined by

$$\begin{aligned}
f^{\Diamond}(x) &= \Diamond c && \text{if } f = c \text{ is a real constant,}\\
f^{\Diamond}(x) &= \Diamond x_l && \text{if } f = \xi_l \text{ is a variable,}
\end{aligned}$$

$$f^{\Diamond}(x) = g^{\Diamond}(x) \mathbin{\Diamond\!\!\!\circ} h^{\Diamond}(x) \quad \text{if } f = (g \circ h), \circ \in \Omega,$$
$$f^{\Diamond}(x) = \varphi^{\Diamond}(g^{\Diamond}(x)) \quad \text{if } f = \varphi(g), \varphi \in \Phi.$$

Again we use NaN as the rounded value of an undefined operation or function, in particular in case of overflow. In computer implementations this has to be indicated by an error flag. It is intuitively clear and will be proved below (Theorem 1.4.1) that $f^{\Diamond}(x)$ is an enclosure for all $f(\tilde{x})$ ($\tilde{x} \in x$). But $f^{\Diamond}(x)$ is not necessarily the tightest representable enclosure; in particular, we note that for $x \in \mathbb{R}^n$ we usually have

$$\Diamond f(x) \neq f^{\Diamond}(x).$$

For example, if $f(\xi) = (1/\xi) * \xi$ then in five-digit decimal floating point arithmetic,

$$\begin{aligned} f^{\Diamond}(3) &= (1 \mathbin{\Diamond\!\!\!/} 3) \mathbin{\Diamond\!\!\!*} 3 = [0.33333, 0.33334) \mathbin{\Diamond\!\!\!*} 3 \\ &= [0.99999, 1.0001], \end{aligned}$$

whereas $\Diamond f(3) = \Diamond 1 = [1, 1]$. As in this example, for thin x the radius of $f^{\Diamond}(x)$ is usually of the order of $\varepsilon_{\mathbb{M}}$ (cf. Corollary 2.1.6). However, for ill-conditioned expressions the outward rounded value may be much wider or even undefined. For example, again in five-digit decimal floating point arithmetic, if

$$f(\xi) = -1 + \text{sqrt}(1 + \xi^2/9 - (\xi/3)^2)$$

(which is a complicated expression for the zero real function) we have

$$\begin{aligned} f^{\Diamond}(308) &= -1 \mathbin{\Diamond\!\!\!+} \text{sqrt}(1 \mathbin{\Diamond\!\!\!+} [10540, 10541] \mathbin{\Diamond\!\!\!-} [102.66, 102.67]^2) \\ &= -1 \mathbin{\Diamond\!\!\!+} \text{sqrt}(10541, 10542] \mathbin{\Diamond\!\!\!-} [10539, 10542]) \\ &= -1 \mathbin{\Diamond\!\!\!+} \text{sqrt}([-1, 3]) = \text{NaN}. \end{aligned}$$

Note, for comparison, that the evaluation of f at $\xi = 308$ in noninterval optimally rounded five-digit decimal floating point arithmetic gives the wrong result -1, due to severe cancellation of digits. Although this example is artificial, such things may happen in badly designed algorithms, and the user should be aware of (and should avoid) it.

Any arithmetical expression $f(\xi)$ in n variables defines two interval functions $f, f^{\Diamond}$: $\mathbb{IR}^n \to \mathbb{IR}^*$, the *interval evaluation* of f which associates with $x \in \mathbb{IR}^n$ its value $f(x)$, and the *rounded interval evaluation* $f^{\Diamond}(x)$. (Here, as in the following, we use the same letter to denote an arithmetical expression and its interval evaluation.) The restriction of f to its *real domain*

$$D_f := \{\tilde{x} \in \mathbb{R}^n \mid f(\tilde{x}) \neq \text{NaN}\}$$

is a real-valued function f_{real}: $D_f \to \mathbb{R}$, the *real evaluation* of f. We now show that f is an inclusion isotonic interval extension of f_{real}, and that both the value and the rounded value of f at $x \in \mathbb{IR}^n$ are enclosures for the range of f_{real} over the rectangular box x.

1.4.1 Theorem (Moore) The interval functions f and $f^{\Diamond}$ associated with an arithmetical expression $f(\xi)$ are inclusion isotone, and we have

$$\{f(\tilde{x}) \mid \tilde{x} \in x\} \subseteq f(x) \subseteq f^{\Diamond}(x) \quad \text{if } x \in \mathbb{I}D_f. \tag{4}$$

Proof First we show that

$$x \subseteq x^0 \Rightarrow f(x) \subseteq f(x^0). \tag{5}$$

Clearly, this is true if f is a constant or a variable. In view of the recursive definition of arithmetical expressions it only remains to show that if (5) holds for g and h in place of f then it holds for $f = g \circ h$ and $f = \varphi(g)$. But if $x \subseteq x^0$, $g(x) \subseteq g(x^0)$, $h(x) \subseteq h(x^0)$ then in the first case

$$\begin{aligned} f(x) = g(x) \circ h(x) &= \square\{\tilde{g} \circ \tilde{h} \mid \tilde{g} \in g(x),\ \tilde{h} \in h(x)\} \\ &\subseteq \square\{\tilde{g} \circ \tilde{h} \mid \tilde{g} \in g(x^0),\ \tilde{h} \in h(x^0)\} = g(x^0) \circ h(x^0) = f(x^0), \end{aligned}$$

and in the second case

$$\begin{aligned} f(x) = \varphi(g(x)) &= \square\{\varphi(\tilde{g}) \mid \tilde{g} \in g(x)\} \\ &\subseteq \square\{\varphi(\tilde{g}) \mid \tilde{g} \in g(x^0)\} = \varphi(g(x^0)) = f(x^0). \end{aligned}$$

This proves (5) and shows that f is inclusion isotone. To show that $f^{\Diamond}$ is inclusion isotone we have to prove that

$$x \subseteq x^0 \Rightarrow f^{\Diamond}(x) \subseteq f^{\Diamond}(x^0). \tag{6}$$

Since $x \subseteq \Diamond x$ (1.3.3a), this holds if f is a constant or a variable. Using the implication $x \subseteq y \Rightarrow \Diamond x \subseteq \Diamond y$ (1.3.3b), (6) follows now inductively as before. Finally, a similar induction argument shows that $f(x) \subseteq f^{\Diamond}(x)$, and (4) follows since, by (5), $\{f(\tilde{x}) \mid \tilde{x} \in x\} \subseteq f(x)$ for $x \in \mathbb{I}D_f$. □

In particular, f is an interval extension of the real function $f_{\text{real}} = f|_{D_f}$. Since many real functions can be represented (and in many different ways) by arithmetical expressions, this yields various inclusion isotone extensions for a large class of real functions. Note that, in view of the examples above, $f^{\Diamond}$ is usually *not* an interval extension of f_{real} since (1) may be violated for $f^{\Diamond}$ in place of f. However, $f^{\Diamond}$ has the advantage that $f^{\Diamond}(x)$ is an enclosure for $f(x)$ which is computable in spite of rounding errors.

The quality of the enclosure in Theorem 1.4.1 may be good or bad, and we shall discuss it in detail in Section 2.1. In this section we restrict ourselves to the demonstration of the problem of *dependence* (or *simultaneity*) which is peculiar to interval arithmetic. Dependence may lead to catastrophic overestimation for apparently simple expressions, and the successful application of interval arithmetic requires that the user is aware of this phenomenon and knows how to avoid it.

1.4.2 Example Consider the arithmetical expression $f(\xi) = 1/(1 - \xi + \xi^2)$. For real ξ, this expression describes a real function with values in $[0, 4/3]$, and the expression is numerically stable for real arguments in the sense that – in the presence of rounding errors – the computed value of $f(\xi)$ for $\xi = \tilde{x}$ is the value of f at a point close to $\tilde{x}$. On the other hand, the attempt to evaluate f at the interval $[0, 2]$ gives the uninformative result

$$f([0,2]) = 1/([-1,1] + [0,4]) = 1/[-1,5] = \text{NaN}.$$

For intervals close to $[0, 1]$ we get

$$f([0,t]) = [1/(1+t^2), 1/(1-t)] \quad \text{for } 0 \le t < 1.$$

Since $\{f(\tilde{x}) \mid \tilde{x} \in [0,t]\} = [1, 4/3]$ for $1/2 \le t \le 1$, the overestimation of the upper bound becomes arbitrarily large for $t \to 1$.

The explanation for this large overestimation is the memoryless nature of interval arithmetic. Once the subexpressions $1 - x = [-1, 1]$ and $x^2 = [0, 4]$ are computed, the arithmetic does not 'remember' that the values represented by the two intervals are strongly correlated. It effectively computes in the next step

$$1 - x + x^2 = [-1, 5] = \square\{1 - \tilde{x}_1 + \tilde{x}_2^2 \mid \tilde{x}_1, \tilde{x}_2 \in x\}$$

instead of

$$[3/4, 3] = \square\{1 - \tilde{x} + \tilde{x}^2 \mid \tilde{x} \in x\}.$$

In such a case we say that the overestimation is due to *dependence* of the variables. Dependence is always present when one or several variables repeatedly occur in an expression; however, there are expressions where dependence does not lead to overestimation. For example, if $f = \xi + \xi$ then $f(x) = \square\{f(\tilde{x}) \mid \tilde{x} \in x\}$ for all $x \in \mathbb{IR}$, and if $f = \xi + \xi^2$ then $f(x) = \square\{f(\tilde{x}) \mid \tilde{x} \in x\}$ for $0 \le x \in \mathbb{IR}$.

In some cases it is possible to transform an expression which suffers from dependence to an equivalent form without dependence. In the above example we could replace f by the expression

$$g(\xi) = 4/(3 + (1 - 2\xi)^2),$$

which is equivalent for real arguments. Now (cf. Corollary 1.4.4 below) the interval evaluation of g gives precisely the range, e.g.,

$$g([0,2]) = 4/(3 + [-3,1]^2) = 4/[3,12] = [1/3, 4/3].$$

The user should make use of such transformations (in particular of 'quadratic completion') whenever possible. If no such transformation is available, more sophisticated techniques must be used to obtain better enclosures for the range; see Chapter 2.

For expressions in which all variables occur only once the dependence problem is absent and we get an optimal enclosure in the following sense:

1.4.3 Theorem (Moore) Let $f(\xi_1, \ldots, \xi_n, \eta_1, \ldots, \eta_m) = f(\xi, \eta)$ be an arithmetical expression in $n + m$ variables and suppose that the variables η_l $(l = 1, \ldots, m)$ occur only once in f. Then

$$\square\{f(\tilde{x}, \tilde{y}) \mid \tilde{x} \in x, \tilde{y} \in y\} = \bigcup_{\tilde{x} \in x} f(\tilde{x}, y) \tag{7}$$

for all $x \in \mathbb{IR}^n$, $y \in \mathbb{IR}^m$ such that $(x, y) \subseteq D_f$.

Proof The right-hand side of (7) contains the left-hand side since $f(\tilde{x}, \tilde{y}) \subseteq f(\tilde{x}, y)$; thus we have equality if we can show that

$$f(\tilde{x}, y) = [f(\tilde{x}, \tilde{y}^1), f(\tilde{x}, \tilde{y}^2)] \quad \text{for suitable } \tilde{y}^1, \tilde{y}^2 \in y. \tag{8}$$

Again this is clear when f is a constant or a variable, and it is sufficient to show (8) for $f = g \circ h$ and $f = \varphi(g)$, assuming that (8) is valid for g and h in place of f. In the first case $(f = g \circ h)$ we have

$$\begin{aligned} f(\tilde{x}, y) &= g(\tilde{x}, y) \circ h(\tilde{x}, y) \\ &= [g(\tilde{x}, \tilde{y}^1), g(\tilde{x}, \tilde{y}^2)] \circ [h(\tilde{x}, \tilde{z}^1), h(\tilde{x}, \tilde{z}^2)] \end{aligned}$$

for suitable $\tilde{y}^1$, $\tilde{y}^2$, $\tilde{z}^1$, $\tilde{z}^2 \in y$. Since $\circ$ is monotone in both arguments, $\inf f(\tilde{x}, y)$ has the form $g(\tilde{x}, \tilde{y}^i) \circ h(\tilde{x}, \tilde{z}^j)$ for suitable $i, j \in \{1, 2\}$. Now each η_l occurs at most once in $g \circ h$; therefore, at least one of g or h does not depend on η_l. This implies that we can choose a vector $\tilde{w}$ such that $\tilde{w}_l \in \{\tilde{y}^i_l, \tilde{z}^j_l\}$ and $g(\tilde{x}, \tilde{w}) = g(\tilde{x}, \tilde{y}^i)$, $h(\tilde{x}, \tilde{w}) = h(\tilde{x}, \tilde{z}^j)$. Therefore $\inf f(\tilde{x}, y) = g(\tilde{x}, \tilde{w}) \circ h(\tilde{x}, \tilde{w}) = f(\tilde{x}, \tilde{w})$ and $\tilde{w} \in \square\{\tilde{y}^i, \tilde{z}^j\} \subseteq y$. By a similar argument, $\sup f(\tilde{x}, \tilde{y}) = f(\tilde{x}, \tilde{y})$ for some $\tilde{y} \in y$, proving (8) for $f = g \circ h$.

In the second case $(f = \varphi(g))$ the continuity of φ implies that $\inf f(\tilde{x}, y) = \inf \varphi(g(\tilde{x}, y)) = \varphi(\tilde{g})$ for some $\tilde{g} \in g(\tilde{x}, y)$. Since g is continuous and (8) holds for g in place of f we have $\tilde{g} \in g(\tilde{x}, \tilde{y})$ for some $\tilde{y} \in y$ so that $\inf f(\tilde{x}, y) = \varphi(g(\tilde{x}, \tilde{y})) = f(\tilde{x}, \tilde{y})$. The same argument applies to $\sup f(\tilde{x}, y)$ so that (8) also holds for $f = \varphi(g)$. Therefore (7) holds. □

1.4.4 Corollary Let $f(\xi)$ be an arithmetical expression in $\xi_1, \ldots, \xi_n$ in which each variable ξ_l $(l = 1, \ldots, n)$ occurs only once. Then

$$f(x) = \square\{f(\tilde{x}) \mid \tilde{x} \in x\} \quad \text{for all } x \in \mathbb{I}D_f. \qquad \square$$

In many applications, some variables in an arithmetical expression have the character of parameters whose values are *interval constants* fixed in advance.

For notational reasons we shall adopt the convention that, if these variables occur only once in an expression, they may be replaced by the constant they represent. Thus

$$f(x) = [1.12, 1.14]\xi^2 + [2.39, 2.40]\xi + 1$$

is short-hand for the arithmetical expression $g(\xi, \eta_1, \eta_2) = \eta_1\xi^2 + \eta_2\xi + 1$, where η_1 and η_2 represent the constants [1.12, 1.14] and [2.39, 2.40], and $f(x) = g(x, [1.12, 1.14], [2.39, 2.40])$ is the value of f at x.

To summarize the discussion of this section: interval arithmetic is capable of producing enclosures for the range of real functions defined by arithmetical expressions over rectangular boxes. However, the quality of the enclosure depends (sometimes crucially) on the way this expression is written. Especially in the case of dependence, where some variable occurs repeatedly in an expression, the amount of overestimation of the range requires closer analysis. With rounded interval arithmetic, enclosures can be calculated on a computer in spite of rounding errors. This implies that, in contrast to classical numerical analysis, the emphasis in the design and analysis of interval algorithms must lie in the consideration of the overestimation properties. Thus we may say that the problem of error analysis of classical numerical methods is now replaced by the problem of overestimation analysis of interval methods. Properly done, this leads, as we shall show, to good and reliable algorithms for the enclosure of the solution to most of the basic problems in numerical analysis with reasonably low overestimation.

1.5 Algebraic properties of interval operations

Only some of the algebraic laws valid for real numbers remain valid for intervals; other laws only hold in a weaker form. There are two general rules:

(i) *Two arithmetical expressions which are equivalent in real arithmetic are equivalent in interval arithmetic when every variable occurs only once on each side.*

This follows from Corollary 1.4.4 since then both sides yield the range of the expression. In particular, the following laws hold for $a, b, c \in \mathbb{IR}$:

$a + b = b + a,$	$ab = ba,$	(commutativity)
$(a + b) \pm c = a + (b \pm c),$	$(ab)c = a(bc),$	(associativity)
$a + 0 = 0 + a = a,$	$1 * a = a * 1 = a,$	(neutral elements)
$a - b = a + (-b) = -b + a,$	$a/b = ab^{-1} = b^{-1}a,$	
$-(a - b) = b - a,$	$a(-b) = (-a)b = -ab,$	
$a - (b \pm c) = (a - b) \mp c,$	$(-a)(-b) = ab.$	

(ii) *If* f, g *are two arithmetical expressions which are equivalent in real arithmetic then the inclusion* $\mathrm{f}(x) \subseteq \mathrm{g}(x)$ *holds if every variable occurs only once in* f.

Indeed, the left-hand side yields the range, and the right-hand side encloses the range. This property implies weak forms of several laws familiar from real arithmetic; in particular, we have for $a, b, c \in \mathbb{IR}$:

$$a(b \pm c) \subseteq ab \pm ac, \qquad (a \pm b)c \subseteq ac \pm bc, \qquad \text{(subdistributivity)}$$

$$\left.\begin{array}{ll} a - b \subseteq (a+c) - (b+c), & a/b \subseteq (ac)/(bc), \\ 0 \in a - a, & 1 \in a/a. \end{array}\right\} \quad \text{(subcancellation)}$$

Usually we have proper inclusion, e.g.,

$$\begin{aligned} [-1, 1]([0, 1] + [-1, 0]) &= [-1, 1] \\ &\subseteq [-2, 2] = [-1, 1][0, 1] + [-1, 1][-1, 0], \\ 0 \in [-1, 1] &= [1, 2] - [1, 2], \\ 1 \in [1/2, 2] &= [1, 2]/[1, 2]. \end{aligned}$$

The failure of the distributive law is a frequent source of overestimation, and appropriate bracketing is advisable. Nevertheless, the subdistributive law in $\mathbb{IR}$, the weak substitute for the distributive law in $\mathbb{R}$, is a useful tool in the theoretical analysis of algorithms. In some special cases of great importance, the distributive law remains valid.

1.5.1 Proposition Let $a, b, c \in \mathbb{IR}$. Then

$$a(b \pm c) = ab \pm ac \quad \text{if } a \text{ is thin}, \tag{1}$$

$$a(b + c) = ab + ac \quad \text{if } b, c \geq 0 \text{ or } b, c \leq 0, \tag{2}$$

$$a(b - c) = ab - ac \quad \text{if } b \geq 0 \geq c \text{ or } b \leq 0 \leq c, \tag{3}$$

Proof (1) holds since for fixed thin a both sides of (1) are arithmetical expressions in which the variables b, c occur only once. By subdistributivity, (2) holds if we can show that $ab + ac \subseteq a(b + c)$. Now every element $\tilde{d} \in ab + ac$ has the form $\tilde{d} = \tilde{a}_1\tilde{b} + \tilde{a}_2\tilde{c}$ with $\tilde{a}_1, \tilde{a}_2 \in a$, $\tilde{b} \in b$ and $\tilde{c} \in c$. Since b and c have the same sign, $\tilde{a} := (\tilde{a}_1\tilde{b} + \tilde{a}_2\tilde{c})/(\tilde{b} + \tilde{c})$ is a convex combination of $\tilde{a}_1$ and $\tilde{a}_2$; hence $\tilde{a} \in a$ and $\tilde{d} = \tilde{a}(\tilde{b} + \tilde{c}) \in a(b + c)$. This implies (2), and (3) follows by substituting $-c$ for c. □

Also a few other rules hold sometimes with equality; e.g., we have for $a, b \in \mathbb{IR}$:

$$\begin{aligned} a + a &= 2a, \\ a - a &= 0 \qquad \text{iff } a \text{ is thin} \end{aligned}$$

$$a * a = a^2 \quad \text{iff } 0 \notin \operatorname{int} a,$$
$$(a/b)b = a, \quad \text{if } b \text{ is thin and nonzero.}$$

Finally we mention some trivial implications which are sometimes useful: if $a, b, c \in \mathbb{IR}$ then

$$\begin{aligned} a + c = b + c &\Leftrightarrow a = b, \\ a + c \subseteq b + c &\Leftrightarrow a \subseteq b, \\ ab = 0 &\Leftrightarrow a = 0 \text{ or } b = 0, \\ 0 \in ab &\Leftrightarrow 0 \in a \text{ or } 0 \in b, \\ ac = bc, 0 \notin c &\Rightarrow a = b. \end{aligned}$$

1.6 Rules for midpoint, radius and absolute value

In this rather technical section we derive a collection of formulae involving midpoints, radii and absolute values of intervals. Most of these rules are needed in later chapters for the derivation or the analysis of convergence and overestimation properties of algorithms. Some of these formulae are used so often that we shall use them without explicit reference.

We begin by noting that

$$\check{a} \in a, \qquad 0 \le \operatorname{rad}(a),$$
$$\underline{a} = \check{a} - \operatorname{rad}(a), \quad \bar{a} = \check{a} + \operatorname{rad}(a),$$

which follows immediately from the definitions.

1.6.1 Proposition (Properties of magnitude and mignitude)

(i) For every $a \in \mathbb{IR}$ there are numbers $\tilde{a}_1, \tilde{a}_2 \in a$ such that

$$|a| = |\tilde{a}_1|, \quad \langle a \rangle = |\tilde{a}_2|.$$

(ii) Let $a, b \in \mathbb{IR}$. Then:

$$a \subseteq b \Rightarrow |a| \le |b| \text{ and } \langle a \rangle \ge \langle b \rangle, \tag{1}$$
$$|a| - \langle b \rangle \le |a \pm b| \le |a| + |b|, \tag{2}$$
$$|a \pm b| = |a| + |b| \quad \text{if } \check{b} = 0 \tag{2a}$$
$$\langle a \rangle - |b| \le \langle a \pm b \rangle \le \langle a \rangle + |b|, \tag{3}$$
$$|ab| = |a|\,|b|, \quad \langle ab \rangle = \langle a \rangle \langle b \rangle, \tag{4}$$
$$|a/b| = |a|/\langle b \rangle, \quad \langle a/b \rangle = \langle a \rangle / |b| \quad \text{if } 0 \notin b, \tag{5}$$
$$|a^{-1}| = \langle a \rangle^{-1} \quad \text{if } 0 \notin a. \tag{6}$$

(iii) Let $a, b \in \mathbb{IR}$, and suppose that $\check{a} = 0$. Then

$$ab = a|b|, \tag{7}$$

$$a/b = a/\langle b\rangle \quad \text{if } 0 \notin b. \tag{8}$$

Proof All properties follow immediately from monotonicity arguments and the fact that $a \circ b = \square\{\tilde{a} \circ \tilde{b} \mid \tilde{a} \in a, \tilde{b} \in b\}$. □

1.6.2 Proposition (Rules relating magnitude and radius) For all $a \in \mathbb{IR}$ we have

$$\text{rad}(a) = |a - \check{a}|, \quad |a| = |\check{a}| + \text{rad}(a), \tag{9}$$

$$\langle a\rangle \geq |\check{a}| - \text{rad}(a), \quad \text{with equality iff } 0 \notin \text{int}(a), \tag{10}$$

$$\text{rad}(a) \leq |a| - \langle a\rangle \leq 2 \cdot \text{rad}(a), \tag{11}$$

$$\text{rad}(a) \leq |a - \tilde{a}| \leq 2 \cdot \text{rad}(a) \quad \text{for all } \tilde{a} \in a. \tag{12}$$

Proof Again, this is straightforward from the definitions. □

1.6.3 Proposition (Inclusion properties) For $a, b \in \mathbb{IR}$ and $\tilde{a}, \tilde{b} \in \mathbb{R}$ we have

$$\tilde{a} \in a \Leftrightarrow |\tilde{a} - \check{a}| \leq \text{rad}(a), \tag{13}$$

$$\tilde{a}, \tilde{b} \in a \Rightarrow |\tilde{a} - \tilde{b}| \leq 2 \cdot \text{rad}(a), \tag{14}$$

$$a \subseteq b \Leftrightarrow |a - \check{b}| \leq \text{rad}(b) \Leftrightarrow |\check{b} - \check{a}| \leq \text{rad}(b) - \text{rad}(a), \tag{15}$$

$$a \subseteq \text{int}(b) \Leftrightarrow |a - \check{b}| < \text{rad}(b) \Leftrightarrow |\check{b} - \check{a}| < \text{rad}(b) - \text{rad}(a), \tag{16}$$

$$a \cap b \neq \emptyset \Leftrightarrow |\check{a} - \check{b}| \leq \text{rad}(a) + \text{rad}(b). \tag{17}$$

Proof (13) and (14) are obvious. (15) holds since $a \subseteq b \Leftrightarrow a - \check{b} \subseteq b - \check{b} = [-\text{rad}(b), \text{rad}(b)] \Leftrightarrow |a - \check{b}| \leq \text{rad}(b) \Leftrightarrow |\check{a} - \check{b}| + \text{rad}(a) \leq \text{rad}(b)$ by (9), and (16) follows in the same way. Finally, if $a \cap b$ is nonempty and $\tilde{a} \in a \cap b$ then $|\check{a} - \check{b}| = |\tilde{a} - \check{b} + \check{a} - \tilde{a}| \leq |\tilde{a} - \check{b}| + |\check{a} - \tilde{a}| \leq \text{rad}(b) + \text{rad}(a)$ by (13). Conversely, if $|\check{a} - \check{b}| \leq \text{rad}(a) + \text{rad}(b)$ then we have $\overline{a} - \underline{b} = \check{a} + \text{rad}(a) - (\check{b} - \text{rad}(b)) = \text{rad}(a) + \text{rad}(b) + (\check{a} - \check{b}) \geq 0$; hence $\overline{a} \geq \underline{b}$ and similarly $\overline{b} \geq \underline{a}$ which implies $a \cap b \neq \emptyset$. Therefore (17) holds. □

1.6.4 Proposition (Properties of the midpoint) Let $a, b \in \mathbb{IR}$. Then

$$c = a \pm b \Rightarrow \check{c} = \check{a} \pm \check{b}, \tag{18}$$

$$c = ab \Rightarrow \check{c} = \check{a}\check{b} \quad \text{if } a \text{ or } b \text{ is thin}, \tag{19}$$

$$c = a/b \Rightarrow \check{c} = \check{a}/\check{b} \quad \text{if } b \text{ is thin and nonzero}. \tag{20}$$

Proof This easily follows from the expressions for $a \circ b$ in terms of the endpoints of a and b. □

Note that, usually, (19) does not hold if a and b are thick (take, e.g., $a = b =$

[1, 3]), and (20) does not hold if b is thick (take $a = 1, b = [1, 3]$). In general, the midpoint of a product depends on both midpoints and radii of the factors. The explicit formula of the midpoint involves the *sign* of a real number $\tilde{x}$, defined as

$$sgn(\tilde{x}) := \begin{cases} 1 & \text{if } \tilde{x} > 0, \\ 0 & \text{if } \tilde{x} = 0, \\ -1 & \text{if } \tilde{x} < 0. \end{cases}$$

1.6.5 Proposition Let $a, b \in \mathbb{IR}$ and $c = ab$. Then:

(i) $\check{c} = \check{a}\check{b} + \operatorname{sgn}(\check{a}\check{b}) \cdot \inf\{\operatorname{rad}(a)|\check{b}|, |\check{a}|\operatorname{rad}(b), \operatorname{rad}(a)\operatorname{rad}(b)\}$,

(ii) $\operatorname{rad}(c) = \sup\{\operatorname{rad}(a)|b|, |a|\operatorname{rad}(b), \operatorname{rad}(a)|\check{b}| + |\check{a}|\operatorname{rad}(b)\}$.

Proof In view of the formulae $ab = (-a)(-b) = -a(-b) = -(-a)b$ we may assume, without loss of generality, that $\check{a} \geq 0$ and $\check{b} \geq 0$. We write $\alpha = \operatorname{rad}(a)$ and $\beta = \operatorname{rad}(b)$, so that c is the hull of the four numbers

$$\begin{aligned}(\check{a} \pm \alpha)(\check{b} \pm \beta) &= \check{a}\check{b} \pm (\alpha\check{b} \pm \check{a}\beta) + \alpha\beta, \\ (\check{a} \pm \alpha)(\check{b} \mp \beta) &= \check{a}\check{b} \pm (\alpha\check{b} - \check{a}\beta) - \alpha\beta,\end{aligned}$$

where in both expressions either the upper or the lower sign is used throughout.

Therefore

$$\begin{aligned}\bar{c} &= \check{a}\check{b} + \alpha\check{b} + \check{a}\beta + \alpha\beta, \\ \underline{c} &= \check{a}\check{b} - \alpha\check{b} - \check{a}\beta - \alpha\beta + 2 \cdot \inf(\alpha\check{b}, \check{a}\beta, \alpha\beta),\end{aligned}$$

so that

$$\begin{aligned}\check{c} &= (\bar{c} + \underline{c})/2 = \check{a}\check{b} + \inf(\alpha\check{b}, \check{a}\beta, \alpha\beta), \\ \operatorname{rad}(c) &= (\bar{c} - \underline{c})/2 = \alpha\check{b} + \check{a}\beta + \alpha\beta - \inf(\alpha\check{b}, \check{a}\beta, \alpha\beta) \\ &= \sup(\check{a}\beta + \alpha\beta, \alpha\check{b} + \alpha\beta, \alpha\check{b} + \check{a}\beta) \\ &= \sup(\bar{a}\beta, \alpha\bar{b}, \alpha\check{b} + \check{a}\beta).\end{aligned}$$

This implies the assertion since $\inf(\alpha\check{b}, \check{a}\beta, \alpha\beta) = 0$ when $\operatorname{sgn}(\check{a}\check{b}) = 0$. □

1.6.6 Corollary If $a, b \in \mathbb{IR}$ and $c = ab$ then $\operatorname{sgn}(\check{c}) = \operatorname{sgn}(\check{a}\check{b})$. □

1.6.7 Proposition (Properties of the radius) Let $a, b \in \mathbb{IR}$. Then

$$a \subseteq b \Rightarrow \operatorname{rad}(a) \leq \operatorname{rad}(b), \tag{21}$$

$$\operatorname{rad}(a \pm b) = \operatorname{rad}(a) + \operatorname{rad}(b), \tag{22}$$

$$|a|\operatorname{rad}(b) \leq \operatorname{rad}(ab) \leq |a|\operatorname{rad}(b) + \operatorname{rad}(a)|\check{b}|, \tag{23}$$

$$\operatorname{rad}(a)|b| \leq \operatorname{rad}(ab) \leq \operatorname{rad}(a)|b| + |\check{a}|\operatorname{rad}(b), \tag{24}$$

$$\operatorname{rad}(a)/\langle b\rangle \le \operatorname{rad}(a/b) \le (\operatorname{rad}(a) + \operatorname{rad}(b)|\check{a}|/|b|)/\langle b\rangle \quad \text{if } 0 \notin b, \tag{25}$$

$$\operatorname{rad}(a^{-1}) = (\operatorname{rad}(a)/|a|)/\langle a\rangle \quad \text{if } 0 \notin a. \tag{26}$$

Proof Expressions (21) and (22) are obvious; (23) and (24) easily follow from Proposition 1.6.5(ii). Since $a^{-1} = [\overline{a}^{-1}, \underline{a}^{-1}]$ and $0 \notin a$ we have

$$\operatorname{rad}(a^{-1}) = \tfrac{1}{2}(\underline{a}^{-1} - \overline{a}^{-1}) = \frac{\overline{a} - \underline{a}}{2\overline{a}\underline{a}} = \frac{\operatorname{rad}(a)}{|a|\langle a\rangle},$$

which implies (26). Finally, (25) follows from (24) if we replace b by b^{-1} and use $ab^{-1} = a/b$ together with (6) and (26) with b in place of a. □

1.6.8 Proposition (Further properties of the radius) Let $a, b \in \mathbb{IR}$.

(i) If $0 \in a$ and $0 \in b$ then $\operatorname{rad}(ab) \le 2 \cdot \operatorname{rad}(a)\operatorname{rad}(b)$.

(ii) With $r^*(b) := \max(0, \operatorname{rad}(b) - |\check{b}|)$ we have

$$\operatorname{rad}(ab) \le \operatorname{rad}(a)|b| + |\check{a}|r^*(b) \quad \text{if } 0 \in a.$$

(iii) If either $\check{a} = 0$, or $\operatorname{rad}(b) = 0$, or $0 \in a$ and $0 \notin \operatorname{int}(b)$ then

$$\operatorname{rad}(ab) = \operatorname{rad}(a)|b|.$$

(iv) If $0 \notin \operatorname{int}(a)$ then $\operatorname{rad}(ab) = |a| \operatorname{rad}(b) + \operatorname{rad}(a)\langle b\rangle$.

(v) If $0 \in a$ and $0 \notin b$ then $\operatorname{rad}(a/b) = \operatorname{rad}(a)/\langle b\rangle$.

Proof If $0 \in a$ then $|\check{a}| \le \operatorname{rad}(a)$ by (13) so that $|a| \le 2 \cdot \operatorname{rad}(a)$ by (9). Now (i) follows directly by applying this to the bound for $\operatorname{rad}(ab)$ given in Proposition 1.6.5. (ii) follows from the same formula since

$$\begin{aligned}|a|\operatorname{rad}(b) &= \operatorname{rad}(a)\operatorname{rad}(b) + |\check{a}|\operatorname{rad}(b) = \operatorname{rad}(a)(|b| - |\check{b}|) + |\check{a}|\operatorname{rad}(b)\\ &\le \operatorname{rad}(a)|b| - |\check{a}|\,|\check{b}| + |\check{a}|\operatorname{rad}(b) \le \operatorname{rad}(a)|b| + |\check{a}|r^*(b),\\ \operatorname{rad}(a)|\check{b}| + |\check{a}|\operatorname{rad}(b) &\le \operatorname{rad}(a)|\check{b}| + \operatorname{rad}(a)\operatorname{rad}(b) = \operatorname{rad}(a)|b|.\end{aligned}$$

(iii) follows from (24) and – in the last case – from (ii) since $r^*(b) = 0$ if $0 \notin \operatorname{int}(b)$. (iv) follows for $0 \in b$ from (iii) (with a, b interchanged), and for $0 \notin b$ from Proposition 1.6.5(ii) since then the right-hand side equals $|a| \operatorname{rad}(b) + \operatorname{rad}(a)(|\check{b}| - \operatorname{rad}(b)) = |\check{a}| \operatorname{rad}(b) + \operatorname{rad}(a)|\check{b}|$.
Finally, (v) follows from (iii) by substituting b^{-1} for b. □

1.6.9 Proposition (Radius of powers and elementary functions) Let $a \in \mathbb{IR}$. Then, for every positive integer n,

$$\operatorname{rad}(a^n) \le |a|^{n-1} \operatorname{rad}(a) \qquad \text{if } 0 \in a, \tag{27}$$

$$\mathrm{rad}(a^n) \le n|a|^{n-1}\,\mathrm{rad}(a) \quad \text{if } 0 \notin a, \tag{28}$$

$$\mathrm{rad}(\mathrm{abs}(a)) \le \mathrm{rad}(a). \tag{29}$$

Moreover, if $\varphi \in \Phi$ is continuously differentiable in a and if φ' is an arithmetical expression for the derivative of φ, then

$$\mathrm{rad}(\varphi(a)) \le |\varphi'(a)|\mathrm{rad}(a). \tag{30}$$

Proof Since $\mathrm{rad}(a^n) = \mathrm{rad}((-a)^n)$ we may assume, without loss of generality, that $\check{a} \ge 0$ so that $|a| = \bar{a}$. If $0 \in a$ then $a^n \subseteq |a|^{n-1}a$ which implies (27). And, if $0 \notin a$ then $\underline{a} > 0$ and $a^n = [\underline{a}^n, \bar{a}^n]$. Thus, the mean value theorem implies the existence of $\tilde{a} \in a$ such that

$$\mathrm{rad}(a^n) = \tfrac{1}{2}(\bar{a}^n - \underline{a}^n) = \tfrac{1}{2}n\tilde{a}^{n-1}(\bar{a} - \underline{a}) = n\tilde{a}^{n-1}\,\mathrm{rad}(a) \le n|a|^{n-1}\,\mathrm{rad}(a).$$

This proves (28). Similarly, since $\varphi(a) = [\varphi(\tilde{a}_1), \varphi(\tilde{a}_2)]$ for suitable $\tilde{a}_1, \tilde{a}_2 \in a$ by continuity of φ, there is a number $\tilde{a} \in \square\{\tilde{a}_1, \tilde{a}_2\} \subseteq a$ such that

$$\mathrm{rad}(\varphi(a)) = \tfrac{1}{2}(\varphi(\tilde{a}_2) - \varphi(\tilde{a}_1)) = \tfrac{1}{2}\varphi'(\tilde{a})(\tilde{a}_2 - \tilde{a}_1) \le |\varphi'(a)|\mathrm{rad}(a).$$

Hence (30) holds. Relation (29) is obvious. □

1.7 Distance and topology

In this section we introduce and discuss a metric on $\mathbb{IR}$ and the corresponding topology.

1.7.1 Proposition Let $a, b \in \mathbb{IR}$. Then the following three definitions are equivalent:

$$q(a,b) := \inf\{q \in \mathbb{R} \mid q \ge 0, a \subseteq b + [-q,q], b \subseteq a + [-q,q]\}, \tag{1}$$

$$q(a,b) := \sup\{|\underline{a} - \underline{b}|, |\bar{a} - \bar{b}|\}, \tag{2}$$

$$q(a,b) := |\check{a} - \check{b}| + |\mathrm{rad}(a) - \mathrm{rad}(b)|. \tag{3}$$

Proof We have $a \subseteq b + [-q,q] = [\underline{b} - q, \bar{b} + q]$ iff $\underline{a} \ge \underline{b} - q$ and $\bar{a} \le \bar{b} + q$, i.e. iff $q \ge \underline{b} - \underline{a}$ and $q \ge \bar{a} - \bar{b}$. Similarly, $b \subseteq a + [-q,q]$ iff $q \ge \underline{a} - \underline{b}$ and $q \ge \bar{b} - \bar{a}$. Therefore both conditions hold simultaneously iff $q \ge |\underline{a} - \underline{b}|$ and $q \ge |\bar{a} - \bar{b}|$. This implies that (1) and (2) are equivalent. By (1.6.15) we also have $a \subseteq b + [-q,q]$ iff $|\check{a} - \check{b}| \le (\mathrm{rad}(b) + q) - \mathrm{rad}(a)$, i.e. iff $q - |\check{a} - \check{b}| \ge \mathrm{rad}(a) - \mathrm{rad}(b)$, and similarly $b \subseteq a + [-q,q]$ iff $q - |\check{a} - \check{b}| \ge \mathrm{rad}(b) - \mathrm{rad}(a)$. Therefore both conditions hold simultaneously iff $q - |\check{a} - \check{b}| \ge |\mathrm{rad}(a) - \mathrm{rad}(b)|$. This shows that (1) and (3) are equivalent, too. □

1.7.2 Proposition The mapping q: $\mathbb{IR} \times \mathbb{IR}$ defined by (1)–(3) is a metric on $\mathbb{IR}$, i.e. if $a, b, c \in \mathbb{IR}$ then

$$q(a, b) \geq 0, \quad \text{with equality iff } a = b, \tag{4}$$

$$q(a, b) = q(b, a), \tag{5}$$

$$q(a, c) \leq q(a, b) + q(b, c). \tag{6}$$

Proof Expressions (4) and (5) are immediate from (2) or (3), and (6) follows from (3) and the triangle inequality. □

We shall call $q(a, b)$ the *distance* of a and b. With this distance $\mathbb{IR}$ is a metric space. In the associated topology a sequence of intervals *converges* iff both the lower and upper bounds converge, equivalently, iff midpoints and radii converge. This follows immediately from (2) and (3). These formulae also imply

$$\lim_{l\to\infty} a_l = \left[\lim_{l\to\infty} \underline{a}_l, \lim_{l\to\infty} \overline{a}_l\right],$$

$$\operatorname{mid}\left(\lim_{l\to\infty} a_l\right) = \lim_{l\to\infty} \operatorname{mid}(a_l), \quad \operatorname{rad}\left(\lim_{l\to\infty} a_l\right) = \lim_{l\to\infty} \operatorname{rad}(a_l).$$

In particular, a sequence of nested intervals

$$a_0 \supseteq a_1 \supseteq \cdots \supseteq a_l \supseteq a_{l+1} \supseteq \cdots$$

always converges to the limit

$$\lim_{l\to\infty} a_l = \bigcap_{l\geq 0} a_l.$$

We note that endpoints, midpoint, radius, magnitude, mignitude, the operations +, −, *, /, ** and the elementary functions are continuous functions of their arguments since they can be expressed as continuous functions of the endpoints of the argument.

For the calculation with distances we mention the following rules.

1.7.3 Proposition Let $a, a', b, b', c \in \mathbb{IR}$. Then

$$q(a, b) = |a - b| \textit{ if } a \text{ or } b \text{ is thin}, \tag{7}$$

$$a \subseteq b \Rightarrow q(a, b) = |b - \check{a}| - \operatorname{rad}(a), \tag{8}$$

$$a \subseteq b \Rightarrow \operatorname{rad}(b) - \operatorname{rad}(a) \leq q(a, b) \leq 2(\operatorname{rad}(b) - \operatorname{rad}(a)), \tag{9}$$

$$a \subseteq b \subseteq c \Rightarrow \max(q(a, b), q(b, c)) \leq q(a, c), \tag{10}$$

$$b \subseteq a + [-1, 1]q(a, b), \tag{11}$$

$$\operatorname{rad}(b) \leq q(a, b) + \operatorname{rad}(a), \tag{12}$$

$$|b| \leq q(a, b) + |a|, \tag{13}$$

$$|a - b| \le q(a, b) + 2 \cdot \mathrm{rad}(a), \tag{14}$$
$$q(a, 0) = |a|, \tag{15}$$
$$q(a + b, a) = |b|, \tag{16}$$
$$q(a + b, a + b') = q(b, b'), \tag{17}$$
$$q(a + a', b + b') \le q(a, b) + q(a', b'), \tag{18}$$
$$q(ca, cb) \le |c| q(a, b), \tag{19}$$
$$q(ac, bc) \le q(a, b)|c|, \tag{20}$$
$$q(a/c, b/c) \le q(a, b)/\langle c\rangle \quad \text{if } 0 \notin c. \tag{21}$$

Moreover, (19)–(21) hold with equality if c is thin.
Proof If a is thin then (1.6.9) and (3) imply $|b - a| = |\mathrm{mid}(b - a)| + \mathrm{rad}(b - a) = |\check{b} - \check{a}| + \mathrm{rad}(b) + \mathrm{rad}(a) = |\check{b} - \check{a}| + \mathrm{rad}(b) - \mathrm{rad}(a) = q(a, b)$; and by symmetry (7) holds. If $a \subseteq b$ then $\mathrm{rad}(a) \le \mathrm{rad}(b)$ so that $q(a, b) = |\check{b} - \check{a}| + \mathrm{rad}(b) - \mathrm{rad}(a) = |b - \check{a}| - \mathrm{rad}(a)$ by (3) and (1.6.9); this implies (8). Expression (9) follows from (3) and (1.6.15). If $a \subseteq b \subseteq c$ then $q(a, b) = |b - \check{a}| - \mathrm{rad}(a) \le |c - \check{a}| - \mathrm{rad}(a) = q(a, c)$ by (8), and $q(b, c) = |c - \check{b}| - \mathrm{rad}(b) \le |c - \check{a}| + |\check{a} - \check{b}| - \mathrm{rad}(b) \le |c - \check{a}| - \mathrm{rad}(a) = q(a, c)$ by (8) and (1.6.15); hence (10) holds. The inclusion (11) directly follows from (1), and by taking radius and absolute value on both sides of (11) we obtain (12) and (13). Equations (14), (15), (16) and (17) immediately follow from (2) or (3), and (18) follows from (3) and the triangle inequality. To prove (20) we put $r = q(a, b)$. Then (11) implies $a \subseteq b + [-r, r]$ so that $ac \subseteq bc + [-r, r]c = bc + [-r, r]|c| = bc + [-r|c|, r|c|]$, and similarly $bc \subseteq ac + [-r|c|, r|c|]$. Application of (1) yields (20), and (19) follows with the commutative law. Relation (21) follows from (20) since $a/c = ac^{-1}$ and $|c^{-1}| = 1/\langle c\rangle$. Finally, for thin a, the equality statements in (19)–(21) immediately follow from (2) or (3). □

1.7.4 Proposition Let $a, b, c \in \mathbb{IR}$ and suppose that $a, b \subseteq c$. Then for every positive integer n,

$$q(a^n, b^n) \le n|c|^{n-1} q(a, b). \tag{22}$$

Moreover, if $\varphi \in \Phi$ is continuously differentiable in c and if φ' is an arithmetical expression for the derivative of φ then

$$q(\varphi(a), \varphi(b)) \le |\varphi'(c)| q(a, b). \tag{23}$$

Finally

$$q(\mathrm{abs}(a), \mathrm{abs}(b)) \le q(a, b). \tag{24}$$

Proof To prove (23) we put $r = q(a, b)$. Then $a \subseteq b + [-r, r]$; so every $\tilde{a} \in a$

can be written as $\tilde{a} = \tilde{b} + \tilde{r}$ with $\tilde{b} \in b$ and $|\tilde{r}| \le r$. By the mean value theorem $\varphi(\tilde{a}) = \varphi(\tilde{b}) + \varphi'(\tilde{c})\tilde{r}$ for a suitable $\tilde{c} \in \square\{\tilde{a}, \tilde{b}\} \subseteq c$. Hence, $\varphi(\tilde{a}) \in \varphi(b) + \varphi'(c)[-r, r] = \varphi(b) + [-s, s]$ with $s = |\varphi'(c)|r$ and therefore $\varphi(a) \subseteq \varphi(b) + [-s, s]$. By symmetry we also have $\varphi(b) \subseteq \varphi(a) + [-s, s]$, so that $q(\varphi(a), \varphi(b)) \le s$ by (1). This implies (23), and the same argument for $\varphi(a) = a^n$ and $\varphi(a) = \text{abs}(a)$ establishes (22) and (24). □

We end this section with a more specialized result required later.

1.7.5 Lemma The mapping β: $\mathbb{IR} \times \mathbb{IR} \to \mathbb{R}$ defined by

$$\beta(x, y) := |x| + q(x, y)$$

has the following properties:

(i) If $x_l, y_l \in \mathbb{IR}$, $x_l \subseteq y_l$ for $l = 1, \ldots, s$ then

$$\beta(x_1 \cdot \ldots \cdot x_s, y_1 \cdot \ldots \cdot y_s) \le \beta(x_1, y_1) \cdot \ldots \cdot \beta(x_s, y_s). \tag{25}$$

(ii) If $x, y \in \mathbb{IR}$, $x \subseteq y$ and $\langle x \rangle > q(x, y)$ then

$$\beta(x^{-1}, y^{-1}) \le (\langle x \rangle - q(x, y))^{-1}. \tag{26}$$

Proof (i) Put $q_l = q(x_l, y_l)$. Then $y_l \subseteq x_l + q_l[-1, 1]$ so that $y_1 y_2 \subseteq (x_1 + q_1[-1, 1])(x_2 + q_2[-1, 1]) \subseteq x_1 x_2 + (|x_1| q_2 + q_1 |x_2| + q_1 q_2)[-1, 1]$. Since $x_1 x_2 \subseteq y_1 y_2$, this implies $q(x_1 x_2, y_1 y_2) \le |x_1| q_2 + q_1 |x_2| + q_1 q_2$. Therefore $\beta(x_1 x_2, y_1 y_2) = |x_1 x_2| + q(x_1 x_2, y_1 y_2) \le |x_1|\,|x_2| + |x_1| q_2 + q_1 |x_2| + q_1 q_2 = (|x_1| + q_1)(|x_2| + q_2) = \beta(x_1, y_1)\beta(x_2, y_2)$. Now (25) follows by induction.

(ii) Put $r = q(x, y)$ and $y' = x + [-r, r]$. By assumption, $\langle y' \rangle \ge \langle x \rangle - r > 0$, so that $0 \notin y'$. Since $x \subseteq y \subseteq y'$ we have $x^{-1} \subseteq y^{-1} \subseteq y'^{-1}$; therefore (1) and (2) imply

$$\begin{aligned} q(x^{-1}, y^{-1}) &\le q(x^{-1}, y'^{-1}) = \max(|\underline{y}'^{-1} - \underline{x}^{-1}|, |\overline{y}'^{-1} - \overline{x}^{-1}|) \\ &= \max(r/|\underline{x}(\underline{x} - r)|, r/|\overline{x}(\overline{x} + r)|) \\ &\le r/(\langle x \rangle(\langle x \rangle - r)) = (\langle x \rangle - r)^{-1} - \langle x \rangle^{-1}. \end{aligned}$$

Therefore

$$\begin{aligned} \beta(x^{-1}, y^{-1}) &= |x^{-1}| + q(x^{-1}, y^{-1}) \\ &= \langle x \rangle^{-1} + q(x^{-1}, y^{-1}) \le (\langle x \rangle - r)^{-1}, \end{aligned}$$

and (26) follows. □

1.8 Appendix. Input/output representation of intervals

A computer implementation of interval arithmetic should contain an input/output capability for comfortable specification of input intervals and easy interpretation of output intervals. In particular, it must be easy to specify and recognize the accuracy of an interval enclosure. We propose the following user friendly scheme:

1.8.1 Syntax

⟨simple number⟩ = ⟨digits⟩{.⟨digits⟩}
                              .⟨digits⟩
⟨number⟩ = {⟨sign⟩}⟨simple number⟩{⟨exponent part⟩}
⟨simple interval⟩ = ⟨simple number⟩
                              ⟨simple number⟩: {⟨digits⟩}
                              [⟨simple number⟩,⟨simple number⟩]
⟨interval⟩ = {⟨sign⟩}⟨simple interval⟩{⟨exponent part⟩}
                              [⟨number⟩,⟨number⟩]
                              NaN

1.8.2 Interpretation

(i) *Rules for simple intervals*. The upper bound is always the number before the :, which must be a syntactically correct number, not ending with a decimal point. After the : only digits are allowed. When there are $k > 0$ such digits, the lower bound is the number obtained by replacing the k last digits of the upper bound by the digits after the :. However, when this would not strictly decrease the number, the $(k + 1)$st last digit is lowered by one (with carry, if it happens to be a zero); and if this is impossible since there is no $(k + 1)$st last digit an error must be signalled.

When there are no digits after the :, the colon is interpreted as if it were replaced by 5:5 or .5:5, depending on whether the number already contains a decimal point or not.

A thin interval must be specified without colon, and *not* by repeating the last digit.

(ii) *Rules for general numbers*. These are treated as if the unsigned number were multiplied by the sign and/or the exponent part.

(iii) *Output*. Standard output should provide two digits after the :, with the upper bound abbreviated such that this accuracy is attained (outward rounding). Thin intervals are given without colon.

Intervals which are too wide to be representable in the : format are given in [,] notation, with lower and upper bound rounded outward to three significant digits; a common exponent and/or sign is taken out of the bracket.

For formatted output the number of digits after the colon should be at the disposal of the programmer.

1.8.3 Examples

1.121:14	=	[1.114,1.121]
−1.121:14	=	[−1.121,−1.114]
1.121:99	=	[1.099,1.121]
1.121:21	=	[1.021,1.121]
1.121	=	[1.121,1.121]
1.121:299	=	[0.299,1.121]
−1.121:299	=	[−1.121,−0.299]
1.121:	=	[1.1205,1.1215]
3.0000:9	=	[2.9999,3.0000]
−3.0000:9	=	[−3.0000,−2.9999]
15.5:3	=	[15.3,15.5]
15.5:5	=	[14.5,15.5]
15:3	=	[13,15]
15:	=	[14.5,15.5]
12:08	=	[8,12]
112:99	=	[99,112]
212:99	=	[199,212]
2:	=	[1.5,2.5]
2.1:4	=	[1.4,2.1]
2.1:14	=	[1.4,2.1]
2.:	=	syntax error
2:.	=	syntax error
2.:15	=	syntax error
2.3:1.5	=	syntax error
15.:	=	syntax error
15.:2	=	syntax error
12:99	=	syntax error
0.12:99	=	syntax error
0.12:12	=	syntax error
1.12:212	=	syntax error

Remarks to Chapter 1

In 1931, Young (1931) developed an arithmetic for calculations with sets of numbers. In 1951, Dwyer (1951) considered the special case of closed intervals (range numbers) in the context of error control in numerical analysis. The first nontrivial applications (to the initial value problem for ordinary differential equations) were given by Moore (1959).

To 1.1 The tolerance problem is further discussed in Deif (1986) and Neumaier (1986c).

To 1.2 It is possible to extend the definition of intervals to the field of *complex numbers*. Two different extensions are in use: rectangular and disc arithmetic. *Rectangular intervals*, sets of the form $x + \mathrm{i}y := \{\tilde{x} + \mathrm{i}\tilde{y} \mid \tilde{x} \in x, \tilde{y} \in y\}$ (where $x, y \in \mathbb{IR}$), are easy to implement but lead to excessive overestimation already for the multiplication of an interval by a sequence of complex numbers. On the other hand, *discs* of the form $\langle c; r\rangle := \{z \in \mathbb{C} \mid |z - c| \leq r\}$ (where $c \in \mathbb{C}$, $0 \leq r \in \mathbb{R}$), introduced by Henrici (1971), are less susceptible to overestimation and have better theoretical properties; but the implementation of the corresponding arithmetic has more difficulties with outward rounding. We refer the interested reader to Alefeld (1970a), Alefeld & Herzberger (1983), and for discs also to Börsken (1978), Hauenschild (1974), Henrici (1974), Krier (1974), Petković (1986), Petković & Petković (1984).

Extensions of the interval concept to include probabilistic or fuzzy aspects are discussed in Dubois & Prade (1987).

To 1.3 The application of interval arithmetic has long been hampered by the difficulty to get access to implementations of rounded interval arithmetic. Due to the efforts of the group around Kulisch (1983) several good implementations with optimal rounding are now publicly available. The language extensions PASCAL-SC (Klatte & Wolff von Gudenberg, 1986; Kulisch, 1987a,b) and FORTRAN-SC (Metzger, 1988; Walter, 1988) allow the user to work with the data type INTERVAL as easily as with other standard data types. Complex (rectangular) interval arithmetic is also provided.

An implementation of interval arithmetic with optimal rounding for arithmetic processors satisfying the IEEE standard for floating point computation is described in Clemmesen (1984); this implementation also supports half-infinite and infinite intervals. More on the IEEE standard and directed rounding (and further references) can be found in Cody (1988).

The characterization of optimal roundings by (1)–(3) is due to Kulisch (1976).

To 1.4 The results of this section are already in the pioneering book by Moore (1966). The symbol NaN is also used in the IEEE floating point standard; see, e.g., Cody (1988) and the references therein. The extension of $\mathbb{IR}$ to $\mathbb{IR}^* = \mathbb{IR} \cup \{\mathrm{NaN}\}$ can be refined in various ways by also allowing half-infinite intervals and complements of open intervals. See, e.g., Kahan (1968) and Ratschek (1988).

To 1.5 A complete analysis of the equality case in the subdistributive law has been given by Ratschek (1971).

To 1.6 Many of the rules discussed here can be found in Kulisch (1969) and Alefeld & Herzberger (1983); they use the *diameter* (also called *width* or *span*) $\mathrm{d}(x) = 2 \cdot \mathrm{rad}(x)$ in place of the radius. The rules for the mignitude $\langle x \rangle$ come from Neumaier (1984a), and Proposition 1.6.5 is taken from Krawczyk & Neumaier (1986). Proposition 1.6.8(iv) is due to Heizmann (personal communication, 1989).

To 1.7 The metric $q(\cdot, \cdot)$ is the restriction to $\mathbb{IR}$ of the Hausdorff metric on sets of numbers; it was introduced by Moore (1966). The equivalence with the definition (1) was proved in Neumaier (1985a). Most properties of the distance can be found in Alefeld & Herzberger (1983). Lemma 1.7.5 is taken from Neumaier (1987a); the improved proof is due to Li You Ming (personal communication, 1987).

To 1.8 At present, unfortunately, none of the publicly available systems for interval calculations has sufficiently flexible input/output facilities.

2

Enclosures for the range of a function

2.1 Analysis of interval evaluation

In this section we take a closer look at the interval evaluation of arithmetical expressions and discuss their asymptotic properties when the radii of the arguments approach zero. Useful general results can be obtained if we avoid the evaluation at intervals in which expressions are not sufficiently smooth, like sqrt([0, 1]). To formulate these results we introduce Lipschitz properties for interval functions and arithmetical expressions.

In the following, $\mathbb{R}^{1\times n}$ and $\mathbb{IR}^{1\times n}$ denote the sets of n-dimensional real and interval row vectors, respectively. We write

$$xy = \sum_{i=1}^{n} x_i y_i$$

for the inner product of $x \in \mathbb{IR}^{1\times n}$ and $y \in \mathbb{IR}^n$. A real function $f: D \subseteq \mathbb{R}^n \rightarrow \mathbb{R}$ is called *Lipschitz continuous* in $D_0 \subseteq D$ if there is a row vector $l^0 \in \mathbb{R}^{1\times n}$ such that

$$|f(x^1) - f(x^2)| \leq l^0|x^1 - x^2| \quad \text{for all } x^1, x^2 \in D_0.$$

By analogy, an interval function $f: \mathbb{IR}^n \rightarrow \mathbb{IR}^*$ is called *Lipschitz continuous* in $x^0 \in \mathbb{IR}^n$ if $f(x^0) \neq \text{NaN}$ and there is a row vector $l^0 \in \mathbb{R}^{1\times n}$ such that

$$q(f(x^1), f(x^2)) \leq l^0 q(x^1, x^2) \quad \text{for all } x^1, x^2 \subseteq x^0.$$

Here the distance function q is used componentwise on vectors (cf. Section 3.3).

We now define a simple condition which guarantees that the interval evaluation of an expression is Lipschitz continuous (Theorem 2.1.1). We call an arithmetical expression f in n variables *Lipschitz* at $x^0 \in \mathbb{IR}^n$ if $f(x^0) \neq \text{NaN}$ and if for all subexpressions g, h of f, the relation $g = h ** \alpha$

$(0 < \alpha < 1)$ implies $h(x^0) > 0$, and the relation $g = \varphi(h)$ with $\varphi \in \Phi$ implies that φ is defined and Lipschitz continuous in a neighborhood of $h(x^0)$, i.e. in an interval containing $h(x^0)$ in its interior. We call f *locally Lipschitz* in $x^0 \in \mathbb{IR}^n$ if f is Lipschitz at every $\tilde{x} \in x^0$.

In the case that Φ consists of the standard set of elementary functions mentioned in Section 1.3, f is Lipschitz at x^0 iff

(i) $f(x^0) \neq \text{NaN}$,
(ii) if f contains some subexpression g of the form $g = \sqrt{h}$ or $g = h ** \alpha$ $(0 < \alpha < 1)$ then h is Lipschitz at x^0 and $h(x^0) > 0$.

The reason is that the square root is Lipschitz continuous in precisely those intervals which are positive, and the other elementary functions in Φ are Lipschitz continuous in every closed interval in which they are continuous. Since (i) already implies that $h(x^0) \geq 0$ when $f = \sqrt{h}$ or $f = h ** \alpha$ $(0 < \alpha < 1)$, we see that the requirement that an expression f is Lipschitz at x^0 is only marginally stronger than the requirement that it is defined at x^0 (i.e. (i) holds). We also see that *it is a trivial matter to check parallel to the evaluation of $f(x^0)$ whether an expression f is Lipschitz at x^0*.

The Lipschitz property excludes certain degeneracies in the interval evaluation. For example, the expression $f_1(\xi) = \text{sqrt}(\xi - \xi)$ is nowhere Lipschitz, and $f_1(x) = \text{NaN}$ for all thick intervals x although its real evaluation is identically zero and hence Lipschitz continuous in $\mathbb{R}$. The expression $f_2(\xi) = 1/(1 - \xi + \xi^2)$ (cf. Example 1.4.2) is not Lipschitz at $[0, 1]$ since $f_2([0, 1]) = \text{NaN}$, but its real evaluation is again Lipschitz continuous in $\mathbb{R}$. The expression $f_3(\xi) = 1 + \text{sqrt}(\xi - 1)$ is not Lipschitz at $x = [1, 10]$, but for subintervals $x_\varepsilon \subseteq x$ converging to the lower bound of x, such as $x_\varepsilon = [1 + \varepsilon^2, 1 + 9\varepsilon^2]$ $(\varepsilon \ll 1)$, the interval $f_3(x_\varepsilon) = [1 + \varepsilon, 1 + 3\varepsilon]$ has radius ε which is much larger than the radius $4\varepsilon^2$ of the argument x_ε itself.

We now show that such a large magnification of the radius cannot occur for subintervals of intervals at which an expression is Lipschitz. This is a consequence of the fact that if f is Lipschitz at x^0 then its interval evaluation (and in particular its real evaluation) is Lipschitz continuous in x^0. To see this for real functions φ which are Lipschitz continuous in an interval $x \in \mathbb{IR}^n$, we define

$$\lambda_\varphi(x) := \sup\left\{\left|\frac{\varphi(\tilde{x}) - \varphi(\tilde{y})}{\tilde{x} - \tilde{y}}\right| \;\middle|\; \tilde{x}, \tilde{y} \in x, \tilde{x} \neq \tilde{y}\right\}.$$

If φ is continuously differentiable then, by the mean value theorem,

$$\lambda_\varphi(x) = \sup\{|\varphi'(\tilde{x})| \mid \tilde{x} \in x\}. \tag{1}$$

2.1.1 Theorem Let f be an arithmetical expression in n variables, and suppose that f is Lipschitz at $x^0 \in \mathbb{IR}^n$. Then the interval evaluation of f is

Table 2.1. *Recursive definition of λ_f as*:

0	if f is a constant
$e^{(j)T}$ (jth row of identity matrix)	if $f = \xi_j$ ($j = 1, \ldots, n$),
$\lambda_g \pm \lambda_h$	if $f = g \pm h$,
$\lvert g(x^0)\rvert\lambda_h + \lvert h(x^0)\rvert\lambda_g$	if $f = g * h$,
$(\lambda_g + \lvert f(x^0)\rvert\lambda_h)/\langle h(x^0)\rangle$	if $f = g/h$,
$\alpha\lvert g(x^0)\rvert^{\alpha-1}\lambda_g$	if $f = g ** \alpha$ ($1 \le \alpha \in \mathbb{R}$),
$\lvert f(x^0)\rvert\left(\frac{\lvert h(x^0)\rvert}{\langle g(x^0)\rangle}\lambda_g + \lvert \ln(g(x^0))\rvert\lambda_h\right)$	if $f = g ** h$,
$\lambda_\varphi(g(x^0)) \cdot \lambda_g$	if $f = \varphi(g)$ ($\varphi \in \Phi$).

Lipschitz continuous in x^0. More specifically, for $x^1, x^2 \in \mathbb{IR}^n$ we have

$$q(f(x^1), f(x^2)) \le \lambda_f(x^0) q(x^1, x^2) \quad \text{if } x^1, x^2 \subseteq x^0, \tag{2}$$

where $\lambda_f \equiv \lambda_f(x^0) \in \mathbb{R}^{1\times n}$ is defined recursively by Table 2.1.

Proof Clearly, (2) is true if f is a constant or a variable. In view of the recursive definition we may therefore assume that $f = g \circ h$ ($\circ \in \Omega$) or $f = \varphi(g)$ ($\varphi \in \Phi$), where (2) holds for g and h in place of f.

If $f = \varphi(g)$ then Proposition 1.7.4 gives

$$\begin{aligned} q(f(x^1), f(x^2)) &= q(\varphi(g(x^1)), \varphi(g(x^2))) \\ &\le \lambda_\varphi(g(x^0)) q(g(x^1), g(x^2)) \\ &\le \lambda_\varphi(g(x^0))\lambda_g q(x^1, x^2) = \lambda_f q(x^1, x^2). \end{aligned}$$

We note in particular that, by (1),

$$\lambda_f = \alpha\lvert g(x^0)\rvert^{\alpha-1}\lambda_g \quad \text{if } f = g ** \alpha \ (1 \le \alpha \in \mathbb{R}), \tag{3a}$$

$$\lambda_f = \lambda_g/\langle g(x^0)\rangle^2 \quad \text{if } f = g^{-1}, \tag{3b}$$

$$\lambda_f = \lambda_g/\langle g(x^0)\rangle \quad \text{if } f = \ln(g), \tag{3c}$$

$$\lambda_f = \lvert f(x^0)\rvert\lambda_g \quad \text{if } f = \exp(g). \tag{3d}$$

If $f = g \pm h$ then

$$\begin{aligned} q(f(x^1), f(x^2)) &= q(g(x^1) \pm h(x^1), g(x^2) \pm h(x^2)) \\ &\le q(g(x^1), g(x^2)) + q(h(x^1), h(x^2)) \\ &\le \lambda_g q(x^1, x^2) + \lambda_h q(x^1, x^2) \\ &= \lambda_f q(x^1, x^2). \end{aligned}$$

If $f = g * h$ then

$$\begin{aligned} q(f(x^1), f(x^2)) &= q(g(x^1)h(x^1), g(x^2)h(x^2)) \\ &\le q(g(x^1)h(x^1), g(x^1)h(x^2)) + q(g(x^1)h(x^2), g(x^2)h(x^2)) \\ &\le |g(x^1)|q(h(x^1), h(x^2)) + q(g(x^1), g(x^2))|h(x^2)| \\ &\le |g(x^0)|\lambda_h q(x^1, x^2) + \lambda_g q(x^1, x^2)|h(x^0)| \\ &= \lambda_f q(x^1, x^2) \end{aligned}$$

If $f = g/h$ then $f = g * h^{-1}$, and (3b) implies (2) with

$$\begin{aligned} \lambda_f = \lambda_{g*h^{-1}} &= |g(x^0)|\lambda_{h^{-1}} + |h^{-1}(x^0)|\lambda_g \\ &= |g(x^0)|\lambda_h/\langle h(x^0)\rangle^2 + \lambda_g/\langle h(x^0)\rangle \\ &= (\lambda_g + |f(x^0)|\lambda_h)/\langle h(x^0)\rangle. \end{aligned}$$

Finally, if $f = g ** h$ then $f = \exp(h * \ln(g))$. Hence (3d) and (3c) imply (2) with

$$\begin{aligned} \lambda_f &= |f(x^0)|\lambda_{h*\ln(g)} \\ &= |f(x^0)|(|h(x^0)|\lambda_{\ln(g)} + |\ln(g(x^0))|\lambda_h) \\ &= |f(x^0)|\left(\frac{|h(x^0)|}{\langle g(x^0)\rangle}\lambda_g + |\ln(g(x^0))|\lambda_h\right). \end{aligned}$$

Thus, (2) holds in all cases. □

Note that (2) implies in particular

$$|f(\tilde{x}^1) - f(\tilde{x}^2)| \le \lambda_f(x^0)|\tilde{x}^1 - \tilde{x}^2| \quad \text{for all } \tilde{x}^1, \tilde{x}^2 \in x^0,$$

so that the real evaluation of f is Lipschitz continuous in every box where f is Lipschitz. We call the row vector $\lambda_f = \lambda_f(x^0)$ defined recursively in Theorem 2.1.1 the *standard Lipschitz constant* for f at x^0. A trivial induction argument shows that for $x, x^0 \in \mathbb{IR}^n$,

$$x \subseteq x^0 \Rightarrow \lambda_f(x) \le \lambda_f(x^0) \tag{4}$$

We note the following implication of Theorem 2.1.1.

2.1.2 Corollary Let f be an arithmetical expression in n variables, and suppose that f is Lipschitz at $x \in \mathbb{IR}^n$. Then

$$f(x) \subseteq f(\check{x}) + \lambda_f(x)(x - \check{x}), \tag{5}$$

$$\operatorname{rad} f(x) \le \lambda_f(x) \cdot \operatorname{rad}(x). \tag{6}$$

Proof For $x^1 = x^0 = x$, $x^2 = \check{x}$, (2) implies $|f(x) - f(\check{x})| \le \lambda_f(x^0)|x - \check{x}|$, which is equivalent with (5). Taking radii in (5) gives (6). □

Our next result shows that under a natural restriction arithmetical expressions can be evaluated at sufficiently narrow intervals.

2.1.3 Theorem Let f be an arithmetical expression in n variables. If f is locally Lipschitz in $x^0 \in \mathbb{IR}^n$ then there is a positive number $\rho > 0$ such that f is defined and Lipschitz in x for all $x \subseteq x^0$ with $\operatorname{rad}(x_i) \leq \rho$ $(i = 1, \ldots, n)$.

Proof Since this is trivial if f is a constant or a variable we may proceed inductively and suppose that the assertion holds when f is replaced by a proper subexpression. Suppose first that $f = \varphi(g)$ with $\varphi \in \Phi$. If the assertion is wrong for f then there is a sequence x^i $(i = 0, 1, 2, \ldots)$ of interval vectors $x^i \subseteq x^0$ with $\lim_{i\to\infty} \operatorname{rad}(x^i) = 0$ such that g is Lipschitz at x^i but f is not. Since x^0 is compact, the midpoints $\check{x}^i$ have an accumulation point $\tilde{x} \in x^0$, and there is a subsequence x^{i_l} $(l = 0, 1, 2, \ldots)$ such that $\lim \check{x}^{i_l} = \tilde{x}$. Since f is locally Lipschitz, φ is Lipschitz continuous in an open neighborhood U of $g(\tilde{x})$. Now $\lim_{l\to\infty} x^{i_l} = \tilde{x}$ and therefore $\lim_{l\to\infty} q(g(x^{i_l}), g(\tilde{x})) = q(g(\tilde{x}), g(\tilde{x})) = 0$. Hence $g(x^{i_l}) \subseteq U$ for sufficiently large l. But this implies that f is Lipschitz at x^{i_l}, a contradiction. Hence the assertion of the theorem holds for $f = \varphi(g)$. The case $f = g \circ h$ $(\circ \in \Omega)$ is treated in the same way. □

Based on Theorem 2.1.3 we can give a simple procedure for obtaining an arbitrarily good enclosure for the range of an expression f over a box x^0 in which f is locally Lipschitz. (This procedure is extremely slow and not suited for practical calculations; see the remarks at the end of this chapter for efficient procedures.) We choose positive integers m_i, partition x^0 into a set $\mathbb{B}_1$ of $b_1 = m_1 \cdot \ldots \cdot m_n$ boxes with sides of length $2\cdot\operatorname{rad}(x_i^0)/m_i \leq 2\rho$ and subdivide each such box further to get a set $\mathbb{B}_k$ of $k^n b_1$ boxes with sides of length $\leq 2\rho/k$. Then we compute the enclosure

$$f_k(x^0) := \bigcup_{x \in \mathbb{B}_k} f(x) \tag{7}$$

for the range

$$f^*(x^0) := \{f(\tilde{x}) \mid \tilde{x} \in x^0\}. \tag{8}$$

If we write γ for the maximum of the sum of the components of $\lambda_f(x)$ over all $x \in \mathbb{B}_1$, then (4) and (6) imply $\operatorname{rad} f(x) \leq \gamma \cdot \max\{\operatorname{rad}(x_i) \mid i = 1, \ldots, n\} \leq \gamma\rho/k$ for all $x \in \mathbb{B}_k$. Since every $f(x)$ contains some element of $f^*(x^0)$, the overestimation is bounded by

$$0 \leq \operatorname{rad} f_k(x^0) - \operatorname{rad} f^*(x^0) \leq \max_{x \in \mathbb{B}_k} (\operatorname{rad} f(x)) \leq \gamma\rho/k. \tag{9}$$

In particular, we have

$$\lim_{k\to\infty} f_k(x^0) = f^*(x^0).$$

2.1.4 Example (cf Example 1.4.2) Let $f(\xi) = 1/(1 - \xi + \xi^2)$. We are interested in the range of f over the interval $x^0 = [0, 1]$. For $0 \leq x \in \mathbb{IR}$ we have

$$f(x) = \left[\frac{1}{1 - \underline{x} + \overline{x}^2}, \frac{1}{1 - \overline{x} + \underline{x}^2}\right] \quad \text{if } \overline{x} < 1 + \underline{x}^2,$$

and $f(x) = \text{NaN}$ otherwise; in particular $f(x^0) = \text{NaN}$. Theorem 2.1.3 holds here, e.g., with $\rho = \frac{1}{2}$. Subdivision of x^0 into the $2k$ intervals $[(i - 1)/2k, i/2k]$ $(i = 1, \ldots, 2k)$ yields

$$f_k(x^0) = \left[\frac{2k}{2k + 1}, \frac{4k}{3k - 3}\right];$$

The infimum is attained for $i = 2k$, the supremum for $i = k + 1$. Since the range is $f^*(x^0) = [1, 4/3]$, we have

$$\text{rad}\,(f_k(x^0)) - \text{rad}\,(f^*(x^0)) = \frac{25k + 2}{6(2k + 1)(3k - 2)} \in \left[\frac{25}{36}, \frac{3}{2}\right]k^{-1},$$

This shows that the order $O(k^{-1})$ in (9) is best possible.

The related expression $\Sigma_{i=1}^{n} 1/(1 - \xi_i + \xi_i^2)$ in n variables, whose range in the unit cube $x^0 = ([0, 1], \ldots, [0, 1])^T$ is wanted shows that, in general, a large number (here 2^n) of subboxes and function evaluations may be required before we even get the first nontrivial enclosure ($\neq$NaN) for the range. This problem disappears, however, in the present example when each term $1/(1 - \xi_i + \xi_i^2)$ is rewritten in the form $4/((2\xi_i - 1)^2 + 3)$ ('quadratic completion'); the real evaluation is unaltered but the interval evaluation now provides the range without overestimation.

To achieve a specified bound ε for the overestimation

$$\text{rad}(f_k(x^0)) - \text{rad}(f^*(x^0)) \leq \varepsilon$$

in (7) we need $k \geq \gamma r/\varepsilon$. Hence an exponential number of at least $(\gamma r/\varepsilon)^n b_1 = O(\varepsilon^{-n})$ interval evaluations is required. This is prohibitively large, and we shall spend the subsequent sections developing refined tools which allow a considerably better approximation of the range.

We conclude the present section by showing that the result of evaluating an arithmetical expression in rounded interval arithmetic approaches the result in exact interval arithmetic when the machine precision tends to infinity, i.e. when the relative accuracy $\varepsilon_{\mathbb{M}}$ tends to zero.

2.1.5 Theorem Let f be an arithmetical expression in n variables. If f is Lipschitz at $x \in \mathbb{IR}^n$ then there are positive numbers c_f and ε_f such that for every rounded interval arithmetic with relative accuracy $\varepsilon_{\mathbb{M}} \leq \varepsilon_f$ the rounded evaluation $f^{\diamond}(x)$ is defined and satisfies

$$q(f^{\diamond}(x), f(x)) \leq c_f \varepsilon_{\mathbb{M}}. \tag{10}$$

Proof By Proposition 1.3.1, this is certainly true if f is a constant or a variable; hence we may proceed inductively and suppose that the theorem holds for all proper subexpressions g and h of f in place of f. Suppose that $f = g \circ h$ with $\circ \in \Omega$ (the case $f = \varphi(g)$ with $\varphi \in \Phi$ is treated in the same way). If $\varepsilon_{\mathbb{M}} \le \min(\varepsilon_g, \varepsilon_h)$ then the induction assumption implies that $g^{\diamond}(x) \to g(x)$ and $h^{\diamond}(x) \to h(x)$ for $\varepsilon_{\mathbb{M}} \to 0$, and, since f is Lipschitz at x, this shows that $z := g^{\diamond}(x) \circ h^{\diamond}(x)$ is defined for sufficiently small $\varepsilon_{\mathbb{M}} > 0$, say for $\varepsilon_{\mathbb{M}} \le \varepsilon_f$. And since $q(g^{\diamond}(x), g(x)) \le c_g \varepsilon_{\mathbb{M}}$ and $q(h^{\diamond}(x), h(x)) \le c_h \varepsilon_{\mathbb{M}}$, an application of Theorem 2.1.1 to the simple expression $\xi_1 \circ \xi_2$ yields

$$q(z, f(x)) \le c\varepsilon_{\mathbb{M}} \tag{11}$$

for some constant c independent of the rounding. Now, using Proposition 1.3.1, $\eta_{\mathbb{M}} \le \varepsilon_{\mathbb{M}}$, and subdistributivity, we have

$$z \subseteq f^{\diamond}(x) \subseteq z + z * e + e \quad \text{with } e = [-\varepsilon_{\mathbb{M}}, \varepsilon_{\mathbb{M}}].$$

Therefore

$$q(f^{\diamond}(x), z) \le |z * e + e| \le (|z| + 1)\varepsilon_{\mathbb{M}} \le c'\varepsilon_{\mathbb{M}}$$

for some constant c'. Together with (11) this implies (10) with $c_f = c + c'$. □

2.1.6 Corollary Let f be an arithmetical expression in n variables. If f is Lipschitz at $\tilde{x} \in \mathbb{R}^n$ then there are positive numbers δ_f, d_f such that for all $\delta \in \mathbb{R}$ with $0 < \delta < \delta_f$, we have

$$\operatorname{rad}(f^{\diamond}(\tilde{x})) \le \delta$$

for every rounded interval arithmetic with relative accuracy $\varepsilon_{\mathbb{M}} \le d_f \delta$.

Proof Take $d_f = c_f^{-1}$, $\delta_f = \varepsilon_f / d_f$ and observe that $\operatorname{rad}(f^{\diamond}(\tilde{x})) \le q(f^{\diamond}(\tilde{x}), f(\tilde{x}))$ since $f(\tilde{x})$ is thin. □

Thus for thin $\tilde{x}$ the radius of $f^{\diamond}(\tilde{x})$ is bounded by a linear function of $\varepsilon_{\mathbb{M}}$. From the proof of Theorem 2.1.5 we see that c_f and d_f can be computed recursively, and thus could be used to find an explicit bound for the relative machine accuracy $\varepsilon_{\mathbb{M}}$ which guarantees a prescribed accuracy δ in the rounded interval evaluation $f^{\diamond}(\tilde{x})$ at thin elements $\tilde{x}$. However, since $\operatorname{rad}(f^{\diamond}(\tilde{x}))$ is usually an almost linear function of $\varepsilon_{\mathbb{M}}$, one can use the simpler rule of thumb: *To reduce* $\operatorname{rad}(f^{\diamond}(\tilde{x}))$ *by a certain factor one has to reduce* $\varepsilon_{\mathbb{M}}$ *by roughly the same factor.*

2.2 Inclusion algebras and recursive differentiation

In this section we introduce the concept of an inclusion algebra, an abstract concept which serves to unify certain recursive methods for the calculation of various kinds of information about functions defined by arithmetical expressions. As a first application we obtain a method for enclosing the range of derivatives of such a function.

Before formal definitions are given, we motivate the concept of an inclusion algebra by an informal discussion. Suppose we are interested in computing enclosures for the range of a real function f and its derivative in an interval $x \in \mathbb{IR}$. The information given by a particular enclosure consists of a pair of intervals (r, d) satisfying

$$f(\tilde{x}) \in r, \quad f'(\tilde{x}) \in d \quad \text{for all } \tilde{x} \in x. \qquad (*)$$

In order to cover the case where it is not possible to compute the enclosures (e.g. when f is not everywhere differentiable) we allow the value NaN for r and d. In general, many functions f will be compatible with a given information pair (r, d), and we may say that f 'belongs to' (r, d), abbreviated to $f \Subset (r, d)$, if the compatibility condition $(*)$ is satisfied. Thus $\mathbb{A} = \mathbb{IR}^* \times \mathbb{IR}^*$ is the set of possible information pairs. For example, a function belongs to the pair $(r, 0)$ iff it is a constant function with value $\in r$. In particular, only the function which has the constant value 1 belongs to the pair $(1, 0)$. This allows us to regard the elements $(r, 0)$ as the constants of $\mathbb{A}$, to write 1 for the constant $(1, 0)$, and to consider the other constants $(r, 0)$ as formal products $r1$.

If the relation $f \Subset (r, d)$ holds for some specific f, the abstract element $(r, d) \in \mathbb{A}$ has found an interpretation as information about f. More generally, we refer to the relation $\Subset$ as an interpretation of the abstract set $\mathbb{A}$.

To be useful, this setting must be filled with life. We show how one can recursively compute information of the form $(*)$ about any function defined by arithmetical expressions, starting from trivial enclosures of the form $(*)$ for linear functions, in particular for the identity function.

Suppose that we already know enclosures $f_1 \Subset (r_1, d_1)$ and $f_2 \Subset (r_2, d_2)$. From this information about f_1 and f_2 we can deduce some information about related functions. By the rules for the differentiation of functions,

$$\begin{aligned}(f \pm g)' &= f' \pm g',\\ (fg)' &= f'g + fg',\\ (f/g)' &= (f'g - fg')/g^2 = (f' - (f/g)g')/g,\end{aligned}$$

we find immediately that

$$\begin{aligned}f_1 \pm f_2 &\Subset (r_1 \pm r_2, d_1 \pm d_2),\\ f_1 f_2 &\Subset (r_1 r_2, d_1 r_2 + r_1 d_2),\\ f_1/f_2 &\Subset (r_1/r_2, (d_1 r_2 - r_1 d_2)/r_2^2),\\ f_1/f_2 &\Subset (r, (d_1 - r d_2)/r_2), \quad \text{where } r = r_1/r_2.\end{aligned}$$

Note that due to the subdistributive law the second enclosure for f_1/f_2 is generally tighter than the first and has the additional advantage that two multiplications have been saved.

We see that the enclosures obtained are independent of the particular functions f_i belonging to (r_i, d_i). This suggests to define operations on $\mathbb{A}$ by

$$\begin{aligned}(r_1, d_1) \pm (r_2, d_2) &:= (r_1 \pm r_2, d_1 \pm d_2),\\ (r_1, d_1) * (r_2, d_2) &:= (r_1 r_2, d_1 r_2 + r_1 d_2),\\ (r_1, d_1)/(r_2, d_2) &:= (r, (d_1 - r * d_2)/r_2), \quad \text{where } r = r_1/r_2.\end{aligned}$$

If we introduce the abbreviation $a_i = (r_i, d_i)$, we can write the above enclosures as

$$f_1 \in a_1, f_2 \in a_2 \Rightarrow f_1 \circ f_2 \in a_1 \circ a_2 \quad \text{for } \circ \in \{+, -, *, /\}.$$

Moreover, putting $d_1 = d_2 = 0$, we find that

$$(r_1 \circ r_2)1 = r_1 1 \circ r_2 1 \quad \text{for all } r_1, r_2 \in \mathbb{IR}^*,$$

so that the constants in $\mathbb{A}$ behave like ordinary intervals.

We have turned $\mathbb{A}$ in what we shall call below an $\mathbb{IR}^*$-algebra. Note that the operations in $\mathbb{A}$ are independent of the interval x over which we consider the range of f and f'. In particular, the same calculation in $\mathbb{A}$ can have different interpretations, given by different relations depending on the choice of x. When an interpretation $\in$ is specified, we have what we shall call an inclusion algebra.

Using the operations in $\mathbb{A}$ we can now recursively compute enclosures for the real evaluation of any rational arithmetical expression $f(\xi)$ in a single variable ξ by starting with the constants, and noting that the identity function $f(\xi) = \xi$ belongs to the element $(x, 1)$ of $\mathbb{A}$. If $f(\xi)$ involves elementary functions, we must extend the arithmetic in $\mathbb{A}$. From the chain rule $(\varphi(f))' = \varphi'(f)f'$ we find

$$f \in (r, d) \Rightarrow \varphi(f) \in (\varphi(r), \varphi'(r)d).$$

Hence a natural consistent extension of φ to $\mathbb{A}$ is given by defining

$$\varphi((r, d)) := (\varphi(r), \varphi'(r)d),$$

so that

$$f \in a \Rightarrow \varphi(f) \in \varphi(a).$$

For example, $\operatorname{sqrt}(r, d) = (\sqrt{r}, d/(2\sqrt{r}))$. Hence, for $x = [0, 3]$, the real evaluation of $f(\xi) = \xi \operatorname{sqrt}(\xi + 1)$ belongs to

$$\begin{aligned}([0, 3], 1)\operatorname{sqrt}(([0, 3], 1) + (1, 0)) &= ([0, 3], 1)\operatorname{sqrt}(([1, 4], 1))\\ &= ([0, 3], 1)\left([1, 2], \left[\frac{1}{4}, \frac{1}{2}\right]\right)\\ &= \left([0, 6], \left[1, \frac{7}{2}\right]\right).\end{aligned}$$

In this example, the range of f is obtained exactly, while the range of f' (namely $[1, 11/4]$) is somewhat overestimated.

We now give a formally precise introduction of inclusion algebras which abstracts the essential features of the above informal discussion.

We call a set $\mathbb{A}$ together with binary operations $+, -, *, /$ on $\mathbb{A}$ an $\mathbb{IR}^*$-*algebra*, if there is a distinguished mapping which associates with each $\alpha \in \mathbb{IR}^*$ an element $\alpha 1 \in \mathbb{A}$ in such a way that

$$(\alpha \circ \beta)1 = \alpha 1 \circ \beta 1 \quad \text{for} \circ \in \{+, -, *, /\}, \alpha, \beta \in \mathbb{IR}^*. \tag{1}$$

The elements $\alpha 1$ ($\alpha \in \mathbb{IR}^*$) are called the *constants* of $\mathbb{A}$. A relation $\Subset$ between the two $\mathbb{IR}^*$-algebras $\mathbb{A}_0$ and $\mathbb{A}$ is called a *covering relation* if for all $a_0, b_0 \in \mathbb{A}_0$ and all $a, b \in \mathbb{A}$, we have

$$a_0 \Subset a, b_0 \Subset b \Rightarrow a_0 \circ b_0 \Subset a \circ b \quad \text{for all} \circ \in \{+, -, *, /\}, \tag{2}$$

and if

$$\alpha 1 \Subset \alpha 1 \quad \text{for all } \alpha \in \mathbb{IR}^*. \tag{3}$$

We may read $a_0 \Subset a$ as 'a_0 is covered by a' or 'a_0 belongs to a'.

For an arbitrary set D, usually a subset of $\mathbb{R}^n$, the set of $\mathcal{F}(D)$ of all functions f: $D \to \mathbb{IR}^*$ becomes an $\mathbb{IR}^*$-algebra by introducing pointwise operations $\circ \in \{+, -, *, /\}$ via

$$(f_1 \circ f_2)(\tilde{x}) = f_1(\tilde{x}) \circ f_2(\tilde{x}) \quad (\tilde{x} \in D),$$

and writing $\alpha 1$ for the function with constant value $\alpha \in \mathbb{IR}^*$. We are interested in $\mathbb{IR}^*$-algebras which capture certain features of the algebra $\mathcal{F}(D)$. An *inclusion algebra* over the set D is a pair $(\mathbb{A}, \Subset)$ consisting of an $\mathbb{IR}^*$-algebra $\mathbb{A}$ and a covering relation $\Subset$ between $\mathcal{F}(D)$ and $\mathbb{A}$, called the *interpretation* of $\mathbb{A}$. If $\Subset$ is clear from the context we simply talk about 'the inclusion algebra' $\mathbb{A}$.

The elementary functions $\varphi \in \Phi$ have a natural meaning for arguments in $\mathcal{F}(D)$ by writing $g = \varphi(f)$ for the function defined by $g(\tilde{x}) := \varphi(f(\tilde{x}))$ for $\tilde{x} \in x$. Usually it is possible to define the elementary function also for arguments which are elements of an inclusion algebra $\mathbb{A}$, sometimes in several natural ways. We shall call any such extension a *consistent extension* of φ (and denote it by the same symbol) if for all $f \in \mathcal{F}(D)$ and $a \in \mathbb{A}$,

$$f \Subset a \Rightarrow \varphi(f) \Subset \varphi(a). \tag{4}$$

We write $(\mathbb{A}, \Subset, \Phi)$ for an inclusion algebra with a fixed set of consistent extensions for the elementary functions $\varphi \in \Phi$. (The case where no consistent extensions are available is covered by putting $\Phi = \emptyset$.)

For each rational arithmetical expression $g(\xi_1, \ldots, \xi_n)$ and arbitrary elements $a_1, \ldots, a_n$ of an $\mathbb{IR}^*$-algebra $\mathbb{A}$, we may define an element $g(a_1, \ldots, a_n)$ recursively by

$$g(a_1, \ldots, a_n) := \alpha 1 \quad \text{if } g(\xi_1, \ldots, \xi_n) = \alpha \in \mathbb{IR}, \tag{5}$$

$$g(a_1, \ldots, a_n) := a_i \quad \text{if } g(\xi_1, \ldots, \xi_n) = \xi_i, \tag{6}$$

$$g(a_1. \ldots, a_n) := g_1(a_1, \ldots, a_n) \circ g_2(a_1, \ldots, a_n) \tag{7}$$
$$\text{if } g = g_1 \circ g_2 \text{ with } \circ \in \{+, -, *, /\}.$$

(Here interval constants α in (5) are interpreted as agreed on in the remark after Corollary 1.4.4.)

If consistent extensions are available for the elementary functions $\varphi \in \Phi$, we may extend this definition to arbitrary arithmetical expressions by adding for each such φ the rule

$$g(a_1, \ldots, a_n) := \varphi(g_0(a_1, \ldots, a_n)) \tag{8}$$

if $g = \varphi(g_0)$ with $\varphi \in \Phi$. In particular, this definition applies to $\mathbb{A} = \mathcal{F}(D)$. A simple induction argument shows that $g(f_1(\tilde{x}), \ldots, f_n(\tilde{x}))$ is the value of $g(f_1, \ldots, f_n)$ at $\tilde{x} \in D$, and hence we are justified to call $g(f_1, \ldots, f_n)$ the *composition* of the arithmetical expression g with the functions $f_1, \ldots, f_n$.

2.2.1 Proposition Let $(\mathbb{A}, \Subset, \Phi)$ be an inclusion algebra over D, and let g be an arithmetical expression in n variables. Then for any $f_1, \ldots, f_n \in \mathcal{F}(D)$ and any $a_1, \ldots, a_n \in \mathbb{A}$ we have

$$f_1 \Subset a_1, \ldots, f_n \Subset a_n \Rightarrow g(f_1, \ldots, f_n) \Subset g(a_1, \ldots, a_n). \tag{9}$$

Proof Straightforward induction. □

This basic observation makes inclusion algebras a useful tool for the calculation of the information described by $\Subset$ about complicated functions in $\mathcal{F}(D)$ from given information about a few simple functions in $\mathcal{F}(D)$.

2.2.2 Example (Piecewise constant enclosures) Define the $\mathbb{IR}^*$-algebra PC_k as the set of k-tuples $a = (a_1, \ldots, a_k)$ of elements $a_i \in \mathbb{IR}^*$, with componentwise operations $a \circ b = (a_1 \circ b_1, \ldots, a_k \circ b_k)$ and constants $\alpha 1 = (\alpha, \ldots, \alpha)$. We can turn PC_k into an inclusion algebra over an arbitrary set D by choosing a partition $D_1, \ldots, D_k$ of D and defining the interpretation

$$f \Subset a :\Leftrightarrow f(\tilde{x}) \subseteq a_i \tag{10}$$

for all $\tilde{x} \in D_i$ $(i = 1, \ldots, k)$. A consistent extension of an elementary function $\varphi \in \Phi$ is given by $\varphi(a) := (\varphi(a_1), \ldots, \varphi(a_k))$.

For a fixed element $a \in PC_k$, the interpretation (10) yields a piecewise constant interval enclosure of $f(\tilde{x})$ when $\tilde{x}$ ranges over D. By choosing k sufficiently large and the partition D_i sufficiently fine we can cover any continuous function $f \in \mathcal{F}(D)$ (defined on a compact topological space D) by an element of PC_k with arbitrary accuracy. This is the principle underlying the simple partition method discussed after Theorem 2.1.3. Note that (10) still is a rather inefficient way of representing information about a given function (cf. Example 2.1.4), and later in this chapter we shall meet more powerful inclusion algebras which represent relevant information about f in a better way.

The algebra PC_k shows that the same $\mathbb{IR}^*$-algebra may be interpreted in several ways as an inclusion algebra. This is a special case ($D' = \{1, \ldots, k\}$) of the following general *substitution principle*.

2.2.3 Proposition Let $(\mathbb{A}, \in, \Phi)$ be an inclusion algebra over D, and let $\mathcal{F}$ be a collection of functions from D to a set D'. Then $(\mathbb{A}, \in', \Phi)$ is an inclusion algebra over D' with the modified interpretation of $\mathbb{A}$ defined as the covering relation $\in'$ between $\mathbb{A}$ and $\mathcal{F}(D')$ defined by

$$f \in' a \Leftrightarrow f(\omega) \in a \quad \text{for all } \omega \in \mathcal{F};$$

here $f(\omega)$: $D \to \mathbb{IR}^*$ is the composition of f and ω.

Proof Trivial. □

Later in this book we shall give several useful inclusion algebras (see Theorems 2.3.8 and 2.3.10). In this section we consider only one nontrivial example of an inclusion algebra. It exploits well-known analytic differentiation techniques to provide enclosures for the first $k + 1$ coefficients of a power series expansion of f given by

$$f(\xi) = \sum_{i \geq 0} \frac{f^{(i)}(0)}{i!} \xi^i.$$

This generalizes the inclusion algebra of the introduction (which is obtained for $k = 1$).

2.2.4 Theorem Let T_k be the $\mathbb{IR}^*$-algebra whose elements are the $(k + 1)$-tuples $a = (a_0, \ldots, a_k)$ with entries $a_i \in \mathbb{IR}^*$ $(i = 0, \ldots, k)$, with constants $\alpha 1 = (\alpha, 0, \ldots, 0)$ for $\alpha \in \mathbb{IR}^*$, and operations defined by $a \circ b = c$, where, for $i = 0, \ldots, k$,

$$c_i = a_i + b_i \qquad \text{if } \circ = +,$$

$$c_i = a_i - b_i \qquad \text{if } \circ = -,$$

$$c_i = \sum_{j=0}^{i} a_j b_{i-j} \qquad \text{if } \circ = *,$$

$$c_i = \left(a_i - \sum_{j=1}^{i} b_j c_{i-j}\right) \Big/ b_0 \quad \text{if } \circ = /.$$

Then T_k is an inclusion algebra over $\mathbb{R}$ for the interpretation $\in$ defined by $f \in a$ if either $a_i = \text{NaN}$ for all i, or f is real valued and k times continuously differentiable in some neighborhood of 0, and

$$\frac{f^{(i)}(0)}{i!} \in a_i \quad \text{for } i = 0, \ldots, k.$$

Proof Clearly, c_i only depends on $c_0, \ldots, c_{i-1}$ and a_j, b_j $(j \le i)$. Hence, the operations are well-defined. To show that $\in$ is a covering relation we write

$$f_i(\xi) := \frac{1}{i!} f^{(i)}(\xi) \quad (i \le k)$$

for the normalized derivatives of a real valued function f, which is defined and k-times continuously differentiable in a neighborhood of 0. Clearly,

$$(f_i)' = (f')_i = (i+1) f_{i+1}. \tag{11}$$

The product rule

$$(fg)_i = \sum_{j=0}^{i} f_j g_{i-j} \tag{12}$$

follows by induction from the trivial case $i = 0$; indeed, if (12) holds for some i then

$$\begin{aligned}(i+1)(fg)_{i+1} = (fg)_i' &= \left(\sum_{j=0}^{i} f_j g_{i-j}\right)' \\ &= \sum_{j=0}^{i} (f_j' g_{i-j} + f_j g_{i-j}') \\ &= \sum_{j=0}^{i} ((j+1) f_{j+1} g_{i-j} + (i-j+1) f_j g_{i-j+1}) \\ &= (i+1) \sum_{j=0}^{i+1} f_j g_{i+1-j},\end{aligned}$$

and division by $i + 1$ yields (12) for the next value of i. By writing $f/g = h$ as $f = gh$, we obtain from (12) the quotient rule

$$h_i = \left(f_i - \sum_{j=1}^{i} g_j h_{i-j}\right) \Big/ g_0 \quad \text{if } h = f/g.$$

Now if $f \in a$ and $g \in b$ then

$$(f \pm g)_i(0) = f_i(0) \pm g_i(0) \in a_i \pm b_i = (a \pm b)_i,$$

$$(f*g)_i(0)=\sum_{j=0}^{i} f_j(0)g_{i-j}(0)\in\sum_{j=0}^{i} a_jb_{i-j}=(a*b)_i,$$

and, if $h=f/g$, then

$$h_i(0)=\left(f_i(0)-\sum_{j=1}^{i} g_i(0)h_{i-j}(0)\right)\Big/g_0(0)$$

$$\in\left(a_i-\sum_{j=1}^{i} b_ih_{i-j}(0)\right)\Big/a_0,$$

so that $h_i(0)\in(a/b)_i$ by induction. Therefore, $f\circ g\Subset a\circ b$ for all operations $\circ\in\{+,-,*,/\}$. Hence $\Subset$ is a covering relation. □

The product rule (12) is the basis for the construction of consistent extensions of a number of elementary functions to T_k.

2.2.5 Proposition The following definitions provide consistent extensions of elementary functions for the inclusion algebra T_k. For $a, b\in T_k$ let

sqr(a) be the element b defined by

$$b_0=\operatorname{sqr}(a_0),\quad b_i=\sum_{j=0}^{i} a_ja_{i-j}\quad(i>0);$$

sqrt(a) be the element b defined by

$$b_0=\operatorname{sqrt}(a_0),\quad b_i=\left(a_i-\sum_{j=1}^{i-1} b_jb_{i-j}\right)\Big/(2b_0)\quad(i>0);$$

ln(a) be the element b defined by

$$b_0=\ln(a_0),\quad b_i=\left(a_i-\frac{1}{i}\sum_{j=1}^{i} jb_ja_{i-j}\right)\Big/b_0\quad(i>0);$$

exp(a) be the element b defined by

$$b_i=c_i*\exp(a_0)\quad(i\ge 0),\text{ with}$$

$$c_0=1,\quad c_i=\frac{1}{i}\sum_{j=1}^{i}(ja_j)c_{i-j}\quad(i>0);$$

a^n be the element b defined by

$$b_i=a_0^{m-i}c_i\qquad\qquad(i\ge 0),\quad\text{with}$$

$$m = \min(n, k),$$
$$c_0 = a_0^{n-m},$$
$$c_i = \frac{1}{i}\sum_{j=1}^{i}(nj + j - i)(a_j a_0^{j-1})c_{i-j} \quad (i > 0);$$

$\sin(a)$ be the element b defined by

$$b_i = c_i \sin(a_0) + s_i \cos(a_0) \quad \text{with}$$

$$\left\{\begin{matrix} c_0 = 1, & c_i = -\frac{1}{i}\sum_{j=1}^{i}(ja_j)s_{i-j} \\ s_0 = 0, & s_i = \frac{1}{i}\sum_{j=1}^{i}(ja_j)c_{i-j} \end{matrix}\right\} \quad (i > 0); \qquad (13)$$

$\cos(a)$ be the element b defined by

$$b_i = c_i \cos(a_0) - s_i \sin(a_0) \quad (i \geq 0), \quad \text{with } c_i, s_i \text{ as in (13).}$$

Proof The consistency of sqr immediately follows from (12). For the other cases support that $f \in a$.

If $g = \text{sqrt}(f)$ then $f = g * g$ so that by (12),

$$f_i = \sum_{j=0}^{i} g_j g_{i-j} = 2g_0 g_i + \sum_{j=1}^{i-1} g_j g_{i-j},$$

$$g_i = \left(f_i - \sum_{j=1}^{i-1} g_j g_{i-j}\right) \Big/ (2g_0) \quad \text{for } i > 0.$$

Since $g_0(0) = g(0) = \text{sqrt}(f(0)) \in \text{sqrt}(a_0) = b_0$ and $f_i(0) \in a_i$, we inductively find that

$$g_i(0) \in \left(a_i - \sum_{j=1}^{i-1} b_j b_{i-j}\right) \Big/ (2b_0) \quad \text{for } i > 0.$$

This proves the consistency of sqrt.

If $g = \ln(f)$ then $g' = f'/f$, and (11) and (12) imply

$$if_i = (f')_{i-1} = (g'f)_{i-1} = \sum_{j=0}^{i-1} (g')_j f_{i-1-j}$$
$$= \sum_{j=0}^{i-1} (j+1) g_{j+1} f_{i-1-j},$$

$$g_i = \left(f_i - \frac{1}{i}\sum_{j=1}^{i-1} jg_j f_{i-j}\right)\bigg/ f_0 \quad \text{for } i > 0.$$

Now the consistency of ln follows as before.

If $g = \exp(f)$ then $g' = f'g$ so that

$$ig_i = (g')_{i-1} = (f'g)_{i-1} = \sum_{j=0}^{i-1} (f')_j g_{i-1-j}$$
$$= \sum_{j=0}^{i-1} (j+1) f_{j+1} g_{i-1-j} = \sum_{j=1}^{i} jf_j g_{i-j}.$$

If we define the numbers

$$\tilde{c}_i := g_i(0)/g_0(0),$$

we find that $\tilde{c}_0 = 1$ and

$$\tilde{c}_i = \frac{1}{i}\sum_{j=1}^{i} jf_j \tilde{c}_{i-j} \quad \text{for } i > 0.$$

Induction shows $\tilde{c}_i \in c_i$ ($i = 0, \ldots, k$), from which one obtains consistency of exp.

If $g = f^n$ then $g' = nf^{n-1}f'$ so that $fg' = nf'g$. Therefore

$$\sum_{j=0}^{i-1} f_j (g')_{i-1-j} = (fg')_{i-1} = n(f'g)_{i-1}$$
$$= n\sum_{j=0}^{i-1} (f')_j g_{i-1-j},$$
$$\sum_{j=0}^{i-1} f_j (i-j) g_{i-j} = n\sum_{j=1}^{i} jf_j g_{i-j}.$$

This yields

$$if_0 g_i = \sum_{j=1}^{i} (nj + j - i) f_j g_{i-j}.$$

If $f(0) \neq 0$ then the numbers $\tilde{c}_i = f(0)^{i-m} g_i(0)$ satisfy $\tilde{c}_0 = a_0^{n-m}$ and

$$\tilde{c}_i = \frac{1}{i}\sum_{j=1}^{i} (nj + j - i) f_j f(0)^{j-1} \tilde{c}_{i-j} \quad \text{for } i > 0,$$

This implies the consistency of a^n. (The excluded case $f(0) = 0$ is handled by a continuity argument.)

Finally, if $g = \sin(f)$ and $h = \cos(f)$ then $g' = f'h$ and $h' = -f'g$. Hence, as before,

$$ig_i = \sum_{j=1}^{i} jf_j h_{i-j}, \quad ih_i = -\sum_{j=1}^{i} jf_j g_{i-j}.$$

Writing $\tilde{s} = g_0(0) = \sin(f(0))$, $\tilde{c} = h_0(0) = \cos(f(0))$, and

$$\tilde{c}_i = g_i(0)\tilde{s} + h_i(0)\tilde{c}, \quad \tilde{s}_i = g_i(0)\tilde{c} - h_i(0)\tilde{s},$$

we find $\tilde{c}_0 = 1$, $\tilde{s}_0 = 0$ and

$$i\tilde{c}_i = -\sum_{j=1}^{i} jf_j \tilde{s}_{i-j}, \quad i\tilde{s}_i = \sum_{j=1}^{i} jf_j \tilde{c}_{i-j}.$$

Since

$$g_i(0) = \tilde{c}_i \tilde{s} - \tilde{s}_i \tilde{c}, \quad h_i(0) = c_i \tilde{c} - \tilde{s}_i \tilde{s},$$

we find (as before) that sin and cos are consistent.   □

Remarks (1) As the proof shows, we could also define $\exp(a)$ as the vector b satisfying

$$b_0 = \exp(a_0), \quad b_i = \frac{1}{i} \sum_{j=1}^{i} (ja_j)b_{i-j} \quad (i > 0).$$

This gives another consistent extension for the exponential function. However, by the subdistributive law, the definition given in the proposition is superior when a_0 is thick. Similar remarks apply to the formulae for the power and the trigonometric functions.

(2) (This remark also applies to other inclusion algebras introduced later.) The operations and elementary functions on T_k can efficiently be implemented in programming languages like PASCAL-SC or ADA which allow the user to define data types and operators between elements of these data types. Also precompilers have been constructed which translate a language extension of FORTRAN with similar features as PASCAL-SC to ordinary FORTRAN.

(3) The element $x^0 := (0, 1, 0, \ldots, 0)$ of T_k covers the identity function. Hence, if f is an arithmetical expression in a single variable and $a = f(x^0)$ then $f \in a$, so that the power series coefficients $f_i = f^{(i)}(0)/i!$ of the real evaluation of f satisfy $f_i \in a_i$ $(i = 0, \ldots, k)$. In exact arithmetic, all a_i are thin (unless one of the a_i has become NaN), so that the exact values of the f_i are

obtained. In rounded arithmetic the properties of rounded interval arithmetic guarantee that reasonably narrow intervals a_i enclosing the f_i will be computed. Since each operation or elementary function in T_k requires at most $O(k^2)$ interval operations, *it takes* $O(k^2N)$ *operations to compute rigorous enclosures for the first k power series coefficients of a function defined by an expression composed of N operations or elementary functions.*

We now apply the substitution principle to obtain other interpretations of T_k which allow the computation of enclosures for the range of derivatives of real valued functions of a real variable.

2.2.6 Theorem For any real number $h > 0$ and any $x \in \mathbb{IR}$, the algebra T_k is also an inclusion algebra over $\mathbb{R}$ for the modified interpretation $\Subset'$ defined by $f \Subset' a$ if either $a_i = \text{NaN}$ for all i, or f is real valued and k-times continuously differentiable in some neighborhood of x, and

$$\frac{h^i}{i!} f^{(i)}(\tilde{x}) \in a_i \quad \text{for } i = 0, \ldots, k \text{ and all } \tilde{x} \in x. \tag{14}$$

The extensions given in Proposition 2.2.5 are also consistent for $(T_k, \Subset')$.

Proof Apply Proposition 2.2.3 with $\mathcal{F} = \{\xi \to \tilde{x} + h\xi \mid \tilde{x} \in x\}$ and note that $g(\xi) = f(\tilde{x} + h\xi)$ implies $g^{(i)}(0) = h^i f^{(i)}(\tilde{x})$ for $i = 0, \ldots, k$. □

Since the identity function is covered by

$$x^\# := (x, h, 0, \ldots, 0) \tag{15}$$

(with respect to the interpretation $\Subset'$ of T_k), (14) implies that we can calculate enclosures

$$f^{(i)}(x) := \frac{i!}{h^i} f(x^\#)_i \quad (i = 0, \ldots, k) \tag{16}$$

for the first k derivatives at $\tilde{x} \in x$ of every function defined by an arithmetical expression f in one variable. The recursions in Theorem 2.2.4 and Proposition 2.2.5 show that $f^{(i)}(x)$ can be considered as the interval evaluation of an arithmetical expression $f^{(i)}$ at x. (This expression could be symbolically computed by the same recursions; however, the resulting formulae are often too messy to be useful. It is one of the advantages of inclusion algebras that one can evaluate higher derivatives recursively without ever constructing explicit expressions for them.) One easily verifies that $f^{(i)}$ is Lipschitz if f is Lipschitz; therefore, if f is Lipschitz in $D \subseteq \mathbb{R}$ then (15) and (16) define Lipschitz continuous interval extensions on D of the first k derivatives of the real evaluation of f. In particular, if x is thin then the $f^{(i)}(x)$ are also thin (or NaN) and yield the precise value of the derivatives. Again, $O(k^2N)$ interval operations are required to compute $f(x), \ldots, f^{(k)}(x)$ for an arithmetical expression involving N operations and elementary functions.

The freedom of choosing $h > 0$ arbitrarily in (14) can be used profitably in cases where a fixed value of h would lead to overflow or underflow: rescaling the a_i to $(h'/h)^i a_i$ with a suitable new h' frequently brings the coefficients back to normal size.

2.3 The mean value form and other centered forms

We have seen that the interval evaluation $f(x)$ of an arithmetical expression f in n variables at a box $x \in \mathbb{IR}^n$ is an enclosure for the range

$$f^*(x) = \{f(\tilde{x}) \mid \tilde{x} \in x\}. \tag{1}$$

However, the overestimation may be large in adverse circumstances. In this section we provide some methods for enclosing the range $f^*(x)$ in such a way that the overestimation remains small for sufficiently narrow boxes x. The resulting formulae are called centered forms since they bound the range by considering the deviation of $f(\tilde{x})$ from a fixed center $\tilde{z} \in x$, often the midpoint $\tilde{z} = \check{x}$. The simplest centered form is the mean value form. It is shown to be inclusion isotone, and a computable bound for the overestimation of the range of a function by the mean value form is derived. This bound allows us in particular to deduce the important quadratic approximation property of the mean value form. The quadratic approximation property extends to more general centered forms defined in terms of linear enclosures and slopes. We show how – using an inclusion algebra – one can recursively determine such linear enclosures and slopes for functions defined by arithmetical expressions. As a special case we find an inclusion algebra for the computation of interval Lipschitz constants and interval extensions of gradients of such functions.

In the classical approach for estimating the variability of $f(\tilde{x})$ for an inaccurately known variable $\tilde{x} \in x$, one represents $\tilde{x}$ as $\tilde{x} = \check{x} + \Delta\tilde{x}$, where $|\Delta\tilde{x}| \leq \Delta x = \text{rad}(x)$. By Taylor development truncated after the linear term, we get

$$f(\tilde{x}) = f(\check{x} + \Delta\tilde{x}) \approx f(\check{x}) + f'(\check{x})\Delta\tilde{x}. \tag{2}$$

so that $f(\tilde{x})$ approximately lies in the interval $f(\check{x}) + f'(\check{x})[-\Delta x, \Delta x]$. In order to get a true bound for the range we may replace the Taylor series in (2) by the mean value theorem, which tells us that

$$f(\tilde{x}) = f(\check{x}) + f'(\xi)(\tilde{x} - \check{x}) \tag{3}$$

for some ξ on the line segment between $\tilde{x}$ and $\check{x}$. Hence $\xi \in x$; if we have arithmetical expressions for the components of f' then $f'(\xi) \in f'(x)$ so that $f(\tilde{x}) \in f(\check{x}) + f'(x)(\tilde{x} - \check{x}) \subseteq f(\check{x}) + f'(x)(x - \check{x})$. The expression

$$f_{\mathrm{m}}(x) := f(\check{x}) + f'(x)(x - \check{x}) \tag{4}$$

is called the *mean value form* of f. Since (4) is independent of $\tilde{x} \in x$ it contains all $f(\tilde{x})$ with $\tilde{x} \in x$; hence the mean value form is an enclosure for the range $f^*(x)$. Note that if f is an expression in several variables then the derivative

$$f'(\xi) = \left(\frac{\partial f}{\partial x_1}(\xi), \ldots, \frac{\partial f}{\partial x_n}(\xi)\right),$$

also called the *gradient* of f, is a row vector and the product in (4) is an inner product. We also note that different expressions f for the same real function do not affect the value of the mean value form, whereas different expressions for the derivative f' may change the value of (4), sometimes drastically.

The requirement that arithmetical expressions for the partial derivatives are available is not very restrictive; indeed if f is Lipschitz at x and only contains elementary functions in the standard set Φ mentioned in Section 1.3 then the gradient can be computed by well-known rules; an inclusion algebra which allows a simple recursive calculation of gradients is described at the end of this section.

In finite precision calculation one has to account for roundoff errors; therefore a rounded mean value form must be computed according to

$$f_{\mathrm{m}}^{\Diamond}(x) := f^{\Diamond}(\check{x}) \oplus f'^{\Diamond}(x) \circledast (x \ominus \check{x});$$

in particular, $f(\check{x})$ must also be evaluated in rounded interval arithmetic. On the other hand, $\check{x}$ need not be the exact midpoint (or an enclosure of it); the enclosure remains valid for every $\tilde{x} \in x$ in place of $\check{x}$. However, in order that the rounded value $\tilde{x}$ of $\check{x}$ is in x, the computation of $\check{x}$ must be done on nonbinary machines according to

$$\check{x} = \underline{x} \oplus 0.5 \circledast (\bar{x} \ominus \underline{x}) \quad \text{or } \check{x} = \bar{x} \ominus 0.5 \circledast (\bar{x} \ominus \underline{x})$$

and not as $\check{x} = 0.5 \circledast (\bar{x} \oplus \underline{x})$; otherwise the computed midpoint of certain narrow intervals (e.g. [0.66666, 0.66666] in five-digit decimal arithmetic) is outside the interval.

2.3.1 Example Consider $f(\xi) = \xi/(\xi - 1)$. The real evaluation is monotonically decreasing, which allows a simple computation of the range. Enclosures for the range over several intervals by interval evaluation and by the mean value form are given in Table 2.2. We notice that the mean value form is much more accurate for narrow intervals, but this advantage is less

Table 2.2. *Comparison of range enclosures for $\xi/(\xi-1)$; correct digits are underlined*

x	$f(x)$	$f_m(x)$
[2, 3]	[1, 3]	[1.66666, 2.166667]
[2.3, 2.4]	[1.642857, 1.846154]	[1.711154, 1.770327]
[2.33, 2.34]	[1.738805, 1.759399]	[1.746237, 1.751891]
[2.333, 2.334]	[1.748875, 1.750938]	[1.749624, 1.750188]
[2, 10]	[0.222222, 10]	[−2.8, 5.2]
[5, 10]	[0.555555, 2.5]	[0.997596, 1.310097]
[1.01, 1.02]	[50.5, 102]	[17.66666, 117.6667]
[1.011, 1.012]	[84.25, 92]	[83.82429, 92.08876]
[1.0111, 1.0112]	[90.276778, 91.09910]	[90.28028, 91.09191]
[1.01111, 1.01112]	[90.92715, 91.00991]	[90.92800, 91.0092]

pronounced when x comes close to the pole. For wide intervals the behavior of the two enclosures is less predictable, and neither enclosure is accurate.

For $x > 1$ an explicit calculation gives the intervals

$$f(x) = \left[\frac{\underline{x}}{\overline{x}-1}, \frac{\overline{x}}{\underline{x}-1}\right], \qquad \operatorname{rad}(f(x)) = \frac{(2\check{x}-1)\operatorname{rad}(x)}{(\overline{x}-1)(\underline{x}-1)},$$

$$f_m(x) = \frac{\check{x}}{\check{x}-1} + \frac{\operatorname{rad}(x)}{(\underline{x}-1)^2}[-1, 1], \qquad \operatorname{rad}(f(x)) = \frac{\operatorname{rad}(x)}{(\underline{x}-1)^2},$$

$$f^*(x) = \left[\frac{\overline{x}}{\overline{x}-1}, \frac{\underline{x}}{\underline{x}-1}\right], \qquad \operatorname{rad}(f^*(x)) = \frac{\operatorname{rad}(x)}{(\overline{x}-1)(\underline{x}-1)}.$$

(Note that Theorem 1.4.3 implies that the transformed expression $1 + 1/(\xi - 1)$ gives the exact range for all $x \in \mathbb{IR}$ with $1 \notin x$.)

The overestimation factor (the quotient of the radius of the enclosure and that of the range) is therefore $2\check{x} - 1 = 1 + 2(\check{x} - 1)$ for the interval evaluation $f(x)$ and $(\overline{x} - 1)/(\underline{x} - 1) = 1 + 2 \cdot \operatorname{rad}(x)/\underline{x} - 1$ for the mean value form $f_m(x)$. Therefore, the interval evaluation is of good quality when x is close to the pole 1 and inaccurate when $x \gg 1$, and the mean value form is of good quality when $\operatorname{rad}(x)$ is small (and very small near the pole). In particular, we see that if $x \subseteq x^0$ and $x^0 > 1$ then the overestimation factor is $1 + O(\operatorname{rad}(x))$. We shall see later that this is typical for mean value forms.

We now investigate the properties of the mean value form.

2.3.2 Theorem (Caprani & Madsen) The mean value form (4) is inclusion isotone, i.e.

$$x \subseteq x^0 \Rightarrow f_m(x) \subseteq f_m(x^0). \tag{5}$$

Proof Let $x \subseteq x^0$. Then, by the mean value theorem,

$$\begin{aligned} f_{\mathrm{m}}(x) &= f(\check{x}) + f'(x)(x - \check{x}) \\ &= f(\check{x}^0) + f'(\xi)(\check{x} - \check{x}^0) + f'(x)(x - \check{x}) \end{aligned}$$

for some $\xi \in x^0$, and since the interval evaluation of f' is inclusion isotone we have

$$\begin{aligned} |f_{\mathrm{m}}(x) - f(\check{x}^0)| &\leq |f'(x^0)|\,|\check{x} - \check{x}^0| + |f'(x^0)|\,|x - \check{x}| \\ &= |f'(x^0)|(|\check{x} - \check{x}^0| + \mathrm{rad}(x)) \\ &\leq |f'(x^0)|\mathrm{rad}(x^0) \end{aligned}$$

by (1.6.15). Since $f(\check{x}^0)$ is thin and $\mathrm{mid}(x^0 - \check{x}^0) = 0$ we have

$$\begin{aligned} \mathrm{mid}(f_{\mathrm{m}}(x^0)) &= f(\check{x}^0) + \mathrm{mid}(f'(x^0)(x^0 - \check{x}^0)) = f(\check{x}^0), \\ \mathrm{rad}(f_{\mathrm{m}}(x^0)) &= \mathrm{rad}(f'(x^0)(x^0 - \check{x}^0)) = |f'(x^0)|\mathrm{rad}(x^0); \end{aligned}$$

hence

$$|f_{\mathrm{m}}(x) - \mathrm{mid}(f_{\mathrm{m}}(x^0))| \leq \mathrm{rad}(f_{\mathrm{m}}(x^0)),$$

and (1.6.15) yields (5). □

Note that the same result holds more generally when $f'(x)$ is replaced by any inclusion isotone interval extension of the derivative in (4), but it becomes false when in (4) the midpoint $\check{x}$ is replaced by an arbitrary $\tilde{x} \in x$.

Our next aim is to find a bound for the overestimation of the range by the mean value form. First we prove the following more general result.

2.3.3 Theorem (Krawczyk & Neumaier) Let $f\colon D \subseteq \mathbb{R}^n \to \mathbb{R}$ be a real function, $x \in \mathbb{I}D$ and $\tilde{z} \in x$. Suppose that $s \in \mathbb{IR}^{1\times n}$ is a row vector such that, for all $\tilde{x} \in x$, we have

$$f(\tilde{x}) = f(\tilde{z}) + \tilde{s}(\tilde{x} - \tilde{z}) \quad \text{for some } \tilde{s} \in s. \tag{6}$$

Then the interval $f(\tilde{z}) + s(x - \tilde{z})$ encloses the range $f^*(x)$, and we have

$$q(f(\tilde{z}) + s(x - \tilde{z}), f^*(x)) \leq 2 \cdot \mathrm{rad}(s)|x - \tilde{z}|. \tag{7}$$

Proof We define two vectors $x_*, x^* \in \mathbb{R}^n$ by

$$\begin{aligned} x_{*i} &= \underline{x}_i, \quad x_i^* = \overline{x}_i \quad \text{if } s_i > 0, \\ x_{*i} &= \overline{x}_i, \quad x_i^* = \underline{x}_i \quad \text{if } s_i < 0, \\ x_{*i} &= \tilde{z}_i, \quad x_i^* = \tilde{z}_i \quad \text{otherwise.} \end{aligned}$$

Clearly $x_*, x^* \in x$. If we denote by $\langle s\rangle$ the row vector with components $\langle s_i\rangle$, then (6) implies that for a suitable $s_* \in s$ we have

$$f(x_*) = f(\tilde{z}) + s_*(x_* - \tilde{z}) = f(\tilde{z}) - \langle s_*\rangle|x - \tilde{z}| \\ \le f(\tilde{z}) - \langle s\rangle|x - \tilde{z}|. \tag{8}$$

By a similar argument we have $f(x^*) \ge f(\tilde{z}) + \langle s\rangle|x - \tilde{z}|$, so that

$$f(\tilde{z}) + [-1, 1]\langle s\rangle|x - \tilde{z}| \subseteq f^*(x). \tag{9}$$

Therefore

$$\begin{aligned} f(\tilde{z}) + s(x - \tilde{z}) &\subseteq f(\tilde{z}) + [-1, 1]|s(x - \tilde{z})| \\ &= f(\tilde{z}) + [-1, 1]|s|\,|x - \tilde{z}| \\ &\subseteq f(\tilde{z}) + [-1, 1](\langle s\rangle + 2\cdot\mathrm{rad}(s))|x - \tilde{z}| \\ &\subseteq f(\tilde{z}) + [-1, 1]\cdot 2\cdot\mathrm{rad}(s)|x - \tilde{z}|. \end{aligned}$$

On the other hand (6) trivially implies the enclosure $f^*(x) \subseteq f(\tilde{z}) + s(x - \tilde{z})$. Now the result follows from (1.7.1) and (1.7.9). □

Applied to the mean value form we obtain

2.3.4 Corollary Let f: $D \subseteq \mathbb{R}^n \to \mathbb{R}$ be a continuously differentiable function, and let f' be a vector of arithmetical expressions for the derivative of f. Then:

(i) The mean value form (4) satisfies

$$0 \le \mathrm{rad}(f_{\mathrm{m}}(x)) - \mathrm{rad}(f^*(x)) \le q(f_{\mathrm{m}}(x), f^*(x)) \\ \le 2\,\mathrm{rad}(f'(x))\mathrm{rad}(x). \tag{10}$$

(ii) If f' is Lipschitz at $x^0 \in \mathbb{IR}^n$ and $0 \notin f'(x^0)$ then there is a constant γ (depending on f' and x^0 only) such that for all $x \subseteq x^0$,

$$1 \le \frac{\mathrm{rad}(f_{\mathrm{m}}(x))}{\mathrm{rad}(f^*(x))} \le 1 + \gamma \max_i (\mathrm{rad}(x_i)). \tag{11}$$

Proof (i) Since (6) holds with $\tilde{z} = \check{x}$ and $s = f'(x)$, we get (10) from (4), (7) and Proposition 1.7.1 (3).

(ii) Let $x \subseteq x^0$ and $\rho = \max_i(\mathrm{rad}(x_i))$. Since $0 \notin f'(x^0)$, the number $\alpha = \min\{\langle f'(x^0)_i\rangle \mid i = 1, \ldots, n\}$ is positive, and since f is inclusion isotone, (9) implies

$$\mathrm{rad}\,(f^*(x)) \ge \langle f'(x)\rangle\mathrm{rad}(x) \ge \langle f'(x^0)\rangle\mathrm{rad}(x) \ge \alpha \sum \mathrm{rad}(x_i).$$

Since f' is Lipschitz at x^0, Corollary 2.1.2 implies that $\operatorname{rad} f'(x)_i \le \beta\rho$ for some constant $\beta > 0$. Hence (10) gives

$$\operatorname{rad}(f_{\mathrm{m}}(x)) \le \operatorname{rad}(f^*(x)) + 2\beta\rho \sum \operatorname{rad}(x_i)$$

$$\le \operatorname{rad}(f^*(x)) + \frac{2\beta\rho}{\alpha} \operatorname{rad}(f^*(x)).$$

This yields the upper bound of (11) with $\gamma = 2\beta/\alpha$. The lower bound in (11) is trivial. □

Formula (10), which expresses the overestimation of the range by a product of two radii, is called the *quadratic approximation property* of the mean value form. If f' is Lipschitz at x^0 and $x \subseteq x^0$ has small radius then Corollary 2.1.2 and (9) imply that

$$q(f_{\mathrm{m}}(x),\ f^*(x)) = \mathrm{O}(r^2) \quad \text{if } \operatorname{rad}(x) = \mathrm{O}(r).$$

This shows that for narrow intervals the mean value form generally gives a much better enclosure for the range than the interval evaluation which only satisfies $q(f(x), f^*(x)) = \mathrm{O}(r)$. It is remarkable that the bound (10) for the overestimation is computable with very little extra effort from the data required for the mean value form; this important observation allows one to check *a posteriori* whether the computed enclosure is indeed narrow enough.

Quite often one can find a natural decomposition

$$f(\tilde{x}) = f(\tilde{z}) + f[\tilde{z}, \tilde{x}](\tilde{x} - \tilde{z}) \tag{12}$$

with a function $f[\cdot,\cdot]\colon D \times D \to \mathbb{R}^{1\times n}$. The row vector $f[\tilde{z}, \tilde{x}]$ is called a *slope* (between $\tilde{z}$ and $\tilde{x}$). Assuming continuity of the slope we have in the univariate case ($n = 1$)

$$f[\tilde{z}, \tilde{x}] = \begin{cases} \dfrac{f(\tilde{x}) - f(\tilde{z})}{\tilde{x} - \tilde{z}} & \text{if } \tilde{x} \ne \tilde{z}, \\ f'(\tilde{z}) & \text{if } \tilde{x} = \tilde{z}; \end{cases}$$

hence $f[\tilde{z}, \tilde{x}]$ is a divided difference of f. In particular, for polynomials in standard power representation, the coefficients of $f[\tilde{z}, \tilde{x}]$ (as a polynomial in $\tilde{x}$) arise as a byproduct of the evaluation of $f(\tilde{z})$ with the Horner scheme. In the multivariate case ($n > 1$), the slope is no longer unique; e.g., if $n = 2$ and $f(\xi) = \xi_1\xi_2$ then both $f[\tilde{z}, \tilde{x}] = (\tilde{x}_2, \tilde{z}_1)$ and $f[\tilde{z}, \tilde{x}] = (\tilde{z}_2, \tilde{x}_1)$ are slopes.

If (12) holds for all $\tilde{x} \in x$ and $f[\zeta, \xi]$ is an arithmetical expression in ζ and ξ then we can enclose the range $f^*(x)$ also by the *slope form* with center $\tilde{z} \in x$, defined as

$$f_s(x, \check{z}) := f(\check{z}) + f[\check{z}, x](x - \check{z}). \tag{13}$$

Theorem 2.3.3 immediately implies the quadratic approximation property

$$q(f_s(x, \check{z}), f^*(x)) \le 2\cdot\text{rad}(f[\check{z}, x])|x - \check{z}|, \tag{14}$$

If $0 \notin f[\check{z}, x^0]$ then, for narrow boxes $x \subseteq x^0$, this again gives an overestimation factor (11) of order $1 + O(r)$ if $\text{rad}(x) = O(r)$. The same observations hold if we replace the slope by the derivative $f'(x)$. Thus, we obtain an enclosure of $f^*(x)$ by the *generalized mean value form*

$$f_m(x, \check{z}) := f(\check{z}) + f'(x)(x - \check{z}), \tag{15}$$

which also has the quadratic approximation property.

The formulae (4), (13), (15), and more generally all formulae of the form $f(\check{z}) + s(x - \check{z})$ (where (6) holds), are called *centered forms* for f with *center* $\check{z}$. All centered forms satisfy the overestimation bound (7), and hence can be checked *a posteriori* for accuracy. General centered forms need not be inclusion isotone; but if the center remains fixed inclusion isotonicity is a trivial consequence of (13).

2.3.5 Example Let $x = [0, 1]$. We consider the arithmetical expressions

$$f(\xi) = \xi^3 - 3\xi^2 + 4\xi + 5,$$
$$f'(\xi) = 3\xi^2 - 6\xi + 4$$
$$f[\zeta, \xi] = \xi^2 + \xi\zeta + \zeta^2 - 3(\xi + \zeta) + 4$$

for a polynomial, its derivative and its slope. Since

$$f(0) = 5, \quad f(0.5) = 6.375, \quad f(1) = 7,$$
$$f(x) = [-3, 10], \quad f'(x) = [-2, 7],$$
$$f[0, x] = [1, 5], \quad f[0.5, x] = [-0.25, 4.25], \quad f[1, x] = [-1, 4],$$

we get the following enclosures for the range $f^*(x) = [5, 7]$:

$$f(x) = [-3, 10],$$
$$f_m(x) = f_m(x, \check{x}) = [2.875, 9.875], \quad f_s(x, \check{x}) = [4.25, 8.5],$$
$$f_m(x, \underline{x}) = [3, 12], \quad f_s(x, \underline{x}) = [5, 10],$$
$$f_m(x, \overline{x}) = [0, 9], \quad f_s(x, \overline{x}) = [3, 8].$$

We notice that the enclosure depends considerably on the choice of the center. In particular, for the same center the slope form consistently gives more accurate results than the generalized mean value form. In fact, we have $f[\check{z}, x] \subseteq f(x)$, and $\text{rad} f[\check{z}, x]$ is roughly one half of $\text{rad} f'(x)$, so that $f_s(x, \check{z})$ has roughly half the radius of $f_m(x, \check{z})$. We shall see later (cf.

Proposition 2.3.12) that this behavior is the general rule for certain canonical expressions for $f'(\xi)$ and $f[\zeta, \xi]$ obtained by using inclusion algebras.

In the present example, the best enclosure is obtained by taking the intersection of $f_s(x, \underline{x})$ and $f_s(x, \overline{x})$, giving the interval $[5, 8]$. On the other hand, if we use the equivalent expression $f'(\xi) = 3(\xi - 1)^2 + 1$ for the derivative, we find $f'(x) = [1, 4] > 0$; this shows that f is monotone on x and $f^*(x) = \square\{f(\underline{x}), f(\overline{x})\} = [5, 7]$.

This example leads to the question whether the center can be chosen in an optimal manner. We can answer this question for generalized mean value forms. To simplify the formulation of an optimality result we need the *cut-off function* cut: $\mathbb{R} \times \mathbb{IR} \to \mathbb{R}$ defined by

$$\operatorname{cut}(\tilde{x}, x) = \begin{cases} \overline{x} & \text{if } \tilde{x} \geq \overline{x}, \\ \underline{x} & \text{if } \tilde{x} \leq \underline{x}, \\ \tilde{x} & \text{otherwise.} \end{cases} \tag{16}$$

2.3.6 Theorem (Baumann) With the definitions

$$p_i = \operatorname{cut}\left(\frac{\operatorname{mid}(f'(x)_i)}{\operatorname{rad}(f'(x)_i)}, [-1, 1]\right) \qquad (i = 1, \ldots, n). \tag{17}$$

$$z_{*i} = \check{x}_i = p_i \operatorname{rad}(x_i), \quad z_i^* = \check{x}_i + p_i \operatorname{rad}(x_i) \quad (i = 1, \ldots, n). \tag{18}$$

we have:

(i) $\inf(f_{\mathrm{m}}(x, \tilde{z}))$ attains its maximum at the center $\tilde{z} = z_*$,
(ii) $\sup(f_{\mathrm{m}}(x, \tilde{z}))$ attains its minimum at the center $\tilde{z} = z^*$,
(iii) $\operatorname{rad}(f_{\mathrm{m}}(x, \tilde{z}))$ attains its minimum at the center $\tilde{z} = \check{x}$.

Proof (i) An arbitrary vector $\tilde{z} \in x$ can be written as $\tilde{z} = \check{x} - \tilde{p} \cdot \operatorname{rad}(x)$ with $\tilde{p} \in (\mathbb{R}^n)^T$ satisfying $|\tilde{p}_i| \leq 1$ for $i = 1, \ldots, n$, and we have

$$x_i - \tilde{z}_i = \operatorname{rad}(x_i)[\tilde{p}_i - 1, \tilde{p}_i + 1] \ni 0 \quad (i = 1, \ldots, n).$$

With the abbreviation $s = f'(x)$, the infimum of (15) takes the form

$$\begin{aligned} \inf(f_{\mathrm{m}}(x, \tilde{z})) &= f(\tilde{z}) + \inf(s(x - \tilde{z})) \\ &= f(\tilde{z}) + \sum \inf(s_i(x_i - \tilde{z}_i)) \\ &= f(\tilde{z}) + \sum \inf(s_i \operatorname{rad}(x_i)[\tilde{p}_i - 1, \tilde{p}_i + 1]) \\ &= f(\tilde{z}) + \sum \operatorname{rad}(x_i)\min(\underline{s}_i(\tilde{p}_i + 1), \overline{s}_i(\tilde{p}_i - 1)). \end{aligned}$$

It remains to show that

$$\alpha_i := \min(\underline{s}_i(\tilde{p}_i + 1), \bar{s}_i(\tilde{p}_i - 1)) \quad (|\tilde{p}_i| \le 1)$$

takes its maximum for $\tilde{p}_i = p_i$. If $s_i \ge 0$ then $\alpha_i \le 0$ (with equality for $\tilde{p}_i = 1 = p_i$), and if $s_i \le 0$ then $\alpha_i \le 0$ (with equality for $\tilde{p}_i = -1 = p_i$). In the remaining case $\underline{s}_i < 0 < \bar{s}_i$ we have $p_i = \check{s}_i/\mathrm{rad}(s_i) = (\bar{s}_i + \underline{s}_i)/(\bar{s}_i - \underline{s}_i)$; hence $\tilde{p}_i \le p_i$ iff $(\bar{s}_i - \underline{s}_i)\tilde{p}_i \le \bar{s}_i + \underline{s}_i$ iff $\bar{s}_i(\tilde{p}_i - 1) \le \underline{s}_i(\tilde{p}_i + 1)$. Therefore

$$\alpha_i = \bar{s}_i(\tilde{p}_i - 1) \le \bar{s}_i(p_i - 1) \quad \text{for } \tilde{p}_i \le p_i,$$
$$\alpha_i = \underline{s}_i(\tilde{p}_i + 1) \le \underline{s}_i(p_i + 1) \quad \text{for } \tilde{p}_i > p_i;$$

and since $\alpha_i = \underline{s}_i(p_i + 1) = \bar{s}_i(p_i - 1)$ for $\tilde{p}_i = p_i$, the maximum of α_i is indeed attained for $\tilde{p}_i = p_i$. Therefore, $\inf f_{\mathrm{m}}(x, \tilde{z})$ attains its maximum at $\tilde{z} = \check{x} - p \cdot \mathrm{rad}(x)$.

(ii) follows by applying (i) with $-f$, $-f'$, $-p_i$ in place of f, f', p_i.

Finally, (iii) holds since $\mathrm{rad}(f_{\mathrm{m}}(x, \tilde{z})) = \mathrm{rad}(s(x - \tilde{z})) \ge |s|\mathrm{rad}(x - \tilde{z}) = |s|\mathrm{rad}(x) = \mathrm{rad}(f_{\mathrm{m}}(x, \check{z}))$ by (15) and (1.6.23). □

Remarks (i) A closer analysis shows that $f_{\mathrm{m}}(x, \tilde{z})$ has minimal radius for all centers $\tilde{z} \in \square\{z_*, z^*\}$.

(ii) While the mean value form has the smallest radius among all generalized mean value forms, a better enclosure for the range is obtained by taking the intersection

$$f_{\mathrm{b}}(x) := f_{\mathrm{m}}(x, z_*) \cap f_{\mathrm{m}}(x, z^*).$$

A further improvement is obtained by intersecting the slope forms with centers z_* and z^*. Of course, these improvements are achieved at the cost of further evaluations of the function (and the slope). We may call the form $f_{\mathrm{b}}(x)$ and the corresponding slope form *bicentered forms*.

In the special case where $0 \notin f'(x)$ so that f is monotone in each variable, we have $p_i \in \{1, -1\}$ for all i; in this case z_* and z^* are corners of the box x, and, as one easily checks, we get

$$f_{\mathrm{m}}(x, z_*) \cap f_{\mathrm{m}}(x, z^*) = \square\{f(z_*), f(z^*)\} = f^*(x).$$

Thus, the bicentered forms give the exact range in an important special case, where the mean value form and other centered forms still overestimate the range by a factor of $1 + \mathrm{O}(\mathrm{rad}(x))$; cf. Example 2.3.1 and Corollary 2.3.4(ii).

In the univariate case there is another possibility for the enclosure of the range in terms of $f(\underline{x}), f(\bar{x})$ and $f'(x)$ motivated by Fig. 2.1. In this figure, the dashed lines indicate the optimal enclosure for $f^*(x)$ with the available

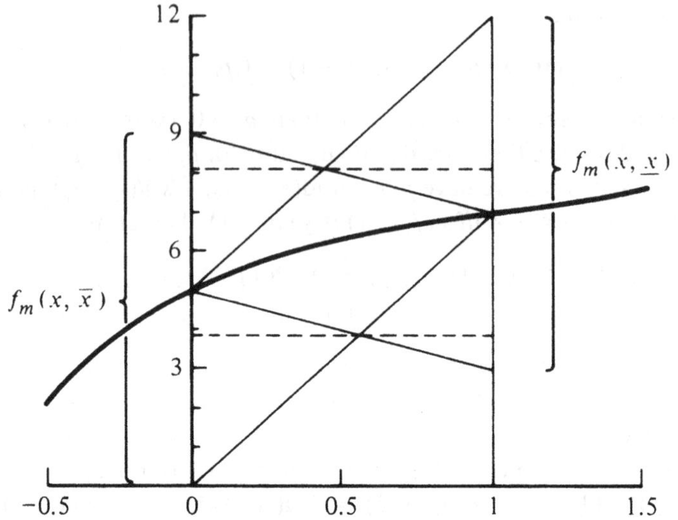

Fig. 2.1. The linear boundary value form.

information. The optimal enclosure for $0 \in \operatorname{int} f'(x)$ is obtained as $[\underline{f}, \bar{f}]$, where

$$\begin{aligned}\underline{f} &= \inf\{\max(f(\underline{x}) + \underline{f'(x)}(\tilde{x} - \underline{x}), f(\overline{x}) + \overline{f'(x)}(\tilde{x} - \overline{x})) \mid \tilde{x} \in x\}\\ &= (\overline{f'(x)}f(\underline{x}) - \underline{f'(x)}f(\overline{x}) + (\overline{x} - \underline{x})\overline{f'(x)}\,\underline{f'(x)})/(\overline{f'(x)} - \underline{f'(x)}),\\ \bar{f} &= \sup\{\min(f(\underline{x}) + \overline{f'(x)}(\tilde{x} - \underline{x}), f(\overline{x}) + \underline{f'(x)}(\tilde{x} - \overline{x})) \mid \tilde{x} \in x\}\\ &= (\overline{f'(x)}f(\overline{x}) - \underline{f'(x)}f(\underline{x}) - (\overline{x} - \underline{x})\overline{f'(x)}\,\underline{f'(x)})/(\overline{f'(x)} - \underline{f'(x)}).\end{aligned}$$

We may call this the *linear boundary value form*.

2.3.7 Example For f as in Example 2.3.5 we get $z_* = 2/9$, $z^* = 7/9$, and (with optimal outward rounding to five decimal digits)

$$\begin{aligned}f_{\mathrm{m}}(x, z_*) &= [4.1961, 11.197]\\ f_{\mathrm{m}}(x, z^*) &= [1.3223, 8.3224],\end{aligned}$$

so that $f_{\mathrm{b}}(x) = [4.1961, 8.3224]$. The linear boundary value form gives a still better enclosure $[13/3, 23/3] \subseteq [4.3333, 7.6667]$. The bicentered slope form gives the best enclosure $[4.7283, 7.6667]$.

We now consider an inclusion algebra which provides linear enclosures of the form (6) for functions defined by arithmetical expressions.

2.3.8 Theorem Let L_n be the $\mathbb{IR}^*$-algebra whose elements are the triples $a = (a_c, a_r, a_s) \in \mathbb{IR}^* \times \mathbb{IR}^* \times (\mathbb{IR}^*)^{1\times n}$, with constants $\alpha 1 = (\alpha, \alpha, 0)$ for $\alpha \in \mathbb{IR}^*$, and operations defined by $a \circ b = d$, where

$$d_c := a_c \circ b_c, \quad d_r := a_r \circ b_r,$$

and

$$\begin{aligned} d_s &:= a_s + b_s && \text{if } \circ = +, \\ d_s &:= a_s - b_s && \text{if } \circ = -, \\ d_s &:= a_r b_s + b_c a_s && \text{if } \circ = *, \\ d_s &:= (a_s - d_c b_s)/b_r && \text{if } \circ = /. \end{aligned}$$

For $x, z \in \mathbb{IR}^n$, let $\in_{x,z}$ be the interpretation defined by $f \in_{x,z} a$ iff, for all $\tilde{z} \in z$, $\tilde{x} \in x$, we have

$$f(\tilde{z}) \in a_c, \quad f(\tilde{x}) \in a_r, \tag{19}$$

$$f(\tilde{x}) - f(\tilde{z}) = \tilde{s}(\tilde{x} - \tilde{z}) \quad \text{for some } \tilde{s} \in a_s. \tag{20}$$

Then, for each interpretation $\in_{x,z}$ $(x, z \in \mathbb{IR}^n)$, L_n is an inclusion algebra over $\mathbb{R}^n$.

Proof We have to show that $\in = \in_{x,z}$ is a covering relation. Hence suppose that $f \in a$, $g \in b$ and $h = f \circ g$, $d = a \circ b$. We show that $h \in d$. Indeed, let $\tilde{z} \in z$, $\tilde{x} \in x$. Then

$$\begin{aligned} h(\tilde{z}) &= f(\tilde{z}) \circ g(\tilde{z}) \in a_c \circ b_c = d_c, \\ h(\tilde{x}) &= f(\tilde{x}) \circ g(\tilde{x}) \in a_r \circ b_r = d_r. \end{aligned}$$

Moreover,

$$\begin{aligned} f(\tilde{x}) - f(\tilde{z}) &= \tilde{s}_f(\tilde{x} - \tilde{z}) \quad \text{with } \tilde{s}_f \in a_s. \\ g(\tilde{x}) - g(\tilde{z}) &= \tilde{s}_g(\tilde{x} - \tilde{z}) \quad \text{with } \tilde{s}_g \in b_s. \end{aligned}$$

For $\circ \in \{+, -\}$, this implies

$$\begin{aligned} h(\tilde{x}) - h(\tilde{z}) &= (f(\tilde{x}) - f(\tilde{z})) \pm (g(\tilde{x}) - g(\tilde{z})) \\ &= \tilde{s}_f(\tilde{x} - \tilde{z}) \pm \tilde{s}_g(\tilde{x} - \tilde{z}) = \tilde{s}(\tilde{x} - \tilde{z}) \end{aligned}$$

with $\tilde{s} = \tilde{s}_f \pm s_g \in a_s \pm b_s = d_s$.

For $\circ = *$ we have

$$\begin{aligned} h(\tilde{x}) - h(\tilde{z}) &= f(\tilde{x})(g(\tilde{x}) - g(\tilde{z})) + (f(\tilde{x}) - f(\tilde{z}))g(\tilde{z}) \\ &= f(\tilde{x}) \cdot \tilde{s}_g(\tilde{x} - \tilde{z}) + \tilde{s}_f(\tilde{x} - \tilde{z}) \cdot g(\tilde{z}) \\ &= \tilde{s}(\tilde{x} - \tilde{z}) \end{aligned}$$

with $\tilde{s} = f(\tilde{x})\tilde{s}_g + g(\tilde{z})\tilde{s}_f \in a_r b_s + b_c a_s = d_s$.

For $\circ = /$ we have

$$\begin{aligned} h(\tilde{x}) - h(\tilde{z}) &= (f(\tilde{x}) - f(\tilde{z}) - h(\tilde{z})(g(\tilde{x}) - g(\tilde{z})))/g(\tilde{x}) \\ &= (\tilde{s}_f(\tilde{x} - \tilde{z}) - h(\tilde{z})\tilde{s}_g(\tilde{x} - \tilde{z}))/g(\tilde{x}) \\ &= \tilde{s}(\tilde{x} - \tilde{z}) \end{aligned}$$

with $\tilde{s} = (\tilde{s}_f - h(\tilde{z})\tilde{s}_g)/g(\tilde{x}) \in (a_s - d_c b_s)/b_r = d_s$.

Hence, in all cases, $h \in d$. Since the axiom for constants is trivially satisfied, the result follows. □

To apply this inclusion algebra for the computation of linear enclosures for specific functions we note that the linear function $h: \mathbb{R}^n \to \mathbb{R}$ defined by $h(\xi) = p + q\xi$ $(p \in \mathbb{R}, q \in \mathbb{R}^{1\times n})$ is covered by $(a_c, a_r, a_s) = (p + qz, p + qx, q)$. Hence we may recursively compute a linear enclosure for any function f defined by

$$f(\xi) := g(p^1 + q^1\xi, \ldots, p^k + q^k\xi), \tag{21}$$

where $p^i \in \mathbb{R}$, $q^i \in \mathbb{R}^{1\times n}$, and g is an arithmetical expression in k variables.

If we write

$$(f(z), f(x), f[z, x]) := g(a^1, \ldots, a^k), \tag{22}$$

where

$$(a^l_c, a^l_r, a^l_s) = (p^l + q^l z, p^l + q^l x, q^l) \quad (l = 1, \ldots, k).$$

then this defines an inclusion isotone interval extension of f and a function $f[\cdot, \cdot]: \mathbb{IR}^n \times \mathbb{IR}^n \to (\mathbb{IR}^*)^{1\times n}$ satisfying the slope property

$$f(\tilde{x}) - f(\tilde{z}) = \tilde{s}(\tilde{x} - \tilde{z}) \quad \text{with } \tilde{s} \in f[z, x] \tag{23}$$

for all $\tilde{x} \in x$, $\tilde{z} \in z$. Moreover, when f is Lipschitz at x^0, Corollary 2.1.2 implies the linear approximation property

$$\operatorname{rad}(f(x)) = O(\operatorname{rad}(x)) \qquad \text{for all } x \subseteq x^0, \tag{24}$$

$$\operatorname{rad} f[z, x] = O(\operatorname{rad}(z) + \operatorname{rad}(x)) \quad \text{for all } x, z \subseteq x^0. \tag{25}$$

Remarks (i) The definitions of the operations can be sharpened by redefining

$$d_r := a_r \circ b_r \cap (d_c + d_s(x - z)). \tag{26}$$

With the interpretation $\in_{x,z}$ the resulting modified $\mathbb{IR}^*$-algebra LI_n again is an inclusion algebra over $\mathbb{R}^n$. When $\operatorname{rad}(z) = 0$ then computations in LI_n generally give sharper enclosures than computations in L_n at the expense of some extra arithmetical work and explicit dependence on the interpretation. The definition (22) now only makes sense when $\operatorname{rad}(z) = 0$ (otherwise center

a_c and range a_r define *different* interval extensions of f). Since (26) is a centered form, Corollary 2.3.4 implies that $f(x)$, computed in LI_n with $\mathrm{rad}(z) = 0$, has the quadratic approximation property

$$\mathrm{rad}(f(x)) - \mathrm{rad}(f^*(x)) = O(\|\mathrm{rad}(x)\|^2) \quad \text{for } z \in x \subseteq x^0 \tag{27}$$

when f is Lipschitz at x^0. The linear approximation property (25) for the slope remains valid without change.

(ii) Note that the definition of multiplication is not commutative; the formula for d_s could be replaced by

$$d_s = a_c b_s + b_r a_s.$$

However, it is not allowed to take the intersection of both formulae for d_s, as the example $f(\xi) = \xi_1 \xi_2$ shows.

2.3.9 Proposition The following definitions provide consistent extensions of elementary functions for the inclusion algebras L_n and LI_n. For $a \in L_n$ (or LI_n) and $\varphi \in \Phi$, let $\varphi(a) = b$ be the element $b \in L_n$ (or LI_n) defined by

$$b_c := \varphi(a_c), \quad b_r := \varphi(a_r),$$

and

$$b_s := (a_c + a_r)a_s \quad \text{if } \varphi = \mathrm{sqr}, \tag{28}$$

$$b_s := a_s/(b_c + b_r) \quad \text{if } \varphi = \mathrm{sqrt}, \tag{29}$$

$$b_s := \varphi[a_c, a_r]a_s \quad \text{otherwise}, \tag{30}$$

where $\varphi[\cdot,\cdot]$ is a weak interval extension for the slope of φ. (Here multiplication and division of a vector by a scalar is interpreted componentwise.)

Proof Suppose that $f \in_{x,z} a$ and $g = \varphi(f)$. Then for $\tilde{x} \in x$ and $\tilde{z} \in z$ we have $f(\tilde{x}) - f(\tilde{z}) = \tilde{s}_f(\tilde{x} - \tilde{z})$ with $\tilde{s}_f \in a_s$. Hence

$$\begin{aligned} g(\tilde{x}) = \varphi(f(\tilde{x})) &= \varphi(f(\tilde{z})) + \varphi[f(\tilde{z}), f(\tilde{x})](f(\tilde{x}) - f(\tilde{z})) \\ &= g(\tilde{z}) + \tilde{s}(\tilde{x} - \tilde{z}) \end{aligned}$$

with $\tilde{s} = \varphi[f(\tilde{z}), f(\tilde{x})]\tilde{s}_f \in \varphi[a_c, a_r]a_s = b_s$ if $\varphi \neq \mathrm{sqr}, \mathrm{sqrt}$. In the two remaining cases we have

$$\varphi[\xi_1, \xi_2] = \xi_1 + \xi_2 \qquad \text{for } \varphi = \mathrm{sqr},$$

$$\varphi[\xi_1, \xi_2] = 1/(\varphi(\xi_1) + \varphi(\xi_2)) \quad \text{for } \varphi = \mathrm{sqrt},$$

so that again $\tilde{s} \in b_s$. Hence, in all cases, $g \in_{x,z} b$; i.e., we have found consistent extensions. □

It remains to find weak interval extensions for the slopes of elementary functions. If φ is continuously differentiable then the mean value theorem implies that

$$\varphi[z, x] := \varphi'(\square(x \cup z)) \tag{31}$$

is such an extension. In particular, for the interpretations $\in_{x,z}$ with $z \subseteq x$ we may assume, without loss of generality, that $a_c \subseteq a_r$; then we get the following simple formulae for b_s:

$$b_s = a_s/a_r \qquad \text{if } \varphi = \ln, \tag{32a}$$
$$b_s = b_r a_s \qquad \text{if } \varphi = \exp, \tag{32b}$$
$$b_s = \cos(a_r)a_s \qquad \text{if } \varphi = \sin, \tag{32c}$$
$$b_s = -\sin(a_r)a_s \qquad \text{if } \varphi = \cos, \tag{32d}$$
$$b_s = a_s/\cos(a_r)^2 \qquad \text{if } \varphi = \tan, \tag{32e}$$
$$b_s = \cosh(a_r)a_s \qquad \text{if } \varphi = \sinh, \tag{32f}$$
$$b_s = \sinh(a_r)a_s \qquad \text{if } \varphi = \cosh, \tag{32g}$$
$$b_s = a_s/\cosh(a_r)^2 \qquad \text{if } \varphi = \tanh. \tag{32h}$$

For the power, we may define $b = a^n$ using

$$b_s = \begin{cases} (n \cdot b_r/a_r)a_s & \text{if } n \le 1, \\ (n \cdot a_r^{n-1})a_s & \text{if } n > 1. \end{cases} \tag{33}$$

For the absolute value we may define $b = \text{abs}(a)$ using

$$b_s = \begin{cases} a_s & \text{if } a_r > 0, \\ -a_s & \text{if } a_r < 0, \\ [-|a_s|, |a_s|] & \text{otherwise.} \end{cases} \tag{34}$$

(A slightly sharper definition is possible if $0 \in \{\underline{a}_r, \overline{a}_r\}$, but (34) is optimal if we insist on inclusion isotonicity.)

The formula (31) may be considered as somewhat crude. It can be replaced by the following formula which follows from the Taylor series with error term:

$$\varphi[z, x] := \varphi'(z) + \frac{1}{2}\varphi''(\square(x \cup z))(x - z). \tag{35}$$

It is also possible to intersect (31) and (35). (Note, however, that there is no advantage in using an intersection formula corresponding to (26) for elementary functions.)

If we use L_n with the interpretation $\in = \in_{x,x}$ (i.e. $z = x$) then $f \in a$ implies the *interval Lipschitz condition*

$$f(\tilde{x}) - f(\tilde{z}) \in a_s(\tilde{x} - \tilde{z}) \quad \text{for all } \tilde{x}, \tilde{z} \in x.$$

In this special case the computations simplify since we may identify a_c and a_r. Hence we get:

2.3.10 Theorem Let LC_n be the $\mathbb{IR}^*$-algebra whose elements are the pairs $a = (a_r, a_s) \in \mathbb{IR}^* \times (\mathbb{IR}^*)^{1\times n}$, with constants $\alpha 1 = (\alpha, 0)$ for $\alpha \in \mathbb{IR}^n$, and operations defined by $a \circ b = d$, where

$$d_r := a_r \circ b_r$$

and

$$\begin{aligned} d_s &:= a_s + b_s && \text{if } \circ = +, \\ d_s &:= a_s - b_s && \text{if } \circ = -, \\ d_s &:= a_r b_s + b_r a_s && \text{if } \circ = *, \\ d_s &:= (a_s - d_r b_s)/b_r && \text{if } \circ = /. \end{aligned}$$

For $x \in \mathbb{IR}^n$, let $\in_x$ be the interpretation defined by $f \in_x a$ if, for all $\tilde{x}, \tilde{z} \in x$, we have

$$f(\tilde{x}) \in a_r \quad \text{and} \quad f(\tilde{x}) - f(\tilde{z}) = \tilde{s}(\tilde{x} - \tilde{z}) \quad \text{for some } \tilde{s} \in a_s. \tag{36}$$

Then, for each interpretation $\in_x$ ($x \in \mathbb{IR}^n$), LC_n is an inclusion algebra over $\mathbb{R}^n$. Moreover, consistent extensions of elementary functions φ are given by defining for $a \in LC_n$ the image $\varphi(a) = b$ by

$$b_r := \varphi(a_r)$$

and

$$\begin{aligned} b_s &:= 2a_r a_s && \text{if } \varphi = \text{sqr}, \\ b_s &:= a_s/(2b_r) && \text{if } \varphi = \text{sqrt}, \end{aligned}$$

together with (32)–(34) for other elementary functions. □

It is not difficult to see that LC_n also provides enclosures for the gradients of differentiable functions given by arithmetical expressions:

2.3.11 Corollary LC_n is an inclusion algebra over $\mathbb{R}^n$ for the interpretation $\in_x'$ defined for $x \in \mathbb{IR}^n$ by

$$f \in_x' a :\Leftrightarrow f(\tilde{x}) \in a_r, f'(\tilde{x}) \in a_s \quad \text{for all } \tilde{x} \in x.$$

(Here the existence of $(\partial/\partial x_i) f(\tilde{x})$ for $\tilde{x} \in x$ is required for all indices i such that $(a_s)_i \neq \text{NaN}$.) Moreover, the extensions of the elementary functions defined above are also consistent for $\in_x'$, except that, in (34), $[-|a_s|, |a_s|]$ must be replaced by a vector of NaNs.

Proof Take the limit $\tilde{z} \to \tilde{x}$ in $f(\tilde{x}) - f(\tilde{z}) \in a_s(\tilde{x} - \tilde{z})$. □

As before, we may recursively compute interval Lipschitz constants and gradients for any function f defined by (21) above. If we write

$$(f(x), f'(x)) := g(a^1, \ldots, a^k), \tag{37}$$

where

$$(a^l_r, a^l_s) = (p^l + q^l x, q^l) \quad (l = 1, \ldots, k),$$

then this defines an inclusion isotone interval extension of f and a function $f': \mathbb{IR}^n \to (\mathbb{IR}^*)^{1\times n}$ satisfying the Lipschitz property

$$f(\tilde{x}) - f(\tilde{z}) = \tilde{s}(\tilde{x} - \tilde{z}) \quad \text{with } \tilde{s} \in f'(x) \tag{38}$$

for all $\tilde{x}, \tilde{z} \in x$. Moreover, if f is continuously differentiable in x^0 then this function is an interval extension of the derivative of f in x^0. Again we have the linear approximation property

$$\begin{aligned} \operatorname{rad}(f(x)) &= O(\operatorname{rad}(x)) \quad \text{for all } x \subseteq x^0, \\ \operatorname{rad}(f'(x)) &= O(\operatorname{rad}(x)) \quad \text{for all } x \subseteq x^0, \end{aligned}$$

when f is Lipschitz at x^0. Note that, in particular, the choice

$$\begin{aligned} p^l &= 0, & q^l &= e^{(l)} \quad \text{for } l = 1, \ldots, n, \\ p^l &= y^{l-n}, & q^l &= 0 \quad \text{for } l = n+1, \ldots, k \end{aligned}$$

yields enclosures for the partial derivatives $\partial_1 g(\tilde{x}, \tilde{y})$ with $\tilde{x} \in x$, $\tilde{y} \in y$.

We end this section by relating the radii of $f[z, x]$ and $f'(x) = f[x, x]$.

2.3.12 Proposition Let f, defined by (21), be Lipschitz at $x^0 \in \mathbb{IR}^n$. Then the slope function derived from the inclusion algebra L_n with consistent extensions derived from (35) satisfies

$$\operatorname{rad}(f[\tilde{z}, x]) = \frac{1}{2}\operatorname{rad}(f[x, x]) + O(\|\operatorname{rad}(x)\|^2)$$

whenever $\tilde{z} \in x \subseteq x^0$.

Proof By induction on the length of the expression of g in (21). The details are left to the reader. □

2.4 Interpolation forms

In this section we improve the methods of the previous section in the case of functions in a single variable. In particular, we derive the parabolic boundary value form and show that it has the cubic approximation property; often it even gives the range without overestimation.

The asymptotic enclosure properties of centered forms were based on the decomposition (2.3.6). In the case of a single variable, this is just Newton's interpolation formula

$$f(\tilde{x}) = f(\tilde{z}) + f[\tilde{z}, \tilde{x}](\tilde{x}, \tilde{z}).$$

This interpretation suggests that perhaps better enclosures can be obtained by starting with Newton's error formula

$$f(\tilde{x}) = p_k(\tilde{x}) + f[\tilde{z}^0, \ldots, \tilde{z}^k, \tilde{x}](\tilde{x} - \tilde{z}^0) \cdots (\tilde{x} - \tilde{z}^k)$$

for the interpolation polynomial p_k of degree $\leq k$ which interpolates f at $k+1$ points $\tilde{z}^0, \ldots, \tilde{z}^k$. Let us write the remainder as $\tilde{c}r_k(\tilde{x})$, where $\tilde{c} = f[\tilde{z}^0, \ldots, \tilde{z}^k, \tilde{x}]$ and $r_k(\tilde{x}) = (\tilde{x} - \tilde{z}^0) \cdots (\tilde{x} - \tilde{z}^k)$. If we know an enclosure c for the coefficients $\tilde{c}$ and bounds for the range of $p_k(\tilde{x}) + \tilde{c}r_k(\tilde{x})$ for $\tilde{x} \in x$ and $\tilde{c} \in c$, then we get an enclosure for $f^*(x)$, a so-called interpolation form for the range. For $k = 0$ this just yields the slope form. For $k > 0$ we get enclosures with superior asymptotic properties: for sufficiently narrow intervals, they overestimate the range only by a term of order $k+2$ in the radius of x.

We now discuss a general setting for an interpolation form. Given $f: D \subseteq \mathbb{R} \to \mathbb{R}$, we require an approximating function $p: \mathbb{R} \to \mathbb{R}$ and a suitable error function $e: \mathbb{R} \to \mathbb{R}$. It is assumed that, for a given interval $x \in \mathbb{I}D$, there is an interval $c \in \mathbb{IR}$ such that, for all $\tilde{x} \in x$, we have

$$f(\tilde{x}) = p(\tilde{x}) + \tilde{c}e(\tilde{x}) \quad \text{for some } \tilde{c} \in c. \tag{1}$$

Clearly, p interpolates f at every point $\tilde{z} \in x$ with $e(\tilde{z}) = 0$. (Although we do not assume this property, the useful error functions indeed have one or several zeros in x.)

The *interpolation form* corresponding to the decomposition (1) is defined as

$$f_{\mathrm{p}}(x) := \square(p^*(x, \underline{c}) \cup p^*(x, \overline{c})), \tag{2}$$

where $p^*(x, \tilde{c})$ is the range of

$$p(\tilde{x}, \tilde{c}) := p(\tilde{x}) + \tilde{c}e(\tilde{x}) \tag{3}$$

over $\tilde{x} \in x$.

2.4.1 Theorem If (1) holds for all $\tilde{x} \in x$ then

$$f^*(x) \subseteq f_{\mathrm{p}}(x) \tag{4}$$

and

$$q(f_{\mathrm{p}}(x), f^*(x)) \leq 2 \cdot \mathrm{rad}(c)|e^*(x)|. \tag{5}$$

Moreover, if $e^*(x) \leq 0$ then

$$f_{\mathrm{p}}(x) = [\inf p^*(x, \overline{c}), \sup p^*(x, \underline{c})]. \tag{6}$$

Proof By (1), $f(\tilde{x})$ can be written as $p(\tilde{x}, \tilde{c})$ for some $\tilde{c} \in c$. Since $p(\tilde{x}, \tilde{c})$ is linear in $\tilde{c}$, $p(\tilde{x}, \cdot)$ takes its maximum in an endpoint of c, so that $f(\tilde{x}) \leq p(\tilde{x}, \check{c})$ for some $\check{c} \in \{\underline{c}, \overline{c}\}$. Hence $f(\tilde{x}) \leq \sup p^*(x, \check{c}) \leq \sup f_{\mathrm{p}}(x)$ by definition of $f_{\mathrm{p}}(x)$, and by a similar argument $f(\tilde{x}) \geq \inf f_{\mathrm{p}}(x)$. This shows (4). In the special case where $e^*(x) \leq 0$, the function $p(\tilde{x}, \tilde{c})$ is decreasing in each component of $\tilde{c}$ so that $p(\tilde{x}, \overline{c}) \leq f(\tilde{x}) \leq p(\tilde{x}, \underline{c})$ which yields (6).

To prove (5) we note that $\sup f_{\mathrm{p}}(x) = \sup p^*(x, c^*)$ for some $c^* \in \{\underline{c}, \overline{c}\}$. Hence $\sup f_{\mathrm{p}}(x) = p(x^*, c^*)$ for some $x^* \in x$. On the other hand, $\sup f^*(x) \geq f(x^*) = p(x^*, \tilde{c})$ for some $\tilde{c} \in c$; therefore

$$\begin{aligned}\sup(f_{\mathrm{p}}(x)) - \sup(f^*(x)) &\leq p(x^*, c^*) - p(x^*, \tilde{c})\\ &= (c^* - \tilde{c})e(x^*)\\ &\leq 2 \cdot \mathrm{rad}(c)|e^*(x)|.\end{aligned}$$

By a similar argument,

$$\inf(f^*(x)) - \inf(f_{\mathrm{p}}(x)) \leq 2 \cdot \mathrm{rad}(c)|e^*(x)|,$$

and (5) follows. □

We now restrict ourselves to the special case where $p(\xi)$ is an interpolation polynomial. Let $f\colon D \subseteq \mathbb{R} \to \mathbb{R}$ be $(k+1)$-times continuously differentiable, $x \in \mathbb{I}D$, and $\tilde{z}^0, \ldots, \tilde{z}^k \in x$. Let

$$e(\xi) := (\xi - \tilde{z}^0) \cdot \ldots \cdot (\xi - \tilde{z}^k),$$

and let $p(\xi)$ be the unique polynomial of degree $\leq k$ such that the function $\gamma\colon D \backslash \{\tilde{z}^0, \ldots, \tilde{z}^k\} \to \mathbb{R}$ defined by

$$\gamma(\xi) := (f(\xi) - p(\xi))/e(\xi)$$

can be continuously extended to D. If the $\tilde{z}^i$ are distinct, $p(\xi)$ is the polynomial of degree $\leq k$ which interpolates f at $\tilde{z}^0, \ldots, \tilde{z}^k$; the general case corresponds to Hermite interpolation involving function and derivative values. If $\tilde{x} \in x$ then, by Newton's interpolation formula,

$$\begin{aligned}\gamma(\tilde{x}) &= f[\tilde{z}^0, \ldots, \tilde{z}^k, \tilde{x}]\\ &= \frac{1}{(k+1)!} f^{(k+1)}(\xi) \quad \text{for some } \xi \in x.\end{aligned} \tag{7}$$

This allows one to find an interval c such that

$$\gamma(\tilde{x}) \in c \quad \text{for all } \tilde{x} \in x; \tag{8}$$

for example, $c = f^{(k+1)}(x)/(k+1)!$ with an interval extension for $f^{(k+1)}$.

Now (8) implies (1) for all $\tilde{x} \in x$, so that the interpolation form (2) is defined. We call $f_p(x) = \square(p^*(x, \underline{c}) \cup p^*(x, \overline{c}))$ the *interpolation form determined by* $\tilde{z}^0, \ldots, \tilde{z}^k$. Since

$$|e(\tilde{x})| = \prod_{i=0}^{k} |\tilde{x} - \tilde{z}^i| \leq (2 \cdot \mathrm{rad}(x))^{k+1} \quad \text{for } \tilde{x} \in x,$$

and $\mathrm{rad}(c) = O(\mathrm{rad}(x))$ if the interval extension of $f^{(k+1)}$ is Lipschitz continuous, we see from (5) that

$$q(f_p(x), f^*(x)) = O(\mathrm{rad}(x)^{k+2}).$$

Thus the interpolation form determined by $k + 1$ points in x has approximation order $k + 2$. This yields enclosures for the range which are asymptotically arbitrarily good. However, in practice, the need to find the precise range of the polynomials $p(\xi, \underline{c})$ and $p(\xi, \overline{c})$ restricts k to a small integer.

We now give explicit formulae for the most important case when $k = 1$ and the endpoints of x serve as interpolation points. In this case $p(\xi, \tilde{c})$ is a parabola, and the resulting interpolation form will be called the *parabolic boundary value form*.

2.4.2 Theorem Let $f\colon D \subseteq \mathbb{R} \to \mathbb{R}$ be a function and $x \in \mathbb{I}D$. Suppose that $c \in \mathbb{I}\mathbb{R}$ satisfies

$$f[\underline{x}, \overline{x}, \tilde{x}] \in c \quad \text{for all } \tilde{x} \in \mathrm{int}(x). \tag{9}$$

Put

$$f^0 = \square\{f(\underline{x}), f(\overline{x})\},$$
$$s = f[\underline{x}, \overline{x}] = (f(\overline{x}) - f(\underline{x}))/(\overline{x} - \underline{x}),$$
$$\underline{d} = \underline{c} \cdot \mathrm{rad}(x) + \left|\frac{s}{2}\right|, \quad \overline{d} = \overline{c} \cdot \mathrm{rad}(x) - \left|\frac{s}{2}\right|.$$

Then the interval $f_p(x)$ defined by $f_p(x) = f(x)$ if x is thin, and otherwise by

$$\underline{f_p(x)} := \begin{cases} \underline{f}^0 & \text{if } \overline{d} \leq 0, \\ \underline{f}^0 - \overline{d}^2/\overline{c} & \text{otherwise;} \end{cases} \tag{10a}$$

$$\overline{f_p(x)} := \begin{cases} \overline{f}^0 & \text{if } \underline{d} \geq 0, \\ \overline{f}^0 - \underline{d}^2/\underline{c} & \text{otherwise,} \end{cases} \tag{10b}$$

satisfies

$$f^*(x) \subseteq f_p(x) \subseteq f^*(x) + 2 \cdot \mathrm{rad}(c)\mathrm{rad}(x)^2[-1, 1]. \tag{11}$$

Proof Without loss of generality assume that x is thick. Let $p(\xi) := f(\underline{x}) + f[\underline{x}, \overline{x}](\xi - \underline{x})$ and $e(\xi) := (\xi - \underline{x})(\xi - \overline{x})$. Then (9) implies (1). Since

$$e^*(x) = [-\text{rad}(x)^2, 0] \le 0,$$

the interpolation form takes the simple form (6), and the overestimation bound (5) yields (11). Hence it only remains to check the closed expression (10) for (6). Now $p(\tilde{x}, \overline{c})$ takes its infimum at an endpoint when $\overline{d} \le 0$; in this case

$$\begin{aligned}\inf p^*(x, \overline{c}) &= \min\{p(\underline{x}, \overline{c}), p(\overline{x}, \overline{c})\}\\ &= \min\{f(\underline{x}), f(\overline{x})\} = \underline{f}^0.\end{aligned}$$

And if $\overline{d} > 0$ then $p(\tilde{x}, \overline{c})$ takes its infimum at $x^* = \check{x} - s/(2\overline{c})$. Hence

$$\begin{aligned}\inf p^*(x, \overline{c}) &= p(x^*, \overline{c}) = p(x^*) + \overline{c}(x^* - \underline{x})(x^* - \overline{x})\\ &= \frac{1}{2}(f(\underline{x}) + f(\overline{x})) - \frac{s^2}{2\overline{c}} + \overline{c}\left(\text{rad}(x) - \frac{s}{2\overline{c}}\right)\left(-\text{rad}(x) - \frac{s}{2\overline{c}}\right)\\ &= \frac{1}{2}(f(\underline{x}) + f(\overline{x})) - \overline{c}\,\text{rad}(x)^2 - \frac{s^2}{4\overline{c}}\\ &= \frac{1}{2}(f(\underline{x}) + f(\overline{x})) - |s|\text{rad}(x) - \frac{\overline{d}^2}{\overline{c}}\\ &= \underline{f}^0 - \overline{d}^2/\overline{c}.\end{aligned}$$

This proves (10a), and (10b) follows in the same way by applying (10a) to $-f$ in place of f. □

Remarks

(i) In the presence of rounding errors one has to take care that the computed $f_p(x)$ contains the true $f_p(x)$. In particular, this requires that one has to compute lower bounds for $\underline{d}$ and $|f[\underline{x}, \overline{x}]|$ and an upper bound for $\overline{d}$. Thus $|s/2|$ is computed from interval enclosures for $f(\overline{x})$ and $f(\underline{x})$ by interval operations as

$$\left|\frac{s}{2}\right| = \text{mig}\left(\frac{f(\overline{x}) - f(\underline{x})}{2(\overline{x} - \underline{x})}\right),$$

and the remaining operations must be performed with directed rounding in the right direction.

(ii) The parabolae $p(\xi, \overline{c})$ and $p(\xi, \underline{c})$ define lower and upper bounds for f in x and thus provide a pointwise enclosure

$$f(\tilde{x}) \in [p(\xi, \overline{c}), p(\xi, \underline{c})] \quad \text{for } \tilde{x} \in x. \tag{12}$$

In geometric terms, the graph of f is enclosed in a sickle-shaped domain if $0 \notin c$ and in a lens-shaped domain if $0 \in c$; see Fig. 2.2.

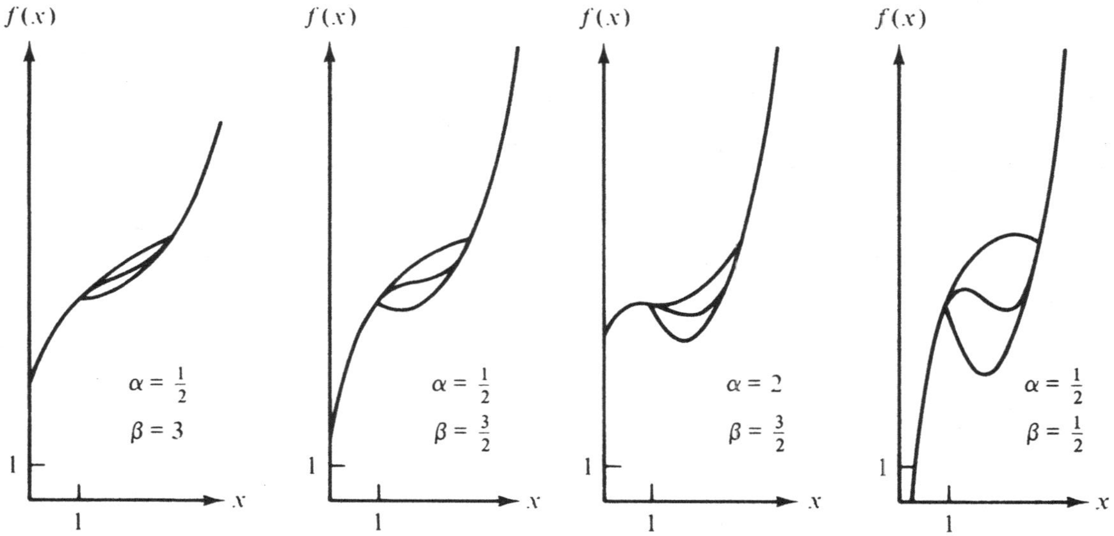

Fig. 2.2. Parabolic enclosures for $f(\xi) = \xi + 5 + \alpha(\xi - \beta)(\xi - 1)(\xi - 3)$ in $x = [1, 3]$; $c = [\alpha(1 - \beta), \alpha(3 - \beta)]$.

(iii) If f is almost linear ($|c|$ small) or sufficiently steep ($|f[\underline{x}, \overline{x}]|$ large), more precisely, if

$$2|c|\mathrm{rad}(x) \leq |f[\underline{x}, \overline{x}]| \tag{13}$$

then both enclosing parabolae have their extrema at the endpoints so that

$$f_{\mathrm{p}}(x) = f^0 = f^*(x);$$

i.e. in this case the range is computed without overestimation.

(iv) If f is twice continuously differentiable in D, and f'' is an interval extension of the second derivative of f then (7) shows that (9) holds with $c = \frac{1}{2}f''(x)$. If f'' is Lipschitz continuous then $\mathrm{rad}(c) = \mathrm{O}(\mathrm{rad}(x))$, and (10) yields

$$0 \leq \mathrm{rad}(f_{\mathrm{p}}(x)) - \mathrm{rad}(f^*(x)) \leq 2 \cdot \mathrm{rad}(c)\mathrm{rad}(x)^2 = \mathrm{O}(\mathrm{rad}(x)^3). \tag{14}$$

This shows that the parabolic boundary value form has *cubic* approximation order. The same holds if c is computed as $c = f[\underline{x}, \overline{x}, x]$ from a Lipschitz continuous interval extension of the second divided difference.

(v) If f is twice continuously differentiable and strictly monotonic in $x^0 \in \mathbb{I}D$ then there are numbers $M, m > 0$ such that $|f''(\xi)| \leq M$, $|f'(\xi)| \geq m$ for all $\xi \in x^0$. Then (9) holds for all $x \subseteq x^0$ with some $c \subseteq [-M/2, M/2]$, and, by the mean value theorem, (13) holds whenever $\mathrm{rad}(x) \leq m/M$. Hence the parabolic boundary value form *gives the range without overestimation for all sufficiently narrow subintervals* of any interval on which f is strictly monotonic and twice continuously differentiable. On the other hand, over large intervals the results are often inferior to simple interval evaluation. For example, if $f(\xi) = \xi^3 - \xi$, $x = [-10, 10]$, then $f^*(x) = [-990, 990] = f^0$ and $f(x) = [-1010, 1010]$, but $f_{\mathrm{p}}(x) = [-1245.025, 1245.025]$. (Here c is calculated as $f[\underline{x}, \overline{x}, x]$ by an extended Horner scheme as described in the next section.)

(vi) The parabolic boundary value form applies to almost arbitrary functions (provided an interval c satisfying (9) can be found); f need not even be continuous. For example, if

$$f(\xi) = \begin{cases} \xi & \text{for } \xi < 0, \\ \xi + \varepsilon & \text{for } \xi \geq 0 \end{cases}$$

(where $0 \leq \varepsilon \leq 2$), and $x = [-1, 1]$ then

$$f[\underline{x}, \overline{x}, x] = \begin{cases} \frac{1}{2}\varepsilon/(1 - \tilde{x}) & \text{if } \tilde{x} < 0, \\ \frac{1}{2}\varepsilon/(-1 - \tilde{x}) & \text{if } \tilde{x} \geq 0; \end{cases}$$

hence (9) holds with $c = [-\varepsilon/2, \varepsilon/2]$. Since $s = 1 + \varepsilon/2$ we have $\underline{d} = (2 - \varepsilon)/4 \geq 0$ and $\overline{d} = (\varepsilon - 2)/4 \leq 0$, so that $f_p(x) = f^0 = f^*(x)$.

(vii) Along the same lines as above, another form with cubic approximation order can be constructed using a Taylor series

$$f(\tilde{x}) = f(\check{x}) + f'(\check{x})(\tilde{x} - \check{x}) + \tfrac{1}{2}f''(\xi)(\tilde{x} - \check{x})^2,$$

namely the interpolation form with respect to $\tilde{z}^0 = \tilde{z}^1 = \check{x}$. Unless f is a linear or quadratic polynomial, such a form will overestimate the range even for monotonic functions and therefore must be considered as inferior to the boundary value form. The details are left to the reader.

2.5 Appendix. The extended Horner scheme

Function values and divided differences of a polynomial

$$f(\xi) = \sum_{i=0}^{n} a_i \xi^i$$

are best computed with the *extended*, $(m+1)$-*fold*, *Horner scheme* defined as

If $m > n$ then $a := 0$ else $a := a_n$;
for $l := 0$ to m do $f_l := a$;
If $m > n$ then $f_n := a_n$;
for $k := n - 1$ down to 0 do
  $\{f_{-1} := a_k$;
  for $l := 0$ to $\min(m, k)$ do $f_l := f_l * x_l + f_{l-1}\}$;

2.5.1 Theorem For $x_0, \ldots, x_m \in \mathbb{IR}$, the extended Horner scheme computes an interval extension of the function value $f_0 = f(x_0)$ and the divided differences $f_l = f[x_0, \ldots, x_l]$ $(l = 1, \ldots, m)$.

Proof We first observe that the polynomials

$$f[\tilde{x}_0, \ldots, \tilde{x}_l, \xi] =: \sum_{i=0}^{n} a_{il} \xi^i \tag{1}$$

have degree $n - l$ so that

$$a_{il} = 0 \quad \text{for } i > n - l. \tag{2}$$

If we insert (1) into the relation

$$f[\tilde{x}_0, \ldots, \tilde{x}_{l-1}, \xi] = f[\tilde{x}_0, \ldots, \tilde{x}_{l-1}, \tilde{x}_l] + f[\tilde{x}_0, \ldots, \tilde{x}_l, \xi](\xi - x_l)$$

and use (2) for $i = n$ we find

$$\sum_{i=0}^{n} a_{il-1}\xi^i = f[\tilde{x}_0, \ldots, \tilde{x}_l] + \sum_{i=0}^{n-1} a_{il}\xi^{i+1} - \sum_{i=0}^{n} a_{il}\tilde{x}_l\xi^i.$$

Comparing the coefficients of ξ^i gives

$$a_{il-1} = a_{i-1l} - a_{il}\tilde{x}_l \quad (i = 0, \ldots, n), \tag{3}$$

where

$$a_{-1l} = f[\tilde{x}_0, \ldots, \tilde{x}_l]. \tag{4}$$

Now (2), (3) and $a_{i0} = a_i$ imply for $i = k - l$ that $f_{kl} := a_{k-ll}$ satisfy

$$f_{nl} = a_n,$$
$$f_{k-1l} = f_{kl}\tilde{x}_l + f_{k-1l-1} \quad (0 \le l \le k \le n).$$

A comparison with the definition of the extended Horner scheme now readily shows that the intervals $f_0, \ldots, f_m$ computed by the recursion satisfy

$$f[\tilde{x}_0, \ldots, \tilde{x}_l] \in f_l \quad \text{whenever } \tilde{x}_i \in x_i \quad (i = 0, \ldots, m). \qquad \square$$

In particular, the extended Horner scheme with $m = 2$ computes in $O(n)$ operations the values $f(\underline{x})$, $s = f[\underline{x}, \overline{x}]$, $c = f[\underline{x}, \overline{x}, x]$, and $f(\overline{x}) = f(\underline{x}) + s(\overline{x} - \underline{x})$ needed for the evaluation of the parabolic boundary value form.

It is possible to generalize the extended Horner scheme to an inclusion algebra which recursively computes the information needed for the parabolic boundary value form for arbitrary arithmetical expressions. The details are left to the reader.

Remarks to Chapter 2

To 2.1 The concept of a Lipschitz expression is new and was chosen to simplify the hypothesis of the statements of this section. Alefeld & Herzberger (1983) prove the conclusion of Theorem 2.1.1 under a different hypothesis. Theorem 2.1.5 is due to Moore (1959) and was rediscovered by Rump (1980).

The subdivision method for the determination of the range is due to Moore (1966) who also suggested numerous improvements. The range determination is a problem of (bound-constrained) *global optimization*, for which a diversity of methods has been devised; see the books by Ratschek & Rokne (1984, 1988). The best methods combine bisection methods with suitable exclusion tests and monotony considerations, and switch locally to an interval method for enclosing a zero of the gradient. Thus the methods resemble bisection methods for zeros of nonlinear equations, treated in

Section 5.6. For details see Baumann (1986), Hansen (1980, 1988), Mohd (1986), and Ratschek & Rokne (1987, 1988), and the introductory paper by Nickel (1986b) which contains a large bibliography. Bisection methods extend to the solution of general *constrained global optimization* problems; see Hansen & Sengupta (1980, 1981), Ichida & Fujii (1979), Fujii, Ichida & Ozasa (1986), Hansen & Walster (1982, 1987, 1990a,b), Mfayokurera (1989), Walster, Hansen & Sengupta (1985). (For deterministic non-interval methods, see, e.g., Pardalos & Rosen, 1987.) For the range of polynomials see also Garloff (1986b), Grassmann & Rokne (1970) and Rokne (1986). *Convex optimization* problems are treated in Bauch *et al.* (1987).

To 2.2 Automatic differentiation has been repeatedly rediscovered as an elegant method of obtaining analytic derivatives. A good account of the method and its implementation is given by Rall (1981, 1983, 1984a, 1986, 1987); see also Corliss (1988). For even more efficient recent methods see Baur & Strassen (1983) and Iri (1984).

The concept of inclusion algebra is new and was chosen to unify the similarities of automatic differentiation and other constructs used in interval analysis. A different attempt of unification has been given by Kaucher & Miranker (1984) with their concept of *functoids*. See also Stetter (1988). For further inclusion algebras see Eckmann & Wittwer (1985), Eiermann (1989a,b).

To 2.3 The concept of a centered form has been introduced by Moore (1966). Theorem 2.3.2 is due to Caprani & Madsen (1980), and Theorem 2.3.3 to Krawczyk & Neumaier (1986) (with a slightly weaker bound); computational aspects in the face of round-off are discussed in Rump (1990). The quadratic approximation property (Corollary 2.3.4) was already known (for historical details see the discussion in Ratschek & Rokne, 1984). Theorem 2.3.6 was found by E. Baumann (1988), and Remark 1 is also due to him (personal communication, 1988). The cut-off function was introduced by Cornelius (1981).

The rules for the inclusion algebra L_n for linear enclosures are taken from Krawczyk & Neumaier (1985), and the corresponding consistent extensions of elementary functions appear in Neumaier (1986a). Similar techniques have been suggested by Hansen (1975) and Lehmann (1985), and for polynomial expressions by Alefeld (1981a).

To 2.4 This section contains an improved version of an idea of Cornelius & Lohner (1984) who gave the first enclosure methods with higher than second order of approximation.

To 2.5 The special cases $m = 1$ and $x_1 = \cdots = x_m = x$ of the extended Horner scheme are treated in many numerical analysis books, usually in the context of finding zeros of polynomials.

For rational functions given as the quotient of two polynomials, Nickel (1968) derived a recursive method for the enclosure of arbitrary divided differences (in the context of error bounds for numerical quadrature). A generalization to multivariate polynomials of the extended Horner scheme is given in Alefeld (1981a).

3

Matrices and sublinear mappings

3.1 Basic facts

In this section we generalize the definitions and rules of interval arithmetic to the matrix case.

An $m \times n$ *interval matrix* (or simply *matrix*) is a rectangular array

$$A = (A_{ik}) = \begin{bmatrix} A_{11} & A_{12} & \cdots & A_{1n} \\ A_{21} & A_{22} & \cdots & A_{2n} \\ \cdot & \cdot & \cdot & \cdot \\ \cdot & \cdot & \cdot & \cdot \\ \cdot & \cdot & \cdot & \cdot \\ A_{m1} & A_{m2} & \cdots & A_{mn} \end{bmatrix}$$

of intervals $A_{ik} \in \mathbb{IR}$. We use the slightly nonstandard notation A_{ik} instead of a_{ik} for the entries of A. This saves us naming conventions when entries of several matrices are used, and is consistent with the formal definition of an $m \times n$ matrix as a mapping A: $\{1, \ldots, m\} \times \{1, \ldots, n\} \to \mathbb{R}$. The set of $m \times n$ interval matrices is denoted by $\mathbb{IR}^{m \times n}$. An interval matrix $A = (A_{ik})$ is interpreted as a set of real $m \times n$ matrices by the convention

$$A = \{\tilde{A} \in \mathbb{R}^{m \times n} \mid \tilde{A}_{ik} \in A_{ik} \quad \text{for } i = 1, \ldots, m; \quad k = 1, \ldots, n\}.$$

An $n \times 1$ interval matrix is just an interval vector, so that $\mathbb{IR}^{n \times 1} = \mathbb{IR}^n$, and a 1×1 interval matrix $A = (A_{11})$ is identified with the 'scalar' interval $A_{11} \in \mathbb{IR}$, so that $\mathbb{IR}^{1 \times 1} = \mathbb{IR}$. In analogy to the 1×1 case we relate certain real matrices to each interval matrix $A \in \mathbb{IR}^{m \times n}$, namely

$$\begin{aligned} \underline{A} = \inf(A) &:= (\underline{A}_{ik}) \quad \overline{A} = \sup(A) := (\overline{A}_{ik}), \\ \check{A} = \operatorname{mid}(A) &:= (\check{A}_{ik}), \qquad \operatorname{rad}(A) := (\operatorname{rad}(A_{ik})), \\ |A| &:= (|A_{ik}|). \end{aligned}$$

For square matrices $A \in \mathbb{IR}^{n \times n}$ we also define the matrix $\langle A \rangle$ with entries

$$\langle A\rangle_{ii} = \langle A_{ii}\rangle, \quad \langle A\rangle_{ik} = -|A_{ik}| \quad \text{for } i \neq k;$$

thus $\langle A\rangle$ has nonnegative diagonal elements and nonpositive off-diagonal elements, cf. Section 3.7. $\langle A\rangle$ is an extension of Ostrowski's comparison matrix of a real matrix to the interval case, and has important applications in the analysis of linear interval equations.

Matrices $A \in \mathbb{IR}^{m\times n}$ with $\text{rad}(A) = 0$ are called *thin*; we identify these matrices with the unique matrix they contain. Matrices with $\text{rad}(A) > 0$ are called *thick*; here, as always in the book, the comparison relations $>, <, \geq, \leq$ are understood componentwise. (In particular, $A > B$ is not equivalent with $A \geq B \neq A$.) We denote the identity matrix of any size by I.

With the partial order $\leq$, $\mathbb{IR}^{m\times n}$ is a locally complete partially ordered space, and an interval matrix $A \in \mathbb{IR}^{m\times n}$ can equivalently be described as the matrix interval

$$A = [\underline{A}, \bar{A}] = \{\tilde{A} \in \mathbb{R}^{m\times n} \quad | \quad \underline{A} \leq \tilde{A} \leq \bar{A}\}.$$

Moreover, if Σ is a bounded set of real $m \times n$ matrices then $\inf(\Sigma)$ and $\sup(\Sigma)$ exist, and the *hull* of Σ,

$$\square\Sigma := [\inf(\Sigma), \sup(\Sigma)]$$

is the tightest interval matrix enclosing Σ. Occasionally, we shall also need the *interior* of a matrix $A \in \mathbb{IR}^{m\times n}$,

$$\text{int}(A) := \{\tilde{A} \in A \quad | \quad \underline{A}_{ik} < \tilde{A}_{ik} < \bar{A}_{ik} \quad \text{whenever } \underline{A}_{ik} \neq \bar{A}_{ik}\}$$

and the boundary

$$\partial A := \{\tilde{A} \in A \quad | \quad \tilde{A}_{ik} = \underline{A}_{ik} \text{ or } \tilde{A}_{ik} = \bar{A}_{ik}\}$$

Note that the interior and boundary of a thin matrix A are A itself; this slightly nonstandard definition will be useful later. In general, $\text{int}(A)$ is not an interval matrix.

We now extend the definitions of addition, subtraction and multiplication to interval matrices. If $A, B \in \mathbb{IR}^{m\times n}$ then we define $A \pm B \in \mathbb{IR}^{m\times n}$ by

$$A + B := \square\{\tilde{A} + \tilde{B} \quad | \quad \tilde{A} \in A, \tilde{B} \in B\},$$
$$A - B := \square\{\tilde{A} - \tilde{B} \quad | \quad \tilde{A} \in A, \tilde{B} \in B\},$$

and if $A \in \mathbb{IR}^{m\times n}$, $B \in \mathbb{IR}^{n\times p}$ then we define $AB \in \mathbb{IR}^{m\times p}$ by

$$AB := \square\{\tilde{A}\tilde{B} \quad | \quad \tilde{A} \in A, \tilde{B} \in B\}.$$

We also define scalar multiplication of $a \in \mathbb{IR}$ and $A \in \mathbb{IR}^{n\times n}$ by

$$aA := \square\{\tilde{a}\tilde{A} \quad | \quad \tilde{a} \in a, \tilde{A} \in A\}.$$

3.1.1 Proposition Let $A, A', B, B', C \in \mathbb{IR}^{m\times n}$. Then

$$A + B = [\underline{A} + \underline{B}, \bar{A} + \bar{B}], \quad A - B = [\underline{A} - \bar{B}, \bar{A} - \underline{B}], \tag{1}$$

$$(A \pm B)_{ik} = A_{ik} \pm B_{ik}, \tag{2}$$

$$A + B = B + A, \quad (A + B) + C = A + (B + C), \tag{3}$$

$$A' \subseteq A, B' \subseteq B \Rightarrow A' \pm B' \subseteq A \pm B, \tag{4}$$

$$A \pm B = \{\tilde{A} \pm \tilde{B} \mid \tilde{A} \in A, \tilde{B} \in B\}. \tag{5}$$

Proof (1) immediately follows from the definition, (2) from (1), and (3), (4) and (5) from (2) using corresponding properties of scalar intervals. □

3.1.2 Proposition Let $A, A' \in \mathbb{IR}^{m\times n}$, $B, B' \in \mathbb{IR}^{n\times p}$. Then

$$(AB)_{ik} = \sum_{j=1}^{n} A_{ij}B_{jk}, \tag{6}$$

$$A' \subseteq A, B' \subseteq B \Rightarrow A'B' \subseteq AB, \tag{7}$$

$$A(B + B') \subseteq AB + AB', \tag{8}$$

with equality if A is thin or if $B, B' \geq 0$,

$$(A + A')B \subseteq AB + AB', \tag{9}$$

with equality if B is thin or if $A, A' \geq 0$.
Proof We have

$$\begin{aligned}(AB)_{ik} &= \square\{(\tilde{A}\tilde{B})_{ik} \mid \tilde{A} \in A, \tilde{B} \in B\} \\ &= \square\left\{\sum_{j=1}^{n} \tilde{A}_{ij}\tilde{B}_{jk} \mid \tilde{A} \in A, \tilde{B} \in B\right\} \\ &= \sum_{j=1}^{n} A_{ij}B_{jk}\end{aligned}$$

since $\Sigma\, \tilde{A}_{ij}\tilde{B}_{jk}$ is an arithmetical expression in which every variable occurs only once. This implies (6), and (7) immediately follows from the definition. Equation (8) holds since by the scalar subdistributive law,

$$\begin{aligned}(A(B \pm B'))_{ik} &= \sum_j A_{ij}(B \pm B')_{jk} = \sum_j A_{ij}(B_{jk} \pm B'_{jk}) \\ &\subseteq \sum_j (A_{ij}B_{jk} \pm A_{ij}B'_{jk}) = (AB)_{ik} \pm (AB')_{ik} \\ &= (AB \pm AB')_{ik},\end{aligned}$$

and equality holds in this argument if A is thin or if $B, B' \geq 0$. A similar argument shows (9). □

Similarly, scalar multiplication satisfies

$$(aA)_{ik} = aA_{ik}, \tag{6a}$$

$$a' \subseteq a, A' \subseteq A \Rightarrow a'A' \subseteq aA, \tag{7a}$$

$$a(B \pm B') \subseteq aB \pm aB', \tag{8a}$$

with equality for thin a, and

$$(a \pm a')B \subseteq aB \pm a'B, \tag{9a}$$

with equality if B is thin. The proofs are left to the reader.

Some other multiplicative properties of scalar intervals do not extend without change to the matrix case.

3.1.3 Example Let

$$A = \begin{pmatrix} 1 & 1 \\ 0 & 1 \end{pmatrix}, \quad B = \begin{pmatrix} 1 & 0 \\ -1 & 1 \end{pmatrix}, \quad C = \begin{pmatrix} [-1, 1] & 0 \\ 0 & [-1, 1] \end{pmatrix}, \quad x = \begin{pmatrix} [-1, 0] \\ [\ 1, 2] \end{pmatrix}.$$

Then

$$(AB)C = \begin{pmatrix} 0 & 1 \\ -1 & 1 \end{pmatrix} C = \begin{pmatrix} 0 & [-1, 1] \\ [-1, 1] & [-1, 1] \end{pmatrix},$$

$$A(BC) = A\begin{pmatrix} [-1, 1] & 0 \\ [-1, 1] & [-1, 1] \end{pmatrix} = \begin{pmatrix} [-2, 2] & [-1, 1] \\ [-1, 1] & [-1, 1] \end{pmatrix},$$

so that the associative law may fail, and

$$A([-1, 1]x) = A\begin{pmatrix} [-1, 1] \\ [-2, 2] \end{pmatrix} = \begin{pmatrix} [-3, 3] \\ [-2, 2] \end{pmatrix},$$

$$[-1, 1](Ax) = [-1, 1]\begin{pmatrix} [0, 2] \\ [1, 2] \end{pmatrix} = \begin{pmatrix} [-2, 2] \\ [-2, 2] \end{pmatrix},$$

so that we may have $A(ax) \neq a(Ax)$ for scalar intervals a. Moreover,

$$\begin{pmatrix} 0 \\ 2 \end{pmatrix} \in Ax = \begin{pmatrix} [0, 2] \\ [1, 2] \end{pmatrix},$$

but $\begin{pmatrix} 0 \\ 2 \end{pmatrix}$ is not of the form $\tilde{A}\tilde{x}$ with $\tilde{A} \in A$, $\tilde{x} \in x$; cf. Fig. 3.1. Hence, we may

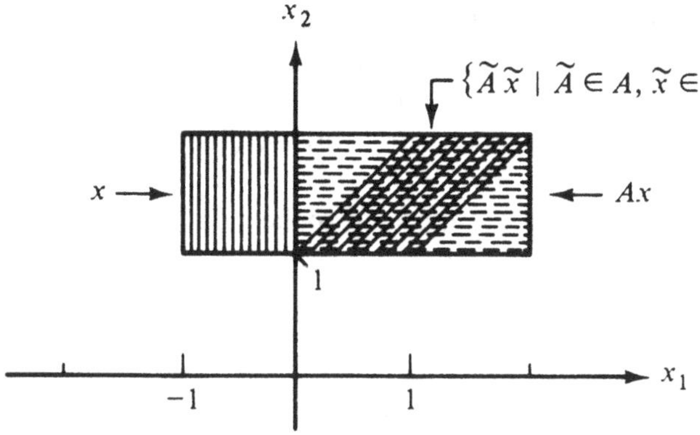

Fig. 3.1. A matrix–vector product.

have $Ax \neq \{\tilde{A}\tilde{x} \mid \tilde{A} \in A, \tilde{x} \in x\}$. However, we are able to prove some weaker results:

3.1.4 Proposition Let $A \in \mathbb{IR}^{m\times n}$, $B \in \mathbb{IR}^{n\times p}$, $C \in \mathbb{IR}^{p\times q}$. Then

$$A\tilde{x} = \{\tilde{A}\tilde{x} \mid \tilde{A} \in A\} \quad \text{for all } \tilde{x} \in \mathbb{R}^n, \tag{10}$$
$$\alpha A = \{\alpha\tilde{A} \mid \tilde{A} \in A\} \quad \text{for all } \alpha \in \mathbb{R}, \tag{10a}$$
$$(AB)C \subseteq A(BC) \quad \text{if } A \text{ is thin, or if } B, C \geq 0, \tag{11a}$$
$$A(BC) \subseteq (AB)C \quad \text{if } C \text{ is thin, or if } A, B \geq 0, \tag{11b}$$
$$A(BC) = (AB)C \quad \text{if } A \text{ and } C \text{ are thin, or if } A, B, C \geq 0, \tag{12}$$
$$A(\alpha B) = \alpha(AB) \quad \text{for all } \alpha \in \mathbb{R}. \tag{13}$$

Proof Let $\tilde{y} \in A\tilde{x}$. Then $\tilde{y}_i \in \Sigma_k A_{ik}\tilde{x}_k$, and, since the right-hand side is an arithmetical expression in which the A_{ik} occur only once, there are $\tilde{A}_{ik} \in A_{ik}$ $(k = 1, \ldots, n)$ such that $\tilde{y}_i = \Sigma \tilde{A}_{ik}\tilde{x}_k$. The matrix $\tilde{A} := (\tilde{A}_{ik})$ then satisfies $\tilde{y} = \tilde{A}\tilde{x}$. Hence, $A\tilde{x} \subseteq \{\tilde{A}\tilde{x} \mid \tilde{A} \in A\}$, and (10) holds by definition of the product. (10a) is immediate. Relation (11a) holds since

$$((AB)C)_{ik} = \sum_j (AB)_{ij}C_{jk} = \sum_j \Big(\sum_l A_{il}B_{lj}\Big)C_{jk}$$

$$\subseteq \sum_j \sum_l (A_{il}B_{lj})C_{jk} \overset{(*)}{=} \sum_j \sum_l A_{il}(B_{lj}C_{jk})$$

$$= \sum_l \sum_j A_{il}(B_{lj}C_{jk}) \overset{(**)}{=} \sum_l A_{il}\Big(\sum_j B_{lj}C_{jk}\Big)$$

$$= \sum_l A_{il}(BC)_{lk} = (A(BC))_{ik}.$$

Here the associative law for scalar interval multiplication was applied in (*), and the distributive law in (**), which holds in this case by the special assumptions (A_{il} thin, or $B_{lj}C_{jk} \geq 0$). Relations (11b) and (13) follow similarly, and (12) is a consequence of (11a) and (11b). □

Note that equation (10) has important applications, cf., e.g., Theorem 3.4.3.

3.1.5 Proposition Let $A \in \mathbb{IR}^{m\times n}$, $B \in \mathbb{IR}^{n\times p}$, $x \in \mathbb{IR}^n$, and suppose that $A \geq 0$. Then

$$Ax = [\underset{\sim}{A}\underline{x}, \tilde{A}\bar{x}] \text{ for suitable } \underset{\sim}{A}, \tilde{A} \in A. \tag{14}$$

In particular, $A \geq 0$ implies

$$AB = [\underline{A}\underline{B}, \bar{A}\bar{B}] \quad \text{if } B \geq 0, \tag{15a}$$

$$AB = [\bar{A}\underline{B}, \bar{A}\bar{B}] \quad \text{if } B \ni 0, \tag{15b}$$

$$AB = [\bar{A}\underline{B}, \underline{A}\bar{B}] \quad \text{if } B \leq 0, \tag{15c}$$

$$AB = [A\underline{B}, A\bar{B}] \quad \text{if } A \text{ is thin.} \tag{15d}$$

Proof Define

$$\underset{\sim}{A}_{ik} := \underline{A}_{ik} \quad \text{if } \underline{x}_k \geq 0, \quad \underset{\sim}{A}_{ik} := \bar{A}_{ik} \text{ otherwise,}$$

$$\tilde{A}_{ik} := \bar{A}_{ik} \quad \text{if } \bar{x}_k \geq 0, \quad \tilde{A}_{ik} := \underline{A}_{ik} \text{ otherwise.}$$

Then

$$(Ax)_i = \sum_k A_{ik}x_k = \sum_k [\underline{A}_{ik}, \bar{A}_{ik}][\underline{x}_k, \bar{x}_k]$$

$$= \sum_k [\underset{\sim}{A}_{ik}\underline{x}_k, \tilde{A}_{ik}\bar{x}_k] = \left[\sum_k \underset{\sim}{A}_{ik}\underline{x}_k, \sum_k \tilde{A}_{ik}\bar{x}_k\right]$$

$$= [(\underset{\sim}{A}\underline{x})_i, (\tilde{A}\bar{x})_i] = [\underset{\sim}{A}\underline{x}, \tilde{A}\bar{x}]_i,$$

and (14) follows. In particular,

$$\underset{\sim}{A} = \underline{A} \quad \text{if } \underline{x} \geq 0, \quad \underset{\sim}{A} = \bar{A} \quad \text{if } \underline{x} \leq 0,$$

$$\tilde{A} = \bar{A} \quad \text{if } \bar{x} \geq 0, \quad \tilde{A} = \underline{A} \quad \text{if } \bar{x} \leq 0,$$

and application to each column $x = Be^{(i)}$ of B implies (15a–c). (Here $e^{(i)}$ is the unit vector with entry 1 in position i.) Equation (15d) follows in the same way. □

We note that analogous formulae hold if the second factor B is nonnegative; in particular we mention

$$AB = [\underline{A}B, \bar{A}B] \quad \text{if } B \geq 0 \text{ is thin.} \tag{16}$$

3.1.6 Proposition Let $A \in \mathbb{IR}^{m\times n}$, $B \in \mathbb{IR}^{n\times p}$, and suppose that $\underline{A} = 0$. Then

$$AB = [\bar{A}\inf\{\underline{B}, 0\}, \bar{A}\sup\{\bar{B}, 0\}]. \tag{17}$$

Proof We have

$$\begin{aligned}(AB)_{ik} &= \sum A_{ij}B_{jk} = \sum [0, \bar{A}_{ij}][\underline{B}_{jk}, \bar{B}_{jk}] \\ &= \sum [\inf\{0, \bar{A}_{ij}\underline{B}_{jk}\}, \sup\{0, \bar{A}_{ij}\bar{B}_{jk}\}] \\ &= \sum [\bar{A}_{ij}\inf\{\underline{B}_{jk}, 0\}, \bar{A}_{ij}\sup\{\bar{B}_{jk}, 0\}] \\ &= \left[\sum \bar{A}_{ij}\inf\{\underline{B}_{jk}, 0\}, \sum \bar{A}_{ij}\sup\{\bar{B}_{jk}, 0\}\right] \\ &= [\bar{A}\cdot\inf\{\underline{B}, 0\}, \bar{A}\cdot\sup(\bar{B}, 0\}]_{ik}.\end{aligned}$$ □

3.1.7 Proposition

$$A \cap B = \begin{cases}[\sup(\underline{A}, \underline{B}), \inf(\bar{A}, \bar{B})] & \text{if } \underline{A} \leq \bar{B} \text{ and } \underline{B} \leq \bar{A}, \\ \emptyset & \text{otherwise;}\end{cases} \tag{18}$$

$$B \subseteq A \Leftrightarrow \underline{A} \leq \underline{B}, \bar{B} \leq \bar{A} \Leftrightarrow |\check{A} - \check{B}| \leq \mathrm{rad}(A) - \mathrm{rad}(B). \tag{19}$$

Proof Consider the entries and use Proposition 1.6.3. □

3.1.8 Proposition Let $A \in \mathbb{IR}^{m\times n}$, and let Σ, Σ' be subsets of $\mathbb{R}^{m\times n}$. Then:

$$\Sigma \subseteq A \Rightarrow \square\Sigma \subseteq A, \tag{20}$$

$$\Sigma' \subseteq \Sigma \Rightarrow \square\Sigma' \subseteq \square\Sigma, \tag{21}$$

$$|\square\Sigma| = \sup\{|\tilde{A}| \mid \tilde{A} \in \Sigma\}. \tag{22}$$

Proof Trivial. □

3.1.9 Proposition

$$A + B \subseteq A + C \Leftrightarrow B \subseteq C, \tag{23}$$

$$A(BC) = (AB)C \quad \text{if } \underline{A} = \underline{B} = 0. \tag{24}$$

Proof Relation (23) is trivial. For (24) suppose that $\underline{A} = \underline{B} = 0$. Then $\overline{AB} = \bar{A}\bar{B}$, and Proposition 3.1.6 implies

$$A(BC) = A[\overline{B} \cdot \inf\{\underline{C}, 0\}, \overline{B} \cdot \sup(\overline{C}, 0\}]$$
$$= [\overline{AB} \cdot \inf\{\underline{C}, 0\}, \overline{AB} \cdot \sup\{\overline{C}, 0\}] = (AB)C. \qquad \square$$

3.1.10 Proposition

$$A' \subseteq A \Rightarrow \mathrm{rad}(A') \leq \mathrm{rad}(A), |A'| \leq |A|, \tag{25}$$
$$|A| = |\check{A}| + \mathrm{rad}(A), \tag{26}$$
$$|A| - |B| \leq |A \pm B| \leq |A| + |B|, \tag{27}$$
$$|AB| \leq |A|\,|B|, \tag{28}$$
$$|A| = |\tilde{A}| \quad \text{for some } \tilde{A} \in A. \tag{29}$$

Proof Apply the corresponding rules for the scalar case. □

3.1.11 Proposition If $A \in \mathbb{IR}^{n\times n}$ then

$$A' \subseteq A \Rightarrow \langle A' \rangle \geq \langle A \rangle, \tag{30}$$
$$\langle A \rangle \geq \langle \check{A} \rangle - \mathrm{rad}(A), \text{ with equality iff } 0 \notin A_{ii} \text{ for } i = 1, \ldots, n, \tag{31}$$
$$\langle A \rangle - |B| \leq \langle A \pm B \rangle \leq \langle A \rangle + |B| \text{ for all } B \in \mathbb{IR}^{n\times n}, \tag{32}$$
$$|AB| \geq \langle A \rangle |B| \text{ for all } B \in \mathbb{IR}^{n\times p}, \tag{33}$$
$$\langle A \rangle = \langle \tilde{A} \rangle \text{ for some } \tilde{A} \in A. \tag{34}$$

Proof Apply the corresponding rules for the scalar case. For (33), observe that

$$|AB|_{ik} = \left|\sum_j A_{ij}B_{jk}\right| \geq |A_{ii}B_{ik}| - \sum_{j\neq i} |A_{ij}B_{jk}|$$
$$= |A_{ii}|\,|B_{ik}| - \sum_{j\neq i} |A_{ij}|\,|B_{jk}| \geq (\langle A \rangle |B|)_{ik}$$

since $|A_{ii}| \geq \langle A_{ii} \rangle$. □

3.1.12 Proposition

$$\mathrm{mid}(A \pm B) = \check{A} \pm \check{B}, \tag{35}$$
$$\mathrm{mid}(AB) = \check{A}\check{B} \text{ if } A \text{ or } B \text{ is thin}, \tag{36}$$
$$\mathrm{rad}(A) = |A - \check{A}| \leq |A|, \tag{37}$$
$$\mathrm{rad}(A \pm B) = \mathrm{rad}(A) + \mathrm{rad}(B), \tag{38}$$
$$\mathrm{rad}(A)|B| \leq \mathrm{rad}(AB) \leq \mathrm{rad}(A)|B| + |\check{A}|\mathrm{rad}(B), \tag{39}$$
$$|A|\mathrm{rad}(B) \leq \mathrm{rad}(AB) \leq |A|\mathrm{rad}(B) + \mathrm{rad}(A)|\check{B}|, \tag{40}$$
$$\mathrm{rad}(AB) = |A|\mathrm{rad}(B) \text{ if } A \text{ is thin or } \check{B} = 0. \tag{41}$$

Proof Apply the corresponding rules for the scalar case. □

3.1.13 Proposition

$$A - \check{A} = [-\mathrm{rad}(A), \mathrm{rad}(A)], \tag{42}$$

$$A[-I, I] = [-I, I]|A| = [-|A|, |A|] = [-1, 1]A, \tag{43}$$

$$AB = |A|B = [-|A|\,|B|, |A|\,|B|] \quad \text{if } \check{B} = 0. \tag{44}$$

Proof Trivial. □

3.2 Norms and spectral radius

In this section we discuss scaled maximum norms and their relations to the spectral radius of square matrices. As an important tool we prove the theorem of Perron and Frobenius on the existence of a nonnegative eigenvector for any nonnegative square matrix.

Scaled maximum norms are a generalization of the l_∞-norm defined for vectors $x \in \mathbb{IR}^n$ and matrices $A \in \mathbb{IR}^{m\times n}$ by

$$\|x\|_\infty := \max\{|x_i| \mid i = 1, \ldots, n\} \qquad \text{(maximum norm)},$$

$$\|A\|_\infty := \max\left\{\sum_{k=1}^{n} |A_{ik}| \mid i = 1, \ldots, m\right\} \quad \text{(row sum norm)}.$$

For a given real scaling vector $u \in \mathbb{R}^n$ with $u > 0$, we define for $x \in \mathbb{R}^n$ and $A \in \mathbb{R}^{n\times n}$ the corresponding *scaled maximum norm* by

$$\|x\|_u := \max\{|x_i|/u_i \mid i = 1, \ldots, n\}, \tag{1}$$

$$\|A\|_u := \|\,|A|u\|_u = \max\left\{\sum_{k=1}^{n} |A_{ik}|u_k/u_i \mid i = 1, \ldots, m\right\}; \tag{2}$$

in the special case where $u = (1, \ldots, 1)^{\mathrm{T}}$ this just reduces to the l_∞-norm. (An extension of scaled maximum norms to rectangular matrices is possible using two scaling vectors $u > 0$, $v > 0$; however, in this book we do not need such an extension.)

The definitions (1) and (2) immediately apply to interval vectors $x \in \mathbb{IR}^n$ and interval matrices $A \in \mathbb{IR}^{n\times n}$. As immediate consequences of the definitions we note that

$$\|x\|_u \le \alpha \Leftrightarrow |x| \le \alpha u,$$

$$\|x\|_u < \alpha \Leftrightarrow |x| < \alpha u,$$

$$\|A\|_u \le \alpha \Leftrightarrow |A|u \le \alpha u,$$

$$\|A\|_u < \alpha \Leftrightarrow |A|u < \alpha u.$$

It is easily verified that $\|\cdot\| = \|\cdot\|_u$ satisfies the rules

$$\begin{aligned}
\|x\| &= \|\,|x|\,\| \geq 0, \text{ with equality iff } x = 0,\\
\|\alpha x\| &= |\alpha|\,\|x\| \text{ for all } \alpha \in \mathbb{R},\\
\|x\| - \|y\| \leq \|x \pm y\| &\leq \|x\| + \|y\|,\\
|x| \leq y &\Rightarrow \|x\| \leq \|y\|,\\
x \subseteq y &\Rightarrow \|x\| \leq \|y\|,
\end{aligned}$$

for all $x, y \in \mathbb{IR}^n$, and

$$\begin{aligned}
&\|A\| = \|\,|A|\,\| \geq 0, \text{ with equality iff } A = 0,\\
&\|Ax\| \leq \|A\|\,\|x\|, \quad \|AB\| \leq \|A\|\,\|B\|,\\
&|B| \leq A \Rightarrow \|B\| \leq \|A\|,\\
&B \subseteq A \Rightarrow \|B\| \leq \|A\|
\end{aligned}$$

for all $A, B \in \mathbb{IR}^{n\times n}$. Moreover,

$$\|I\| = 1.$$

3.2.1 Lemma Let $A \in \mathbb{R}^{n\times n}$ and $0 < u \in \mathbb{R}^n$. Then

$$|\lambda| \leq \|A\|_u \text{ for every eigenvalue } \lambda \text{ of } A.$$

Proof Let x be an eigenvector associated with λ, i.e. $Ax = \lambda x$, $x \neq 0$. We may scale x such that $\|x\|_u = 1$. Then

$$\begin{aligned}
|\lambda| = |\lambda|\,\|\,|x|\,\|_u = \|\,|\lambda|\,|x|\,\|_u = \|\,|Ax|\,\|_u &\leq \|\,|A|\,|x|\,\|_u\\
&\leq \|\,|A|\,\|_u\|\,|x|\,\|_u = \|A\|_u.
\end{aligned}$$ □

The maximal absolute value of the eigenvalues of A,

$$\rho(A) := \max\{|\lambda| \mid \lambda \in \mathbb{C}, Ax = \lambda x \text{ for some } x \in \mathbb{C}^n\backslash\{0\}\},$$

is called the *spectral radius* of $A \in \mathbb{R}^{n\times n}$. Lemma 3.2.1 immediately implies that

$$\rho(A) \leq \|A\|_u \quad \text{if } A \in \mathbb{R}^{n\times n}, u > 0. \tag{3}$$

The spectral radius $\rho(A)$ is a continuous function of A since the eigenvalues of A continuously depend on A. The following result on the spectral radius is fundamental.

3.2.2 Theorem (Perron, Frobenius) If $A \in \mathbb{R}^{n\times n}$ is nonnegative then the spectral radius $\rho(A)$ is an eigenvalue of A, and there is a real, nonnegative eigenvector $x \neq 0$ with $Ax = \rho(A)x$.

Proof Put $\Omega := \{x \geq 0 \mid \|x\|_\infty = 1\}$. First we treat the special case $A > 0$. Since $Ax > 0$ for all $x \in \Omega$, the number

$$\rho_x := \sup\{\alpha \geq 0 \mid Ax \geq \alpha x\} = \inf\left\{\sum A_{ik}x_k/x_i \mid i = 1, \ldots, n, x_i \neq 0\right\}$$

is defined (and nonnegative) for all $x \in \Omega$. Now Ω is compact, and since $A > 0$, ρ_x is continuous as a function of x; therefore ρ_x attains its supremum $\rho = \sup\{\rho_x \mid x \in \Omega\}$ for some vector $u \in \Omega$. We show that $Au = \rho u$. Indeed, $w := Au - \rho u$ is nonnegative by construction, and $Aw = A \cdot Au - \rho Au = \|Au\|_\infty(Ax - \rho x)$, where $x = \|Au\|_\infty^{-1}Au \in \Omega$. If $w \neq 0$ then $Aw > 0$ since $A > 0$, so that $Ax > \rho x$, $\rho_x > \rho$ against the maximality of ρ. Therefore $w = 0$, i.e. $Au = \rho u$. In particular, ρ is an eigenvalue of A, and $\rho < \rho(A)$. On the other hand, $u = \rho^{-1}Au > 0$ so that $\rho(A) \leq \|A\|_u = \rho$; therefore $\rho = \rho(A)$. Thus, the theorem holds when $A > 0$.

In the general case, let E be an arbitrary positive $n \times n$ matrix and put $A(t) := A + tE$, $\rho(t) := \rho(A + tE)$. Since $A(t) > 0$ for $t > 0$, there is a nonnegative vector $u(t) \in \Omega$ with $A(t)u(t) = \rho(t)u(t)$. Since Ω is compact, there is an accumulation point $u \in \Omega$ of $u(t)$ for $t \to +0$. Since $A(t) \to A$ and $\rho(t) \to \rho(A)$ for $t \to +0$, any such accumulation point satisfies $Au = \rho(A)u$. □

A real nonnegative eigenvector $x \neq 0$ with $Ax = \rho(A)x$ is called a *Perron vector* of the nonnegative matrix $A \in \mathbb{R}^{n \times n}$.

3.2.3 Corollary If $0 \leq A \in \mathbb{R}^{n\times n}$ and $0 < \alpha \in \mathbb{R}$ then

$$\rho(A) = \inf\{\|A\|_u \mid u > 0\}, \tag{4}$$

$$\rho(A) < \alpha \Leftrightarrow \exists u > 0: Au < \alpha u, \tag{5}$$

$$\rho(A) \geq \alpha \Leftrightarrow \exists u > 0: Au \geq \alpha u \neq 0. \tag{6}$$

In particular

$$\rho(A) < 1 \Rightarrow \lim_{l\to\infty} A^l = 0. \tag{7}$$

Proof By (3) we have $\rho(A) \leq \inf\{\|A\|_u \mid u > 0\}$, and the backward implication in (5). Let the notation be as in the last proof. If $\rho(A) < \alpha$ then $\rho(t) < \alpha$ for sufficiently small $t > 0$, and $u = u(t)$ satisfies $|A|u = Au \leq A(t)u = \rho(t)u < \alpha u$. Thus, the forward implication of (5) holds; moreover $\|A\|_u < \alpha$, and therefore $\inf\{\|A\|_u \mid u > 0\} \leq \inf\{\alpha \mid \rho(A) < \alpha\} = \rho(A)$. This proves (4) and (5).

The forward implication in (6) follows by choosing u as a Perron vector of A. To show the backward implication we assume that $u \geq 0$ satisfies $Au \geq \alpha u \neq 0$, and, w.l.o.g., $\|u\|_\infty = 1$. If $A > 0$ then $u \in \Omega$, $\alpha \leq \rho_u \leq \rho(A)$; in the general case we have $A(t)u \geq \alpha u$ for all $t > 0$ so that similarly $\alpha \leq \rho(t)$ for all $t > 0$. In the limit $t \to 0$ this again yields $\alpha \leq \rho(A)$.

Finally, if $\rho(A) < 1$ then by (4) we have $\beta = \|A\|_u < 1$ for some $u > 0$. Hence $\|A^i\|_u \le \|A\|_u^i = \beta^i$, which shows that $\|A^i\|_u \to 0$ for $i \to \infty$. This implies (7). □

We note two further properties of the spectral radius.

3.2.4 Proposition Let $A, B \in \mathbb{R}^{n\times n}$. Then

$$|A| \le B \Rightarrow \rho(A) \le \rho(B), \tag{8}$$

$$\rho(AB) = \rho(BA). \tag{9}$$

Proof Let λ be an eigenvalue of A with $|\lambda| = \rho(A)$, and x an associated eigenvector. Then $u := |x| \ne 0$, and $\rho(A)u = |\lambda|\,|x| = |\lambda x| = |Ax| \le |A|\,|x| \le B|x| = Bu$, and by (6) we get $\rho(B) \ge \rho(A)$. Hence (8) holds. To show (9) we suppose that $\rho(AB) < \rho(BA)$. Let λ be an eigenvalue of BA with $|\lambda| = \rho(BA) > 0$, and x an associated eigenvector. Then $BAx = \lambda x \ne 0$; in particular $x' := Ax$ is nonzero and satisfies $ABx' = ABAx = \lambda Ax = \lambda x'$. But then λ is an eigenvalue of AB, $\rho(AB) \ge |\lambda| = \rho(BA)$, contradiction. Thus $\rho(AB) \ge \rho(BA)$, and (9) follows by symmetry. □

3.2.5 Proposition Let $A \in \mathbb{R}^{n\times n}$ satisfy $\rho(A) < 1$. Then $I - A$ is nonsingular. Moreover, if $A \ge 0$ then $(I - A)^{-1} \ge 0$.

Proof Since 1 cannot be an eigenvalue of A, $I - A$ is nonsingular. Now suppose that $A \ge 0$. By (5), there is a vector $u > 0$ with $Au < u$. Put $B := (I - A)^{-1}$ and $v_k := \min\{B_{ik}/u_i \mid i = 1, \ldots, n\}$. Then $B \ge uv^T$, and for each index k there is an index i such that $u_i v_k = B_{ik}$. Hence $0 \le (B - AB)_{ik} \le (u - Au)_i v_k$. This implies $v_k \ge 0$ and therefore $B \ge uv^T \ge 0$. □

Remark The standard argument for $(I - A)^{-1} \ge 0$ is based on the convergence of the Neumann series

$$(I - A)^{-1} = \sum_{i=0}^{\infty} A^i \ge 0.$$

3.3 Distance and topology

In this section we discuss the concept of a matrix-valued distance and the properties of a corresponding topology. As an application we prove the important contraction mapping theorem for interval functions.

3.3.1 Proposition Let $A, B \in \mathbb{IR}^{m\times n}$. Then the following three definitions are equivalent:

$$q(A, B) := \inf\{Q \in \mathbb{R}^{m\times n} \mid Q \geq 0, A \subseteq B + [-Q, Q], B \subseteq A + [-Q, Q]\}, \quad (1)$$

$$q(A, B) := \sup\{|\underline{A} - \underline{B}|, |\overline{A} - \overline{B}|\}, \quad (2)$$

$$q(A, B) := |\check{A} - \check{B}| + |\mathrm{rad}(A) - \mathrm{rad}(B)|. \quad (3)$$

Proof We have $A \subseteq B + [-Q, Q] = [\underline{B} - Q, \overline{B} + Q]$ iff $\underline{A} \geq \underline{B} - Q$ and $\overline{A} \leq \overline{B} + Q$, i.e. iff $Q \geq \underline{B} - \underline{A}$ and $Q \geq \overline{A} - \overline{B}$. Similarly $B \subseteq A + [-Q, Q]$ iff $Q \geq \underline{A} - \underline{B}$ and $Q \geq \overline{B} - \overline{A}$. Therefore both conditions hold simultaneously iff $Q \geq |\underline{A} - \underline{B}|$ and $Q \geq |\overline{A} - \overline{B}|$. This implies that (1) and (2) are equivalent. By Proposition 3.1.7 we also have $A \subseteq B + [-Q, Q]$ iff $|\check{A} - \check{B}| \leq Q + \mathrm{rad}(B) - \mathrm{rad}(A)$, i.e. iff $Q - |\check{A} - \check{B}| \geq \mathrm{rad}(A) - \mathrm{rad}(B)$, and similarly $B \subseteq A + [-Q, Q]$ iff $Q - |\check{A} - \check{B}| \geq \mathrm{rad}(B) - \mathrm{rad}(A)$. Therefore, both conditions hold simultaneously iff $Q - |\check{A} - \check{B}| \geq |\mathrm{rad}(A) - \mathrm{rad}(B)|$. This implies that (1) and (3) are equivalent. □

3.3.2 Proposition Let $A, B, C \in \mathbb{IR}^{m\times n}$. Then

$$q(A, B) \geq 0, \text{ with equality iff } A = B, \quad (4)$$

$$q(A, B) = q(B, A), \quad (5)$$

$$q(A, C) \leq q(A, B) + q(B, C). \quad (6)$$

Proof Expressions (4) and (5) are immediate from (2) or (3), and (6) follows from the triangle inequality applied to (3). □

Thus, $q(A, B)$ has properties analogous to a metric; therefore it is called the *distance* of A and B. Note that $q(A, B)$ is a real, nonnegative matrix of the same size as A and B.

By combining the distance with the l_∞-norm, we obtain with

$$d(A, B) := \|q(A, B)\|_\infty$$

a metric in $\mathbb{IR}^{m\times n}$. The axioms of a metric immediately follow from (4), (5), (6) and the triangle inequality. In the associated topology a sequence of matrices *converges* iff lower and upper bounds converge, equivalently, iff midpoints and radii converge. This immediately follows from (2) and (3), which imply

$$\lim_{l\to\infty} A_l = \left[\lim_{l\to\infty} \underline{A}_l, \lim_{l\to\infty} \overline{A}_l\right],$$

$$\mathrm{mid}\left(\lim_{l\to\infty} A_l\right) = \lim_{l\to\infty} \mathrm{mid}(A_l), \quad \mathrm{rad}\left(\lim_{l\to\infty} A_l\right) = \lim_{l\to\infty} \mathrm{rad}(A_l).$$

This also implies that convergence is componentwise, i.e. the (i, k)-entry of $\lim_{l\to\infty} A_l$ is the limit of the (i, k)-entries of A_l.

Endpoints, midpoint, radius, absolute value, sum, difference and product of matrices A, B are continuous functions of A and B since they can be continuously expressed in terms of the endpoints of A and B. Also, as we already observed in the scalar case, it is obvious that a sequence

$$A_0 \supseteq A_1 \supseteq \cdots \supseteq A_l \supseteq A_{l+1} \supseteq \cdots$$

of nested interval matrices always converges to the limit

$$\lim_{l\to\infty} A_l = \bigcap_{l\geq 0} A_l.$$

Another important convergence result is the following *contraction mapping theorem* for interval mappings.

3.3.3 Theorem (Schröder) Let Φ: $\mathbb{IR}^n \to \mathbb{IR}^n$ be a mapping such that

$$q(\Phi(x), \Phi(y)) \leq Pq(x, y) \quad \text{for all } x, y \in \mathbb{IR}^n, \tag{7}$$

where $P \in \mathbb{R}^{n\times n}$ is a nonnegative matrix with spectral radius $\rho(P) < 1$. Then, for all $x^0 \in \mathbb{IR}^n$, the sequence defined by the iteration

$$x^{l+1} := \Phi(x^l) \quad (l = 0, 1, 2, \ldots)$$

converges to the unique fixed point of Φ, i.e. the solution x^* of the equation $x^* = \Phi(x^*)$.

Proof By Proposition 3.2.5, $I - P$ is nonsingular, and the equation $(I - P)w = q(x^0, x^1)$ has a nonnegative solution $w \in \mathbb{R}^n$. By induction on l, we show the inequality

$$q(x^l, x^m) \leq P^m w \quad \text{for } l \geq m \geq 0, \tag{8}$$

which is obvious for $l = 0$. If (8) holds for some l then for $m \geq 1$ we get $q(x^{l+1}, x^m) = q(\Phi(x^l), \Phi(x^{m-1})) \leq Pq(x^l, x^{m-1}) \leq PP^{m-1}w = P^m w$. For $m = 0$ we get $q(x^{l+1}, x^0) \leq q(x^{l+1}, x^1) + q(x^1, x^0) \leq Pw + (I - P)w = w$. So (8) holds generally.

Since $\rho(P) < 1$, (3.2.7) implies $P^l \to 0$ for $l \to \infty$; therefore $q(x^l, x^m) \to 0$ for $l, m \to \infty$. By (2), this implies that $\underline{x}^l$ $(l = 0, 1, 2, \ldots)$ and $\overline{x}^l$ $(l = 0, 1, 2, \ldots)$ are Cauchy sequences; hence they are convergent. Therefore, the interval sequence x^l $(l = 0, 1, 2, \ldots)$ has a limit x^*, and since (7) implies the continuity of Φ we have $x^* = \Phi(x^*)$.

To show uniqueness of x^* we suppose that $x \neq x^*$ satisfies $x = \Phi(x)$. Then $u := q(x, x^*)$ is a nonzero, nonnegative vector satisfying $u = q(\Phi(x), \Phi(x^*)) \leq Pq(x, x^*) = Pu$, contradicting (3.2.6). Hence the fixed point is unique. □

Remark More generally, the above theorem holds with the same proof when $\mathbb{IR}^n$ is replaced by a partially ordered vector space V with a *hypermetric*, i.e. a vector-valued distance relation $q: V \times V \to \mathbb{R}^n$. Later we shall need the case when $V = \mathbb{R}^n$ and $q(x, y) = |x - y|$.

For the calculation with distances we note the following rules.

3.3.4 Lemma Let $A, B \in \mathbb{IR}^{m\times n}$. Then $Q = q(A, B)$ has the entries

$$Q_{ik} = q(A_{ik}, B_{ik}) \quad (i = 1, \ldots, m; k = 1, \ldots, n).$$

Proof Apply (1.7.2) to the entries of (2). □

3.3.5 Proposition Let $A, A', B, B' \in \mathbb{IR}^{m\times n}$, $C, C' \in \mathbb{IR}^{n\times p}$. Then

$$q(A, B) = |A - B| \text{ if } A \text{ or } B \text{ is thin,} \tag{9}$$

$$A \subseteq B \Rightarrow q(A, B) = |B - \check{A}| - \operatorname{rad}(A), \tag{10}$$

$$A \subseteq B \Rightarrow \operatorname{rad}(B) - \operatorname{rad}(A) \le q(A, B) \le 2(\operatorname{rad}(B) - \operatorname{rad}(A)), \tag{11}$$

$$A \subseteq B \subseteq B' \Rightarrow \max\{q(A, B), q(B, B')\} \le q(A, B'), \tag{12}$$

$$B \subseteq A + [-I, I]q(A, B), \tag{13}$$

$$\operatorname{rad}(B) \le q(A, B) + \operatorname{rad}(A), \tag{14}$$

$$|B| \le q(A, B) + |A|, \tag{15}$$

$$|A - B| \le q(A, B) + 2 \cdot \operatorname{rad}(A), \tag{16}$$

$$q(A, 0) = |A|, \tag{17}$$

$$q(A + B, A) = |B|, \tag{18}$$

$$q(A + B, A + B') = q(B, B'), \tag{19}$$

$$q(A + A', B + B') \le q(A, B) + q(A', B'), \tag{20}$$

$$q(\alpha A, \alpha B) = |\alpha| q(A, B) \quad \text{for } \alpha \in \mathbb{R}, \tag{21}$$

$$q(AC, BC) \le q(A, B)|C|, \tag{22}$$

$$q(AC, AC') \le |A| q(C, C'). \tag{23}$$

Proof Combine the previous lemma with the rules stated in Proposition 1.7.3. □

3.4 Linear interval equations

In this section we define linear interval equations and consider some of their basic aspects. We characterize the interval matrices which do not contain a singular matrix, and show how to compute the hull of the solution set of linear interval equations with only two variables.

A *linear interval equation* with coefficient matrix $A \in \mathbb{IR}^{m\times n}$ and right-hand side $b \in \mathbb{IR}^m$ is defined as the family of linear equations

$$\tilde{A}\tilde{x} = \tilde{b} \quad (\tilde{A} \in A, \tilde{b} \in b). \tag{1}$$

Thus we consider linear systems of equations in which the coefficients are unknown numbers ranging in certain intervals. We are interested in enclosures for the *solution set* of (1), given by

$$\Sigma(A, b) := \{\tilde{x} \in \mathbb{R}^n \mid \tilde{A}\tilde{x} = \tilde{b} \text{ for some } \tilde{A} \in A, \tilde{b} \in b\}.$$

We show that the solution set $\Sigma(A, b)$ usually is not an interval vector, and need not even be convex; in general, $\Sigma(A, b)$ has a very complicated structure.

3.4.1 Example Consider the system (1) with

$$A = \begin{pmatrix} [\ 2, 4] & [-1, 1] \\ [-1, 1] & [\ 2, 4] \end{pmatrix}, \quad b = \begin{pmatrix} [-3, 3] \\ 0 \end{pmatrix}.$$

Then $\Sigma(A, b)$ is the butterfly-shaped region drawn in Fig. 3.2. It is quite difficult to verify the figure from the definition given. A simpler description of $\Sigma(A, b)$ is provided by Corollary 3.4.4 below which implies that, for the present example,

$$\Sigma(A, b) = \{x \in \mathbb{R}^2 \mid 2|x_2| \le |x_1|, 2|x_1| \le 3 + |x_2|\}.$$

In order to guarantee that the solution $\Sigma(A, b)$ of (1) is bounded we require that the matrix A is *regular*, i.e. that every matrix $\tilde{A} \in A$ has rank n. In this case, standard results from linear algebra imply that for each $\tilde{A} \in A, \tilde{b} \in b$, the equation $\tilde{A}\tilde{x} = \tilde{b}$ has at most one solution $\tilde{x}$, and $\tilde{A}^T\tilde{A}$ is a nonsingular $n \times n$ matrix. In particular, the solution $\tilde{x}$, if it exists, can be written as

$$\tilde{x} = \tilde{A}^+\tilde{b}, \quad \text{where } \tilde{A}^+ = (\tilde{A}^T\tilde{A})^{-1}\tilde{A}^T, \tag{2}$$

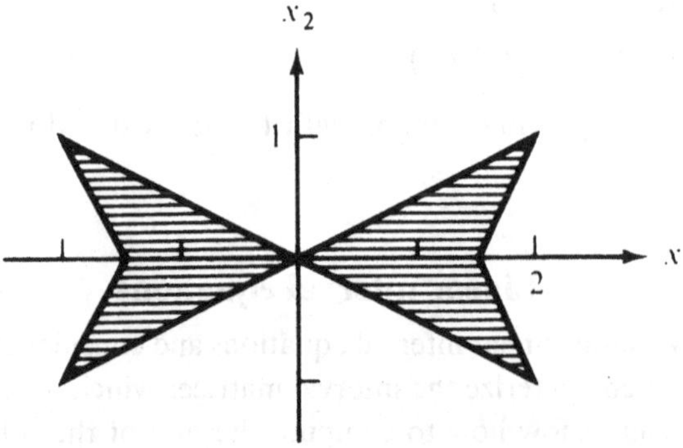

Fig. 3.2. Solution set of a linear interval equation.

and thus continuously depends on $\tilde{A}$ and $\tilde{b}$. Since $\tilde{A}$ and $\tilde{b}$ vary over bounded intervals, it follows that $\Sigma(A, b)$ is bounded, and the *hull of the solution set*, which we denote as

$$A^{H}b := \square\Sigma(A, b), \tag{3}$$

is defined. From the definition we immediately get the relation

$$A' \subseteq A, b' \subseteq b \Rightarrow A'^{H}b' \subseteq A^{H}b. \tag{4}$$

If $A \in \mathbb{IR}^{n\times n}$ is a regular square matrix, then for each $\tilde{A} \in A$, $\tilde{b} \in b$, the equation $\tilde{A}\tilde{x} = \tilde{b}$ has a unique solution $\tilde{x} = \tilde{A}^{-1}\tilde{b}$ so that we can express $A^{H}b$ as

$$A^{H}b = \square\{\tilde{A}^{-1}\tilde{b} \mid \tilde{A} \in A, \tilde{b} \in b\} \quad \text{for } b \in \mathbb{IR}^{n}. \tag{5}$$

In particular, (5) defines a mapping A^{H}: $\mathbb{IR}^{n} \to \mathbb{IR}^{n}$ which we call the *hull inverse* of A. We can also define a *matrix inverse* of a regular, square matrix $A \in \mathbb{IR}^{n\times n}$ by

$$A^{-1} := \square\{\tilde{A}^{-1} \mid \tilde{A} \in A]. \tag{3a}$$

From (3.1.10) we get the inclusion

$$A^{H}b \subseteq A^{-1}b, \text{ with equality if } A \text{ is thin.} \tag{6}$$

(Note that the left-hand side denotes the result of the mapping A^{H} applied to b, whereas the right-hand side denotes the multiplication of the matrix A^{-1} with b.) Another case of equality in (6) occurs when A is a regular diagonal matrix, i.e. $A_{ii} \not\ni 0$, $A_{ik} = 0$ for $i \neq k$. In this case we have

$$A^{H}b = A^{-1}b = c \quad \text{with } c_i = b_i/A_{ii} \quad (i = 1, \ldots, n). \tag{6a}$$

It is a remarkable fact that in general $A^{H}b$ cannot be written as a product of b with an interval matrix A' independent of b. Indeed, if $A^{H}b = A'b$ for all $b \in \mathbb{IR}^{n}$ then, by choosing the unit vectors $e^{(1)}, \ldots, e^{(n)}$ for b we find that the columns of A' and A^{-1} agree. Therefore we must have equality in (6). But this is not always the case, as the following example shows.

3.4.2 Example Take

$$A := \begin{pmatrix} 2 & [-1, 0] \\ [-1, 0] & 2 \end{pmatrix}, \quad b := \begin{pmatrix} 1.2 \\ -1.2 \end{pmatrix}.$$

Then $\tilde{A} \in A$ iff

$$\tilde{A} := \begin{pmatrix} 2 & -\alpha \\ -\beta & 2 \end{pmatrix}$$

with $\alpha, \beta \in [0, 1]$. By Cramer's rule we have

$$\Sigma(A, b) = \left\{ \begin{pmatrix} 1.2(2 - \alpha)/(4 - \alpha\beta) \\ 1.2(\beta - 2)/(4 - \alpha\beta) \end{pmatrix} \;\middle|\; \alpha, \beta \in [0, 1] \right\},$$

cf. Fig. 3.3.

Hence,

$$A^{\mathrm{H}} b = \square\Sigma(A, b) = \begin{pmatrix} [\ 0.3, & 0.6] \\ [-0.6, & -0.3] \end{pmatrix}.$$

On the other hand,

$$A^{-1} = \square\left\{ \frac{1}{4 - \alpha\beta} \begin{pmatrix} 2 & \alpha \\ \beta & 2 \end{pmatrix} \;\middle|\; \alpha, \beta \in [0, 1] \right\}$$
$$= \begin{pmatrix} [1/2, 2/3] & [\ 0\ , 1/3] \\ [\ 0\ , 1/3] & [1/2, 2/3] \end{pmatrix}$$

so that

$$A^{-1} b = \begin{pmatrix} [\ 0.2, 0.8] \\ [-0.8, 0.2] \end{pmatrix} \neq A^{\mathrm{H}} b.$$

Since A is symmetric, it is conceivable that A stands for an unknown *symmetric* matrix

$$\tilde{A} = \begin{pmatrix} 2 & -\alpha \\ -\alpha & 2 \end{pmatrix}.$$

In this case the modified solution set

$$\Sigma_{\mathrm{sym}}(A, b) = \{\tilde{x} \in \mathbb{R}^n \;\mid\; \tilde{A}\tilde{x} = \tilde{b} \text{ for some } \tilde{A} \in A, \tilde{b} \in b \text{ with } \tilde{A}^{\mathrm{T}} = \tilde{A}\}$$

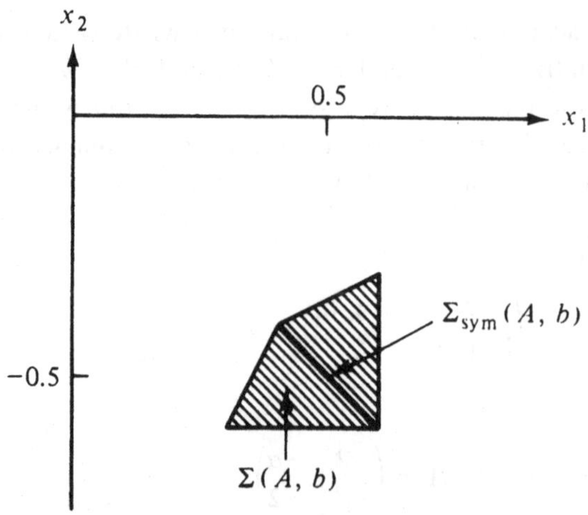

Fig. 3.3. Solution set $\Sigma(A, b)$ and symmetric solution set $\Sigma_{\mathrm{sym}}(A, b)$.

is smaller, namely

$$\Sigma_{\text{sym}}(A, b) = \left\{ \begin{pmatrix} (2.4 - 1.2\alpha)/(4 - \alpha^2) \\ (1.2\alpha - 2.4)/(4 - \alpha^2) \end{pmatrix} \;\middle|\; \alpha \in [0, 1] \right\}$$

(indicated by a heavy line in Fig. 3.3), and its hull is

$$\Box\Sigma_{\text{sym}}(A, b) = \begin{pmatrix} [\ \ 0.4, & 0.6] \\ [-0.6, & -0.4] \end{pmatrix} \neq A^{\text{H}}b.$$

Thus the hull inverse is not adapted to the optimal treatment of symmetric matrices. However, at present, no special methods have been devised for this case, and we shall content ourselves with the unsymmetric treatment of symmetric matrices.

There are some special cases where equality holds in (6). This is immediate if A is a regular diagonal matrix, and follows from equation (3.1.10) if A is thin.

If $A \in \mathbb{IR}^{m \times n}$ is a rectangular matrix and $m < n$ then $\Sigma(A, b)$ is either empty or unbounded and $A^{\text{H}}b$ is not defined. If $m \geq n$ and A is regular then $\Sigma(\text{A}, b)$ may still be empty, so that $A^{\text{H}}b$ is not necessarily an interval vector. However, it is sometimes more appropriate to consider, in place of (1), the family of linear least squares problems

$$\|\tilde{A}\tilde{x} - \tilde{b}\|_2 = \min! \quad (\tilde{A} \in A, \tilde{b} \in \text{b}). \tag{1a}$$

It is well known that when $\tilde{A}$ has rank n the solution of the least squares problem is given by (2); thus it makes sense to consider the *least squares hull*

$$A^{\text{L}}b := \Box\{\tilde{A}^{+}\tilde{b} \;\mid\; \tilde{A} \in A, \tilde{b} \in b\} \quad \text{for } b \in \mathbb{IR}^m. \tag{5a}$$

Clearly,

$$A^{\text{H}}b \subseteq A^{\text{L}}b,$$

The converse is not true since, e.g., $A^{\text{H}}b$ may be empty whereas $A^{\text{L}}b \in \mathbb{IR}^n$ for all $b \in \mathbb{IR}^m$. However, in the square case, we have $A^{\text{H}}b = A^{\text{L}}b$; in particular, the previous remarks on writing $A^{\text{H}}b$ as a matrix–vector product apply to A^{L}, too.

The solution set $\Sigma(A, b)$ of (1) has the following neat characterization.

3.4.3 Theorem (Beeck) Let $A \in \mathbb{IR}^{m \times n}$, $b \in \mathbb{IR}^m$. Then

$$\Sigma(A, b) = \{\tilde{x} \in \mathbb{R}^n \;\mid\; A\tilde{x} \cap b \neq 0\} = \{\tilde{x} \in \mathbb{R}^n \;\mid\; 0 \in A\tilde{x} - b\}.$$

Proof If $\tilde{x} \in \Sigma(A, b)$ then $\tilde{A}\tilde{x} = \tilde{b}$ for some $\tilde{A} \in A$, $\tilde{b} \in b$; hence $\tilde{b} \in A\tilde{x} \cap b$. Conversely, if $A\tilde{x} \cap b \neq \emptyset$ then $A\tilde{x} \cap b$ contains some $\tilde{b} \in \mathbb{R}^m$; clearly $\tilde{b} \in b$, and by equation (3.1.10), $\tilde{b} = \tilde{A}\tilde{x}$ for some $\tilde{A} \in A$. Therefore $\tilde{x} \in \Sigma(A, b)$. Since $A\tilde{x} \cap b \neq \emptyset$ iff $0 \in A\tilde{x} - b$, the theorem follows. □

3.4.4 Corollary (Oettli & Prager) Let $A \in \mathbb{IR}^{m \times n}$, $b \in \mathbb{IR}^m$. Then

$$\tilde{x} \in \Sigma(A, b) \Leftrightarrow |\check{A}\tilde{x} - \check{b}| \leq \text{rad}(A)|\tilde{x}| + \text{rad}(b).$$

Proof Put $a := A\tilde{x}$. Then

$$A\tilde{x} \cap b \neq \emptyset \Leftrightarrow a \cap b \neq \emptyset \Leftrightarrow |\check{a} - \check{b}| \leq \text{rad}(a) + \text{rad}(b).$$

Since $\check{a} = \check{A}\tilde{x}$ and $\text{rad}(a) = \text{rad}(A)|\tilde{x}|$, the result follows. □

3.4.5 Corollary For $A \in \mathbb{IR}^{m \times n}$, the following statements are equivalent:

(i) A is regular,
(ii) $\tilde{x} \in \mathbb{R}^n$, $0 \in A\tilde{x} \Rightarrow \tilde{x} = 0$,
(iii) $\tilde{x} \in \mathbb{R}^n$, $|\check{A}\tilde{x}| \leq \text{rad}(A)|\tilde{x}| \Rightarrow \tilde{x} = 0$.

Proof A is regular iff every equation $\tilde{A}\tilde{x} = 0$ $(\tilde{A} \in A)$ has the trivial solution $\tilde{x} = 0$ only, i.e. iff $\Sigma(A, 0) = \{0\}$. Now apply Theorem 3.4.3 and Corollary 3.4.4. □

3.4.6 Example We determine all values of t $(0 \leq t \in \mathbb{R})$ for which the matrix $A \in \mathbb{IR}^{n \times n}$ defined by

$$A_{ii} = t, \quad A_{ik} = [0, 2] \quad \text{for } i \neq k \tag{7}$$

is regular. Suppose that A is not regular. Then $0 \in A\tilde{x}$ for some $\tilde{x} \neq 0$. By symmetry of the entries of A we may assume that $\tilde{x}_1 \leq \cdots \leq \tilde{x}_l \leq 0 \leq \tilde{x}_{l+1} \leq \cdots \leq \tilde{x}_n$ for some index l. Since $A \geq tI$ we have $\tilde{x}_1 < 0 < \tilde{x}_n$. The condition $0 \in A\tilde{x}$ now becomes

$$t\tilde{x}_n + 2\sum_{i \leq l} \tilde{x}_i \leq 0 \leq t\tilde{x}_1 + 2\sum_{i > l} \tilde{x}_i$$

and hence

$$t \leq \min\left\{-2\sum_{i \leq l} \tilde{x}_i/\tilde{x}_n, -2\sum_{i > l} \tilde{x}_i/\tilde{x}_1\right\}.$$

The right-hand side becomes largest for the choice $\tilde{x}_i = (n - l)^{1/2}$ for $i \leq l$, and $\tilde{x}_i = l^{1/2}$ for $i > l$. Optimization of l now shows that the matrix A given by (7) is regular iff $t > n$ (for n even) or $t > (n^2 - 1)^{1/2}$ (for n odd).

The solution set of $Ax = b$, $b_i = [-1, 1]$ consists in the regular case of the union of two cubes and $2^n - 2$ spikes which grow infinitely long when t approaches the bound n or $(n^2 - 1)^{1/2}$. See Fig. 3.4 for the case $n = 3$, $t = 3.5$.

The decision whether an interval matrix is regular or not, and the computation of good enclosures of $A^{\text{H}}b$ and $A^{\text{L}}b$, are, in general, very difficult problems. For a subclass of matrices, these problems are discussed in detail in Chapter 4.

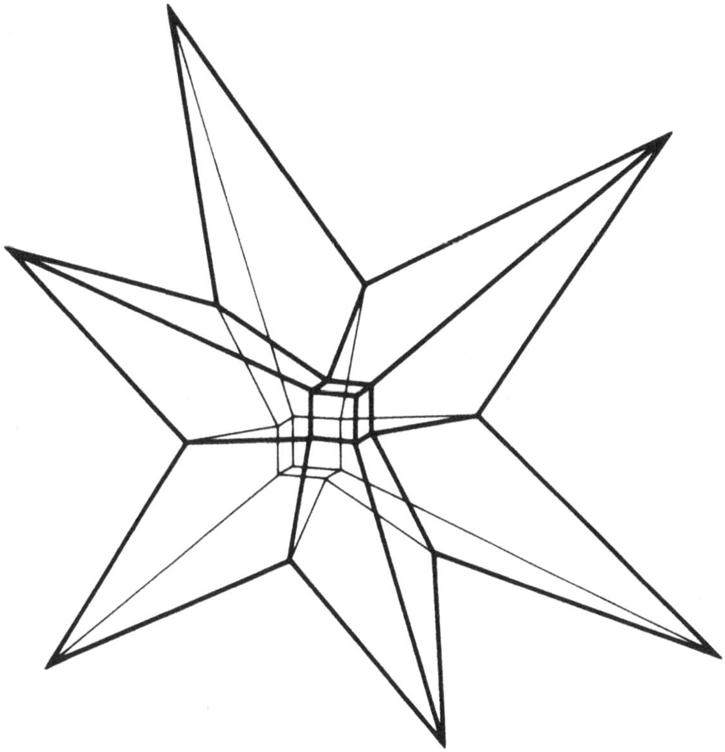

Fig. 3.4. A projection of a three-dimensional solution set.

The computation of $A^{\mathrm{H}}b$ for square A is treated in Section 6.2. However, for $n \leq 2$, simpler procedures exist. Indeed if $n = 1$ then $A^{\mathrm{H}}b = b/A$; and for $n = 2$, Cramer's rule can be adapted to interval problems so that it yields $A^{\mathrm{H}}b$.

3.4.7 Example Let $A \in \mathbb{IR}^{2\times 2}$, $b \in \mathbb{IR}^2$. Choose indices i, j, k, l such that $\{i, j\} = \{k, l\} = \{1, 2\}$ and $0 \notin A_{ik}$. (If A is regular, this can be achieved with an arbitrary choice for l.) Then any $\tilde{x} \in \Sigma(A, b)$ satisfies

$$\begin{pmatrix} \tilde{A}_{ik} & \tilde{A}_{il} \\ \tilde{A}_{jk} & \tilde{A}_{jl} \end{pmatrix} \begin{pmatrix} \tilde{x}_k \\ \tilde{x}_l \end{pmatrix} = \begin{pmatrix} \tilde{b}_i \\ \tilde{b}_j \end{pmatrix} \quad \text{for some } \tilde{A} \in A, \tilde{b} \in b.$$

Now $\tilde{A}$ is regular iff $0 \neq \pm \tilde{A}_{ik}^{-1} \det(\tilde{A}) = \tilde{A}_{jl} - \tilde{A}_{il} \cdot \tilde{A}_{jk}/\tilde{A}_{ik}$. Since the right-hand side contains each variable only once we find

$$A \text{ regular} \Leftrightarrow 0 \notin A_{jl} - A_{il}q_l, \quad \text{where } q_l = A_{jk}/A_{ik}. \tag{8}$$

In this case, Cramer's rule gives

$$\tilde{x}_l = (\tilde{b}_j - \tilde{b}_i\tilde{q}_l)(\tilde{A}_{jl} - \tilde{A}_{il}\tilde{q}_l),$$

where $\tilde{q}_l = \tilde{A}_{jk}/\tilde{A}_{ik} \in q_l$. Since the formula for $\tilde{x}_l$ is monotone in $\tilde{q}_l$, and since the remaining variables occur only once, the optimal enclosure $x := A^{\mathrm{H}}b$ of the solution set $\Sigma(A, b)$ satisfies

$$x_l = \square\left(\frac{b_j - b_i\underline{q}_l}{A_{jl} - A_{il}\underline{q}_l} \cup \frac{b_j - b_i\overline{q}_l}{A_{jl} - A_{il}\overline{q}_l}\right).$$

3.5 Sublinear mappings

In this section we define sublinear mappings and discuss their properties. Sublinear mappings are an important abstract tool for the study of iterative methods for the solution of linear and nonlinear systems of equations. Their properties are related to those of linear mappings in vector spaces in a similar way as interval matrix–vector operations are related to the corresponding operations in vector spaces.

A well-known result in linear algebra states that, for a real matrix $\tilde{A} \in \mathbb{R}^{m\times n}$, the mapping which maps a real vector $x \in \mathbb{R}^n$ to the matrix product $\tilde{A}x$ is a linear mapping from $\mathbb{R}^n$ to $\mathbb{R}^m$, and, conversely, every such linear mapping is obtained in this way from a suitable matrix $\tilde{A} \in \mathbb{R}^{m\times n}$. We should like to extend this correspondence to the matrix–vector multiplication mappings A^{M}: $\mathbb{IR}^n \to \mathbb{IR}^m$ defined by

$$A^{\mathrm{M}}x := Ax \quad (x \in \mathbb{IR}^n), \tag{1}$$

where now $A \in \mathbb{IR}^{m\times n}$ is an interval matrix and the argument is interval vector valued. However, since the distributive law is not valid, the linearity axiom must be weakened if we want to cover all these mappings. Moreover, the example of the hull inverse A^{H}, mentioned in Section 3.2, is not of the form (1), but shares many properties of matrix–vector multiplication, and should also be covered. The following definition generalizes the most relevant properties of A^{M} and A^{H}, and has proved to be very useful.

A mapping S: $\mathbb{IR}^n \to \mathbb{IR}^m$ is called *sublinear* if the axioms

(S1) $x \subseteq y \Rightarrow Sx \subseteq Sy$ (inclusion isotonicity),
(S2) $\alpha \in \mathbb{R} \Rightarrow S(\alpha x) = \alpha(Sx)$ (homogeneity), and
(S3) $S(x + y) \subseteq Sx + Sy$ (subadditivity)

are valid for all $x, y \in \mathbb{IR}^n$; it is called *linear* if (S1), (S2) and

(S3a) $S(x + y) = Sx + Sy$ (additivity)

hold for all $x, y \in \mathbb{IR}^n$. We emphasize that (S2) need not be valid for scalar intervals $\alpha \in \mathbb{IR}$.

3.5.1 Examples

(i) The mapping $S\colon \mathbb{IR}^n \to \mathbb{IR}^n$ defined by $Sx = x - x$ is sublinear.

(ii) For every $A \in \mathbb{IR}^{m\times n}$, the mapping A^{M} defined by (1) is sublinear; it is linear if A is thin.

(iii) If A is regular then the mapping A^{L} which maps $x \in \mathbb{IR}^m$ to the least square hull $A^{\mathrm{L}}x \in \mathbb{IR}^n$ is also sublinear. (S1) and (S2) are obvious, and (S3) is shown as follows. Let $\tilde{A} \in A$, $\tilde{b} \in x + y$. Then $\tilde{b} = \tilde{x} + \tilde{y}$ for suitable $\tilde{x} \in x$, $\tilde{y} \in y$ by equation (3.1.5), so that $\tilde{A}^+\tilde{b} = \tilde{A}^+\tilde{x} + \tilde{A}^+\tilde{y} \in A^{\mathrm{L}}x + A^{\mathrm{L}}y$. Therefore $A^{\mathrm{L}}(x+y) = \square\{\tilde{A}^+\tilde{b} \mid \tilde{A} \in A, \tilde{b} \in x + y\} \subseteq A^{\mathrm{L}}x + A^{\mathrm{L}}y$.

If A is regular and thin then A^{L} is linear since, by equation (3.1.10), $A^{\mathrm{L}}x = A^+x$ and A^+ is thin.

(iv) In particular, if A is regular and square then the hull inverse A^{H} is a sublinear mapping, and if, in addition, A is thin then A^{H} is linear. We shall call any sublinear mapping $S\colon \mathbb{IR}^n \to \mathbb{IR}^n$ satisfying

$$A^{\mathrm{H}}b \subseteq Sb \text{ for all } b \in \mathbb{IR}^n$$

an *inverse* of $A \in \mathbb{IR}^{n\times n}$.

Every sublinear mapping $S\colon \mathbb{IR}^n \to \mathbb{IR}^m$ can be extended to matrix arguments $A \in \mathbb{IR}^{n\times p}$ by applying S to each column $Ae^{(i)}$ of A separately:

$$SA := (S(Ae^{(1)}), \ldots, S(Ae^{(p)})).$$

In other words, SA is the $m \times p$ interval matrix satisfying

$$(SA)e^{(k)} = S(Ae^{(k)}) \quad (k = 1, \ldots, p). \tag{2}$$

For the matrix–vector multiplication mapping, this definition just amounts to matrix multiplication:

$$A^{\mathrm{M}}B = AB \quad \text{for } A \in \mathbb{IR}^{m\times n},\ B \in \mathbb{IR}^{n\times p}.$$

Some properties of matrix multiplication immediately extend to arbitrary sublinear mappings:

3.5.2 Proposition Let $S\colon \mathbb{IR}^n \to \mathbb{IR}^n$ be sublinear and $A, B \in \mathbb{IR}^{n\times p}$. Then

$$A \subseteq B \Rightarrow SA \subseteq SB, \tag{3}$$

$$S(\alpha A) = \alpha(SA) \quad \text{for } \alpha \in \mathbb{R}, \tag{4}$$

$$S0 = 0, \tag{4a}$$

$$S(A \pm B) \subseteq SA \pm SB, \tag{5}$$

$$S(A\tilde{C}) \subseteq (SA)\tilde{C} \quad \text{for } \tilde{C} \in \mathbb{R}^{p\times q}, \tag{6}$$

$$\check{A} = 0 \Rightarrow \operatorname{mid}(SA) = 0. \tag{7}$$

If S is linear then (5) and (6) hold with equality.
Proof Expressions (3) and (4) follow by applying (S1) and (S2) columnwise, and putting $\alpha = 0$ we get (4a). Expression (5) follows for the + sign by applying (S3) columnwise, and it holds for the − sign since

$$S(A - B) = S(A + (-1)B) \subseteq SA + (-1)SB = SA - SB.$$

To show (6) we first note that for $\tilde{x} \in \mathbb{R}^p$ we have

$$(A\tilde{x})_i = \sum_k A_{ik}\tilde{x}_k = \sum_k \tilde{x}_k(Ae^{(k)})_i$$

so that

$$A\tilde{x} = \sum_k \tilde{x}_k(Ae^{(k)}). \tag{8}$$

Since S is subadditive and homogeneous, this implies

$$\begin{aligned} S(A\tilde{x}) &\subseteq \sum_k S(\tilde{x}_k(Ae^{(k)})) = \sum_k \tilde{x}_k(S(Ae^{(k)})) \\ &= \sum_k \tilde{x}_k((SA)e^{(k)}) = (SA)\tilde{x} \end{aligned}$$

by (2) and (8), with SA in place of A. Hence $S(A\tilde{x}) \subseteq (SA)\tilde{x}$, and application to each column $\tilde{x} = \check{C}e^{(j)}$ of $\check{C}$ shows (6).

If $\check{A} = 0$ then $A = -A$. Therefore $SA = S((-1)A) = (-1)SA = -SA$ so that $\text{mid}(SA) = -\text{mid}(SA)$. This implies (7). Finally, if S is linear then all arguments in the proof hold with equality. □

We now extend the notion of *absolute value* from matrices to sublinear mappings $S\colon \mathbb{IR}^n \to \mathbb{IR}^m$ by defining

$$|S| := |S[-I, I]|.$$

Clearly, $|S|$ is a nonnegative real $m \times n$ matrix and $|A^{\mathrm{M}}| = |A|$.

3.5.3 Proposition Let $S\colon \mathbb{IR}^n \to \mathbb{IR}^m$ be sublinear and $A, B \in \mathbb{IR}^{n\times p}$. Then

$$S[-I, I] = [-|S|, |S|], \tag{9}$$

$$q(SA, SB) \leq |S|q(A, B), \tag{10}$$

$$|SA| \leq |S|\,|A|. \tag{11}$$

In particular, every sublinear mapping is continuous.
Proof Let $B := S[-I, I]$. Then $|B| = |S|$ by definition, and $\check{B} = 0$ by (7). Hence $S[-I, I] = B = [-|B|, |B|] = [-|S|, |S|]$ by relation (3.1.42), and (9)

holds. To prove (10) we note that $B \subseteq A + [-I, I]q(A, B)$ by (3.3.11) so that

$$\begin{aligned} SB &\subseteq S(A + [-I, I]q(A, B)) && \text{by (3)} \\ &\subseteq SA + S([-I, I]q(A, B)) && \text{by (5)} \\ &\subseteq SA + (S[-I, I])q(A, B) && \text{by (6)} \\ &= SA + [-|S|, |S|]q(A, B) && \text{by (9)}. \end{aligned}$$

With $Q := |S|q(A, B)$ we get $SB \subseteq SA + [-Q, Q]$, and by symmetry $SA \subseteq SB + [-Q, Q]$. By Proposition 3.3.1 this implies (10) and hence the continuity of S. Finally, (11) follows by putting $B = 0$ in (10). □

In order to be able to characterize the linear mappings we need a further concept: the *core* of a sublinear mapping $S: \mathbb{IR}^n \to \mathbb{IR}^m$, defined as

$$\operatorname{cor}(S) := SI = (Se^{(1)}, \ldots, Se^{(n)}).$$

For example, $\operatorname{cor}(A^{\mathrm{M}}) = A$.

3.5.4 Theorem Let $S: \mathbb{IR}^n \to \mathbb{IR}^m$ be sublinear. Then

$$|\operatorname{cor}(S)| \le |S|, \tag{12}$$

$$Sx \subseteq \operatorname{cor}(S)\check{x} + |S|(x - \check{x}) \quad \text{for all } x \in \mathbb{IR}^n. \tag{13}$$

If S is linear then (13) holds with equality and $\operatorname{cor}(S)$ is thin; in particular, a linear mapping is uniquely determined by its core and absolute value.

Proof By (S1), $\operatorname{cor}(S) = SI \subseteq S[-I, I]$. Hence $|\operatorname{cor}(S)| \le |S[-I, I]| = |S|$ and (12) holds. To prove (13) we note that $x = \check{x} + [-I, I]\operatorname{rad}(x)$ so that

$$\begin{aligned} Sx &= S(\check{x} + [-I, I]\operatorname{rad}(x)) \subseteq S\check{x} + S([-I, I]\operatorname{rad}(x)) \\ &\subseteq S\check{x} + (S[-I, I])\operatorname{rad}(x) = S(I\check{x}) + [-|S|, |S|] \cdot \operatorname{rad}(x) \\ &\subseteq (SI)\check{x} + |S|[-\operatorname{rad}(x), \operatorname{rad}(x)] = \operatorname{cor}(S)\check{x} + |S|(x - \check{x}) \end{aligned}$$

by (S3), (6), (9) and again (6). This implies (13). If S is linear then (6) holds with equality so that we have equality in (13), i.e. $Sx = \operatorname{cor}(S)\check{x} + |S|(x - \check{x})$ is determined for all $x \in \mathbb{IR}^n$ by $\operatorname{cor}(S)$ and $|S|$. For linear S, $\operatorname{cor}(S)$ is thin since $2 \cdot \operatorname{rad}(\operatorname{cor}(S)) = |\operatorname{cor}(S) - \operatorname{cor}(S)| = |SI - SI| = |S(I - I)| = 0$. □

If S is not linear, then S is not necessarily determined by its core and absolute value. In particular, we need not have equality in (12).

3.5.5 Example Let $A \in \mathbb{IR}^{n \times n}$ be regular. Then

$$\begin{aligned} &\operatorname{cor}(A^{\mathrm{H}}) = A^{\mathrm{H}}I = \square\{\tilde{A}^{-1} \mid \tilde{A} \in A\} = A^{-1}, \\ &|A^{\mathrm{H}}| = |A^{\mathrm{H}}[-I, I]| \le |A^{-1}[-I, I]| = |A^{-1}| = |\operatorname{cor}(A^{\mathrm{H}})| \le |A^{\mathrm{H}}|. \end{aligned}$$

Therefore

$$\operatorname{cor}(A^{\mathrm{H}}) = A^{-1}, \quad |A^{\mathrm{H}}| = |A^{-1}|. \tag{14}$$

Hence A^{H} and $(A^{-1})^{\mathrm{M}}$ have the same core and absolute value. For the specific example

$$A := \begin{pmatrix} [-1,7] & -1 \\ 3 & 1 \end{pmatrix}, \quad b := \begin{pmatrix} [-1,1] \\ [-1,1] \end{pmatrix}$$

we get, using the method of Example 3.4.7,

$$A^{-1} = \begin{pmatrix} [\ \ 0.1, \ \ 0.5] & [\ \ 0.1, 0.5] \\ [-1.5, -0.3] & [-0.5, 0.7] \end{pmatrix}, \quad A^{\mathrm{H}}b = \begin{pmatrix} [-1,1] \\ [-2,2] \end{pmatrix}.$$

Since $\check{b} = 0$ equality in (13) would imply $A^{\mathrm{H}}b = |A^{\mathrm{H}}|b = |A^{-1}|b$ and $A^{\mathrm{H}} = (A^{-1})^{\mathrm{M}}$ would imply $A^{\mathrm{H}}b = A^{-1}b = |A^{-1}|b$, too. But this is not the case since

$$|A^{-1}|b = \begin{pmatrix} [-1,1] \\ [-2.2, 2.2] \end{pmatrix}.$$

Hence A^{H} and $(A^{-1})^{\mathrm{M}}$ are distinct sublinear mappings with the same core and absolute value.

The axioms defining a sublinear mapping do not suffice to prove the following generalization of the matrix relation (3.1.40):

(S4) $\operatorname{rad}(Sx) \geq |S|\operatorname{rad}(x)$ for all $x \in \mathbb{IR}^n$.

Hence we consider this property as an additional axiom and call a sublinear mapping S: $\mathbb{IR}^n \to \mathbb{IR}^m$ *normal* if (S4) holds. Clearly, the matrix multiplication mapping A^{M} is normal; on the other hand, the hull inverse A^{H} need not be normal; e.g., for the previous example we have

$$\operatorname{rad}(A^{\mathrm{H}}b) = \begin{pmatrix} 1 \\ 2 \end{pmatrix} \ngeq |A^{\mathrm{H}}|\operatorname{rad}(b) = \begin{pmatrix} 1 \\ 2.2 \end{pmatrix}.$$

3.5.6 Proposition Let S: $\mathbb{IR}^n \to \mathbb{IR}^m$ be sublinear and $x \in \mathbb{IR}^n$.

(i) If S is linear then

$$\operatorname{rad}(Sx) = |S|\operatorname{rad}(x); \tag{15}$$

in particular, every linear mapping is normal.

(ii) If S is normal then

$$\check{x} = 0 \Rightarrow Sx = |S|x. \tag{16}$$

Proof (i) If S is linear then (13) holds with equality and $\operatorname{cor}(S)$ is thin, so that

$$\begin{aligned}\operatorname{rad}(Sx) &= \operatorname{rad}(|S|(x-\check{x})) = \operatorname{rad}([-|S|\operatorname{rad}(x), |S|\operatorname{rad}(x)])\\ &= |S|\operatorname{rad}(x).\end{aligned}$$

(ii) If S is normal and $\check{x} = 0$ then

$$|\,|S|x| \le |S|\,|x| = |S|\operatorname{rad}(x) \le \operatorname{rad}(Sx),$$

and by (13) and (7) we obtain

$$Sx \subseteq |S|x \subseteq [-\operatorname{rad}(Sx), \operatorname{rad}(Sx)] = Sx.$$

Hence $Sx = |S|x$. □

The set of normal sublinear mappings is closed under addition and composition.

3.5.7 Proposition

(i) If $S, T\colon \mathbb{IR}^n \to \mathbb{IR}^m$ are sublinear mappings and $a, b \in \mathbb{R}$ then the mapping $aS + bT\colon \mathbb{IR}^n \to \mathbb{IR}^m$ defined by

$$(aS + bT)x := a(Sx) + b(Tx)$$

is sublinear; it is normal (linear) if S and T are normal (linear).

(ii) If $S\colon \mathbb{IR}^n \to \mathbb{IR}^m$ and $T\colon \mathbb{IR}^p \to \mathbb{IR}^n$ are sublinear mappings then the mapping $ST\colon \mathbb{IR}^p \to \mathbb{IR}^m$ defined by

$$(ST)x := S(Tx)$$

is sublinear and satisfies

$$|ST| \le |S|\,|T|; \tag{17}$$

it is normal (linear) if S and T are normal (linear); in this case, (17) holds with equality.

Proof Trivial. □

Note that in general $A^M B^M \ne (AB)^M$, due to lack of associativity.

We now introduce another important property of sublinear mappings. A sublinear mapping $S\colon \mathbb{IR}^n \to \mathbb{IR}^m$ is called *regular* if

$$\tilde{x} \in \mathbb{R}^n, \quad 0 \in S\tilde{x} \Rightarrow \tilde{x} = 0. \tag{18}$$

By Corollary 3.4.5, multiplication by an interval matrix $A \in \mathbb{IR}^{m\times n}$ is regular iff A itself is regular. As a sufficient condition for regularity we have the following:

3.5.8 Proposition Let $S: \mathbb{IR}^n \to \mathbb{IR}^m$ be sublinear.

(i) If the matrix $S\tilde{A}$ is regular for some nonsingular matrix $\tilde{A} \in \mathbb{R}^{n\times n}$ then S is regular.
(ii) If $m = n$ and $Sx \subseteq D^{-1}(x - Tx)$ for some nonsingular diagonal matrix $D \in \mathbb{IR}^n$ and some sublinear mapping $T: \mathbb{IR}^n \to \mathbb{IR}^n$ with $\rho(|T|) < 1$ then S is regular.

Proof (i) Since $\tilde{A}$ is nonsingular we can write any $\tilde{x} \in \mathbb{R}^n$ as $\tilde{x} = \tilde{A}\tilde{y}$ for some $\tilde{y} \in \mathbb{R}^n$. Hence $0 \in S\tilde{x}$ implies $0 \in S(\tilde{A}\tilde{y}) \subseteq (S\tilde{A})\tilde{y}$ by (6), so that $\tilde{y} = 0$ and $\tilde{x} = \tilde{A}\tilde{y} = 0$.

(ii) Since D is diagonal, $0 \in S\tilde{x}$ implies $0 \in \tilde{x} - T\tilde{x}$, hence $\tilde{x} \in T\tilde{x}$. But then $|\tilde{x}| \le |T\tilde{x}| \le |T|\,|\tilde{x}|$, and since $\rho(|T|) < 1$ this implies $|\tilde{x}| = 0$, hence $\tilde{x} = 0$. □

Note that if $A \in \mathbb{IR}^{n\times n}$ is regular then A^{H} need not be regular.

3.5.9 Example Let

$$A = \begin{pmatrix} 3 & [-2,2] & 0 \\ 0 & 3 & [-2,2] \\ [-2,2] & 0 & 3 \end{pmatrix}, \quad b = \begin{pmatrix} 6 \\ 6 \\ 6 \end{pmatrix}.$$

A is regular (strictly diagonal dominant). Since

$$\begin{pmatrix} 3 & 2 & 0 \\ 0 & 3 & -1.5 \\ 0 & 0 & 3 \end{pmatrix} \begin{pmatrix} 0 \\ 3 \\ 2 \end{pmatrix} = \begin{pmatrix} 6 \\ 6 \\ 6 \end{pmatrix}$$

we have $(0, 3, 2)^{\mathrm{T}} \in A^{\mathrm{H}}b$, and by symmetry, also $(2, 0, 3)^{\mathrm{T}}$, $(3, 2, 0)^{\mathrm{T}} \in A^{\mathrm{H}}b$. Hence $([0, 3], [0, 3], [0, 3])^{\mathrm{T}}$, the hull of the three points, is contained in $A^{\mathrm{H}}b$. Thus $0 \in A^{\mathrm{H}}b$, and since b is thin, A^{H} is not regular.

3.6 *M*-matrices and inverse positive matrices

In this section we study the classes of inverse positive matrices, i.e. regular square interval matrices with nonnegative inverse. Such matrices naturally arise in operations research and in the numerical treatment of ordinary and elliptic partial differential equations. The most important easily recognizable subclass consists of the so-called M-matrices. In the context of interval calculations M-matrices play a distinguished role since they behave particularly well in algorithms for the solution of linear interval equations discussed in Chapter 4. (Most results of this section hold with only trivial changes for matrices which can be scaled by thin diagonal matrices to M-matrices or inverse positive matrices. The formulation of these generalizations is left to the reader.)

We motivate the concept of an M-matrix by the following observation.

3.6.1 Lemma Let $P \in \mathbb{R}^{n\times n}$ be a nonnegative matrix with spectral radius $\rho(P) < \alpha$. Then $\alpha I - P$ is regular, $Q := (\alpha I - P)^{-1} \geq 0$, and $\rho(Q) = (\alpha - \rho(P))^{-1}$.

Proof Proposition 3.2.5, applied with $A = \alpha^{-1}P$, shows that $\alpha I - P$ is regular and $Q \geq 0$. If z is an eigenvector of $P, Pz = \lambda z$, then $Qz = (\alpha - \lambda)^{-1}z$, and conversely. Now $|\lambda| \leq \rho(P)$ so that $|(\alpha - \lambda)^{-1}| \leq (\alpha - \rho(P))^{-1}$, and since equality holds for the Perron eigenvalue we have $\rho(Q) = (\alpha - \rho(P))^{-1}$. □

By relation (3.2.5), there is a vector $u > 0$ such that $Pu < \alpha u$. Hence the matrix $A = \alpha I - P$ satisfies $Au > 0$ and $A_{ik} \leq 0$ for $i \neq k$. This motivates the following definition.

An *M-matrix* is a square matrix $A \in \mathbb{IR}^{n\times n}$ such that $A_{ik} \leq 0$ for all $i \neq k$, and $Au > 0$ for some positive vector $u \in \mathbb{R}^n$. For instance, the matrix in Example 3.4.2 is an M-matrix $\left(\text{take } u = \begin{pmatrix}1\\1\end{pmatrix}\right)$.

3.6.2 Example The tridiagonal $n \times n$ matrix

$$A = \begin{bmatrix} 2 & -1 & & & \\ -1 & \cdot & & \cdot & 0 \\ & \cdot & \cdot & \cdot & \\ & 0 & \cdot & \cdot & -1 \\ & & & -1 & 2 \end{bmatrix}$$

is an M-matrix. Indeed, if u is the vector with $u_i = (i+1)(n+1-i)$ then $u > 0$ and $Au = (n+3, 2, 2, \ldots, 2, n+3)^{\mathrm{T}} > 0$. For any diagonal matrix $D \in \mathbb{IR}^{n\times n}$ with $D_{ii} \geq 0$, the matrix

$$A + D = \begin{bmatrix} 2 + D_{11} & -1 & & & \\ -1 & \cdot & & \cdot & 0 \\ & \cdot & \cdot & \cdot & \\ 0 & & \cdot & \cdot & -1 \\ & & & -1 & 2 + D_{nn} \end{bmatrix}$$

is an M-matrix, too, since $(A + D)u \geq Au > 0$. This example is relevant in the context of finite difference methods for the solution of two-point boundary value problems.

For the determination of u in more general instances see the Remark after Proposition 3.7.3 below. We now note some useful facts for real M-matrices.

3.6.3 Proposition Let $A, B \in \mathbb{R}^{n \times n}$ and suppose that A is an M-matrix and $B \geq 0$. Then

(i) A is regular, $A_{ii} > 0$ for all i, and $A^{-1} \geq 0$,
(ii) $0 < x \in \mathbb{R}^n \Rightarrow A^{-1}x > 0$,
(iii) $A - B$ is an M-matrix iff $\rho(A^{-1}B) < 1$.

Proof Let $u > 0$ be such that $Au > 0$. Since $A_{ik} \leq 0$ for $i \neq k$ we have

$$A_{ii}u_i \geq \sum_{k=1}^{n} A_{ik}u_k = (Au)_i > 0 \qquad (1)$$

so that $A_{ii} > 0$. In particular, $\alpha := \max\{A_{ii} \mid i = 1, \ldots, n\} > 0$. Now the matrix $P := \alpha I - A$ is nonnegative and satisfies $Pu = \alpha u - Au < \alpha u$; hence $\rho(P) < \alpha$ by relation (3.2.5). Since $A = \alpha I - P$, the preceding lemma now implies (i).

To prove (ii) we note that, since $A^{-1}A = I$, no row of A^{-1} consists of zeros only. Therefore, if $x > 0$, each component of $A^{-1}x$ is a sum of nonnegative products with at least one positive term. Hence $A^{-1}x > 0$, and (ii) holds.

To prove (iii) let $A - B$ be an M-matrix. Then $(A - B)u > 0$ for some $u > 0$, and since $A^{-1} \geq 0$ we get $A^{-1}Bu < u$. Hence $\rho(A^{-1}B) < 1$ by relation (3.2.5). Conversely, if $\rho(A^{-1}B) < 1$ then $I - A^{-1}B$ is nonsingular by Lemma 3.6.1, and (with an arbitrary $v > 0$)

$$u := (I - A^{-1}B)^{-1}A^{-1}v > 0.$$

Hence $(A - B)u = A(I - A^{-1}B)u = v > 0$. Since $(A - B)_{ik} \leq A_{ik} \leq 0$ for $i \neq k$, this implies that $A - B$ is an M-matrix. □

3.6.4 Corollary Let $A \in \mathbb{R}^{n\times n}$ be a matrix with $A_{ik} \le 0$ for all $i \ne k$. Then the following statements are equivalent:

(i) A is an M-matrix,
(ii) $A^{-1} \ge 0$,
(iii) $u \ge 0, Au \le 0 \Rightarrow u = 0$.

Proof By the previous proposition, (i) implies (ii). If (ii) holds and $Au \le 0$ then multiplication with $A^{-1} \ge 0$ gives $u \le 0$. Therefore (ii) implies (iii). Now suppose that (iii) holds but A is not an M-matrix. With $\alpha := \max\{1, A_{11}, \ldots, A_{nn}\} \ge 1$, we have $B := I - \alpha^{-1}A \ge 0$, and $I - B = \alpha^{-1}A$ is not an M-matrix. Proposition 3.6.3 applied with the identity in place of A shows that $\rho(B) \ge 1$. By relation (3.2.6) there is a vector $u \ge 0$ with $Bu \ge u \ne 0$, violating (iii). Therefore (iii) implies (i). □

The first part of Proposition 3.6.3 extends to interval M-matrices.

3.6.5 Theorem

(i) If $A \in \mathbb{IR}^{n\times n}$ is an M-matrix and $B \subseteq A$ then B is an M-matrix; in particular, all $\tilde{A} \in A$ are M-matrices.
(ii) $A \in \mathbb{IR}^{n\times n}$ is an M-matrix iff $\underline{A}$ and $\bar{A}$ are M-matrices.
(iii) Every M-matrix $A \in \mathbb{IR}^{n\times n}$ is regular, and

$$A^{-1} = [\bar{A}^{-1}, \underline{A}^{-1}] \ge 0. \tag{2}$$

Proof (i) immediately follows from the definition. If $\underline{A}$ and $\bar{A}$ are M-matrices then $A_{ik} \le \bar{A}_{ik} \le 0$ for $i \ne k$, and $Au \ge \underline{A}u \ge 0$ for some $u > 0$. Hence A is an M-matrix, and (ii) holds in view of (i). Finally, by (i) and Proposition 3.6.3(i), each $\tilde{A} \in A$ is regular and $\tilde{A}^{-1} \ge 0$. Now $\underline{A} \le \tilde{A} \le \bar{A}$ implies $\tilde{A}^{-1}\underline{A} \le I \le \tilde{A}^{-1}\bar{A}$, hence $\tilde{A}^{-1} \le \underline{A}^{-1}$ and $\bar{A}^{-1} \le \tilde{A}^{-1}$. Therefore $A^{-1} = \square\{\tilde{A}^{-1} \mid \underline{A} \le \tilde{A} \le \bar{A}\} \subseteq [\bar{A}^{-1}, \underline{A}^{-1}]$, and (2) follows since the bounds belong to A^{-1}. □

If we assume that the boundary matrices have a nonnegative inverse, we can extend this still further.

3.6.6 Proposition (Kuttler) Let $A \in \mathbb{IR}^{n\times n}$. If $\underline{A}$, $\bar{A}$ are regular and $\underline{A}^{-1}, \bar{A}^{-1} \ge 0$ then A is regular and

$$A^{-1} = [\bar{A}^{-1}, \underline{A}^{-1}] \ge 0.$$

Proof Pick $v > 0$. Then $u := \underline{A}^{-1}v > 0$ since $\underline{A}^{-1}$ cannot have a row of zeros. Now, if $\tilde{A} \in A$ then $\underline{A} \le \tilde{A} \le \bar{A}$ so that

$$\bar{A}^{-1}\tilde{A} \leq I \leq \underline{A}^{-1}\tilde{A}. \tag{3}$$

Hence $\tilde{B} := \bar{A}^{-1}\tilde{A}$ satisfies $\tilde{B} \leq I$ and $\tilde{B}u = \bar{A}^{-1}\tilde{A}u \geq \bar{A}^{-1}\underline{A}u = \bar{A}^{-1}v > 0$. Therefore $\tilde{B}$ is an M-matrix. In particular, $\tilde{B}$ and $\tilde{A} = \bar{A}\tilde{B}$ are regular. Now $\tilde{A}^{-1} = \tilde{B}^{-1}\bar{A}^{-1} \geq 0$; hence (3) implies $\bar{A}^{-1} \leq \tilde{A}^{-1} \leq \underline{A}^{-1}$. Since the bounds are attained for $\tilde{A} = \bar{A}$ and $\tilde{A} = \underline{A}$, the assertion follows. □

We call an interval matrix A *inverse positive* if A is regular and $A^{-1} \geq 0$; equivalently, if the assumptions of Proposition 3.6.6 hold. Every M-matrix is inverse positive, and indeed the M-matrices are the only known class of inverse positive matrices for which this property can be checked without inverting the endpoint matrices. For inverse positive matrices not only A^{-1} but also $A^{\mathrm{H}}b$ can be described explicitly.

3.6.7 Theorem Let $A \in \mathbb{IR}^{n\times n}$ be inverse positive. Then

$$x := A^{\mathrm{H}}b = [\tilde{A}^{-1}\underline{b}, \underset{\sim}{A}^{-1}\bar{b}] \tag{4}$$

where $\tilde{A}, \underset{\sim}{A} \in A$ are defined by

$$\tilde{A}_{ik} = \bar{A}_{ik} \quad \text{if } \underline{x}_k \geq 0 \text{ and } \tilde{A}_{ik} = \underline{A}_{ik} \text{ otherwise},$$

$$\underset{\sim}{A}_{ik} = \bar{A}_{ik} \quad \text{if } \bar{x}_k \leq 0 \text{ and } \underset{\sim}{A}_{ik} = \underline{A}_{ik} \text{ otherwise}.$$

In particular,

$$A^{\mathrm{H}}b = A^{-1}b = \begin{cases} [\bar{A}^{-1}\underline{b}, \underline{A}^{-1}\bar{b}] & \text{if } b \geq 0, \\ [\underline{A}^{-1}\underline{b}, \underline{A}^{-1}\bar{b}] & \text{if } b \ni 0, \\ [\underline{A}^{-1}\underline{b}, \bar{A}^{-1}\bar{b}] & \text{if } b \leq 0. \end{cases} \tag{5}$$

Proof For $\tilde{x} \in \mathbb{R}^n$ the matrix $A(\tilde{x}) \in \mathbb{R}^{n\times n}$ with entries

$$A(\tilde{x})_{ik} = \bar{A}_{ik} \quad \text{if } \tilde{x}_k \geq 0,$$
$$A(\tilde{x})_{ik} = \underline{A}_{ik} \quad \text{otherwise},$$

has the property that $A(\tilde{x}) \in A$ and

$$A(\tilde{x})\tilde{x} \geq \tilde{A}\tilde{x} \quad \text{for all } \tilde{A} \in A. \tag{6}$$

We show that

$$A(\tilde{x})\tilde{x} = \underline{b} \Leftrightarrow \tilde{x} = \inf(A^{\mathrm{H}}b). \tag{7}$$

To see this we first show that there is a vector $\tilde{x} \in \mathbb{R}^n$ with $A(\tilde{x})\tilde{x} = \underline{b}$. Define $\tilde{x}^0 := \bar{A}^{-1}\underline{b}$ and $\tilde{x}^{l+1} := A(\tilde{x}^l)^{-1}\underline{b}$ $(l = 0, 1, 2, \ldots)$. Then (6) implies $A(\tilde{x}^0)\tilde{x}^0 \geq \bar{A}\tilde{x}^0 = \underline{b}$ and $A(\tilde{x}^l)\tilde{x}^l \geq A(\tilde{x}^{l-1})\tilde{x}^l = \underline{b}$ for $l \geq 1$ so that $\tilde{x}^l \geq A(\tilde{x}^l)^{-1}\underline{b} = \tilde{x}^{l+1}$ for all $l \geq 0$. Thus the $\tilde{x}^l$ form a decreasing sequence and since $A(\tilde{x})$ only changes when some entry $\tilde{x}_k$ changes its sign there is an index $l \le n-1$ with $\tilde{x}^l = \tilde{x}^{l+1}$. Hence $\tilde{x} := \tilde{x}^l$ satisfies $A(\tilde{x})\tilde{x} = A(\tilde{x}^l)\tilde{x}^{l+1} = \underline{b}$. On the other hand, let $\tilde{x}$ be any solution of $A(\tilde{x})\tilde{x} = \underline{b}$. If $\tilde{z} = \tilde{A}^{-1}\tilde{b}$ ($\tilde{A} \in A$, $\tilde{b} \in$ b) then $\tilde{A}\tilde{z} = \tilde{b} \ge \underline{b} = A(\tilde{x})\tilde{x} \ge \tilde{A}\tilde{x}$ by (6), so that multiplication with $\tilde{A}^{-1} \ge 0$ gives $\tilde{z} \ge \tilde{x}$. Hence $A^{\mathrm{H}}b \ge \tilde{x}$, and since $\tilde{x} = A(\tilde{x})^{-1}\underline{b} \in A^{\mathrm{H}}b$ we have $\tilde{x} = \inf(A^{\mathrm{H}}b)$. In particular, $\tilde{x}$ is unique so that (7) holds. But clearly (7) implies that $\inf(A^{\mathrm{H}}b) = \tilde{A}^{-1}\underline{b}$ with $\tilde{A} = A(\underline{x})$. Replacing b by $-b$ we also find $\sup(A^{\mathrm{H}}b) = -\inf(A^{\mathrm{H}}(-b)) = -\underset{\sim}{A}^{-1}(-\overline{b}) = \underset{\sim}{A}^{-1}\overline{b}$ with $\underset{\sim}{A} = A(-\overline{x})$. This implies (4), and (5) is an obvious consequence. □

The proof gives a constructive method for obtaining $A^{\mathrm{H}}b$; one has to solve at most $2n$ linear systems to find $\tilde{x}^l = \inf(A^{\mathrm{H}}b)$ and similarly $\sup(A^{\mathrm{H}}b)$. With a suitable implementation (using rank one modifications of the inverse) the total effort can be held at $O(n^3)$ operations. In the presence of roundoff errors the decision whether $\tilde{x}_k \ge 0$ may be difficult, so care has to be taken to get the proper bounds.

We end the section with some remarks about regularity. The inverse matrix $A^{-1} = [\overline{A}^{-1}, \underline{A}^{-1}]$ of an inverse positive matrix need not be regular. For example, if

$$A = \begin{pmatrix} 2 & -1 \\ [-6, -2] & 3.5 \end{pmatrix}$$

then

$$A^{-1} = \begin{pmatrix} [0.7, 3.5] & [0.2, 1.0] \\ [0.4, 6.0] & [0.4, 2.0] \end{pmatrix}$$

contains the singular matrix $\begin{pmatrix} 1 & 1 \\ 1 & 1 \end{pmatrix}$. However, we have:

3.6.8 Theorem Let $A \in \mathbb{IR}^{n \times m}$ be inverse positive. Then the hull inverse A^{H} is regular.

Proof Suppose $0 \in A^{\mathrm{H}}\tilde{b}$. Then, by (4), there are $\underset{\sim}{A}$, $\tilde{A} \in A$ such that $\underset{\sim}{A}^{-1}\tilde{b} \le 0 \le \tilde{A}^{-1}\tilde{b}$. Since $\underline{A}^{-1}\underset{\sim}{A} \ge I$, $\underline{A}^{-1}\tilde{A} \ge I$, this implies $0 \le \underline{A}^{-1}\tilde{b} \le 0$, so that $\underline{A}^{-1}\tilde{b} = 0$. Hence $\tilde{b} = 0$. □

3.7 *H*-matrices

H-matrices are a generalization of *M*-matrices obtained by lifting the restrictive sign condition in the definition of an *M*-matrix. It turns out that one important property of *M*-matrices, their regularity, carries over to *H*-matrices, whereas other properties, e.g. explicit formula for the inverse matrix, do not generalize. However, the inverse of an H-matrix A can at least be bounded by the inverse of a related matrix, namely Ostrowski's comparison matrix $\langle A\rangle$. We therefore begin this section by studying the properties of the comparison operator; then we apply them to obtain information about H-matrices. We also show how to verify the H-matrix (and M-matrix) property for numerically given matrices.

For any square matrix $A \in \mathbb{IR}^{n\times n}$ we can force upon A the sign condition of M-matrices by introducing the *comparison matrix* $\langle A\rangle$ with entries

$$\langle A\rangle_{ii} := \langle A_{ii}\rangle, \quad \langle A\rangle_{ik} := -|A_{ik}| \quad \text{for } i \neq k.$$

It is immediately clear that for a real M-matrix $A \in \mathbb{R}^{n\times n}$ we have $\langle A\rangle = A$, whereas for an interval M-matrix $A \in \mathbb{IR}^{n\times n}$ we have $\langle A\rangle = \underline{A}$. In general, the comparison matrix has the following properties.

3.7.1 Proposition Let $A, B \in \mathbb{IR}^{n\times n}$ and $C = \mathbb{IR}^{n\times p}$. Then

$$\langle A\rangle = \langle \tilde{A}\rangle \quad \text{for some } \tilde{A} \in A, \tag{1}$$

$$\langle A\rangle \geq \langle \check{A}\rangle - \operatorname{rad}(A), \tag{2}$$

with equality if $0 \notin A_{ii}$ for $i = 1, \ldots, n$,

$$\langle B\rangle \geq \langle A\rangle - q(A, B), \tag{3}$$

$$\langle A \pm B\rangle \geq \langle A\rangle - |B|, \tag{4}$$

$$|A \pm B| \geq \langle A\rangle - \langle B\rangle, \tag{5}$$

$$|AC| \geq \langle A\rangle |C|, \tag{6}$$

$$B \subseteq A \Rightarrow \langle B\rangle \geq \langle A\rangle. \tag{7}$$

Proof Define $\tilde{A}$ on the diagonal by $\tilde{A}_{ii} = \underline{A}_{ii}$ if $0 < A_{ii}$, $\tilde{A}_{ii} = \overline{A}_{ii}$ if $0 > A_{ii}$, $\tilde{A}_{ii} = 0$ if $0 \in A_{ii}$, and off the diagonal by $\tilde{A}_{ik} = \overline{A}_{ik}$ if $0 \leq \check{A}_{ik}$, $\tilde{A}_{ik} = \underline{A}_{ik}$ if $0 > \check{A}_{ik}$. Then $\langle \tilde{A}\rangle = \langle A\rangle$ and (1) holds.

Relation (2) holds since $\langle A\rangle_{ii} = \langle A_{ii}\rangle \geq \langle \check{A}_{ii}\rangle - \operatorname{rad}(A_{ii}) = \langle \check{A}\rangle_{ii} - \operatorname{rad}(A)_{ii}$ by formula (1.6.10) and $\langle A\rangle_{ik} = -|A_{ik}| = -|\check{A}_{ik}| - \operatorname{rad}(A_{ik}) = -|\check{A}|_{ik} - \operatorname{rad}(A)_{ik}$ for $i \neq k$. If $0 \notin A_{ii}$ for $i = 1, \ldots, n$ then equality holds throughout the argument so that in this case (2) holds with equality.

Relation (4) holds since $\langle A \pm B\rangle_{ii} = \langle A_{ii} \pm B_{ii}\rangle \geq \langle A_{ii}\rangle - |B_{ii}| = \langle A\rangle_{ii} - |B|_{ii}$ by expression (1.6.3) and $\langle A \pm B\rangle_{ik} = -|A \pm B|_{ik} \geq -|A|_{ik} - |B|_{ik} = \langle A\rangle_{ik} - |B|_{ik}$ for $i \neq k$, and (5) follows in the same way as (4), using (1.6.2). The inequality (3) follows from relation (1.7.13) off the diagonal, and a similar relation on the diagonal.

Relation (6) holds since

$$|AC|_{ik} = \left|\sum_j A_{ij}C_{jk}\right| \geq |A_{ii}C_{ik}| - \sum_{j\neq i} |A_{ij}C_{jk}|$$

$$= |A_{ii}|\,|C_{ik}| - \sum_{j\neq i} |A_{ij}|\,|C_{jk}|$$

$$\geq \langle A_{ii}\rangle|C_{ik}| - \sum_{j\neq i} |A_{ij}|\,|C_{jk}| = (\langle A\rangle|C|)_{ik}$$

Finally, if $B \subseteq A$ then $\langle B\rangle_{ii} = \langle B_{ii}\rangle \geq \langle A_{ii}\rangle = \langle A\rangle_{ii}$ by (1.6.1), and $\langle B\rangle_{ik} = -|B_{ik}| \geq -|A_{ik}| = \langle A\rangle_{ik}$ for $i \neq k$, so that (7) holds. □

An *H-matrix* is a square matrix $A \in \mathbb{IR}^{n\times n}$ such that its comparison matrix $\langle A\rangle$ is an *M*-matrix; equivalently, A is an *H*-matrix iff

$$\langle A\rangle u > 0 \quad \text{for some } u > 0. \tag{8}$$

(Note that this forces $0 \notin A_{ii}$ for $i = 1, \ldots, n$.) In particular, every *M*-matrix is an *H*-matrix. In the special case when $u = (1, \ldots, 1)^{\mathrm{T}}$, the condition $\langle A\rangle u > 0$ can be written as

$$\langle A_{ii}\rangle > \sum_{k\neq i} |A_{ik}| \quad \text{for } i = 1, \ldots, n. \tag{9}$$

A matrix $A \in \mathbb{IR}^{n\times n}$ satisfying (9) is called a *strictly diagonally dominant matrix*. For example, the matrix in Example 3.4.1 is diagonally dominant, and hence an *M*-matrix. For many *H*-matrices (8) holds with

$$u = (\langle A_{11}\rangle^{-1}, \ldots, \langle A_{nn}\rangle^{-1})^{\mathrm{T}}.$$

It is a very important fact that, if a matrix is sufficiently close to the identity matrix, then it is strictly diagonally dominant, and hence an *H*-matrix. More specifically we have the following:

3.7.2 Proposition Let $A \in \mathbb{IR}^{n\times n}$. If

$$\beta := \|I - A\|_u < 1 \quad \text{for some } u > 0 \tag{10}$$

then A is an *H*-matrix and $\langle A\rangle u \geq (1-\beta)u > 0$. In particular, if

$$\|I - A\|_\infty < 1 \tag{11}$$

then A is strictly diagonally dominant.

Proof If (10) holds then $\beta u \geq |I - A|u \geq (\langle I\rangle - \langle A\rangle)u = u - \langle A\rangle u$ by (5), so that $\langle A\rangle u \geq (1-\beta)u > 0$. Hence A is an *H*-matrix. The second part follows by putting $u = (1, \ldots, 1)^{\mathrm{T}}$. □

Relations (10) and (11) are only sufficient criteria since there are strictly diagonally dominant *H*-matrices (like $A = -I$) which do not satisfy (10) or (11). As a necessary and sufficient condition we prove the following:

3.7.3 Proposition For $A \in \mathbb{IR}^{n\times n}$ the following conditions are equivalent:

(i) A is an H-matrix;
(ii) $\langle A\rangle$ is regular, and $\langle A\rangle^{-1}e > 0$, where $e = (1, \ldots, 1)^T$;
(iii) $0 \le u \in \mathbb{R}^n$, $\langle A\rangle u \le 0 \Rightarrow u = 0$.

Proof If (i) holds then $\langle A\rangle$ is an M-matrix. Hence, by Proposition 3.6.3, $\langle A\rangle$ is regular, and $\langle A\rangle^{-1} \ge 0$, $\langle A\rangle^{-1}e > 0$; moreover, if $\langle A\rangle u \le 0$ then $u = \langle A\rangle^{-1}\langle A\rangle u \le 0$. Thus (i) implies (ii) and (iii). If (ii) holds then the vector $u := \langle A\rangle^{-1}e$ satisfies $u > 0$ and $\langle A\rangle u = e > 0$ so that A is an H-matrix. Finally, if (iii) holds then $\alpha := 1 + \max\{\langle A_{ii}\rangle \mid i = 1, \ldots, n\} \ge 1$ and $B := I - \alpha^{-1}\langle A\rangle \ge 0$. If $\rho(B) \ge 1$ then by relation (3.2.6) we have $Bu \ge u \ne 0$ for some $u \ge 0$, and $\langle A\rangle u = \alpha(u - Bu) \le 0$, contradicting (iii). Therefore $\rho(B) < 1$ and by relation (3.2.5) we have $Bu < u$ for some $u > 0$, i.e. $\langle A\rangle u = \alpha(u - Bu) > 0$. Therefore A is an H-matrix. □

Remark Condition (ii) is useful for the construction of a vector u with $\langle A\rangle u > 0$. For an H-matrix A, an approximation $\tilde{u}$ to the solution of the linear system $\langle A\rangle u = e$ satisfies $\tilde{u} \approx u = \langle A\rangle^{-1}e > 0$ and $\langle A\rangle\tilde{u} \approx \langle A\rangle u = e > 0$ so that, for sufficiently good approximations $\tilde{u}$, we still have $\tilde{u} > 0$ and $\langle A\rangle\tilde{u} > 0$. On the other hand (iii) can be used to show that a matrix A is not an H-matrix by finding a nonzero vector $u \ge 0$ with $\langle A\rangle u \le 0$.

3.7.4 Corollary Every regular (lower or upper) triangular matrix $A \in \mathbb{IR}^{n\times n}$ is an H-matrix.
Proof Without loss of generality let A be a lower triangular matrix. Then the equation $\langle A\rangle u = e$ implies

$$\langle A_{ii}\rangle u_i = 1 + \sum_{k<i} |A_{ik}|u_k \quad (i = 1, \ldots, n). \tag{12}$$

Since A is regular we have $0 \notin A_{ii}$, i.e. $\langle A_{ii}\rangle > 0$ for all i. Solving (12) for $i = 1, \ldots, n$ gives $u_i > 0$ for all i. Hence $\langle A\rangle^{-1}e = u > 0$, and A is an H-matrix. □

Theorem 3.6.5 extends to H-matrices in the following modified form.

3.7.5 Theorem

(i) If $A \in \mathbb{IR}^{n\times n}$ is an H-matrix and $B \subseteq A$ then B is an H-matrix; in particular, all $\tilde{A} \in A$ are H-matrices.
(ii) $A \in \mathbb{IR}^{n\times n}$ is an H-matrix iff $\check{A}$ is an H-matrix and

$$\rho(\langle\check{A}\rangle^{-1}\,\mathrm{rad}(A)) < 1; \tag{13}$$

in this case

$$\langle A\rangle = \langle\check{A}\rangle - \mathrm{rad}(A). \tag{14}$$

(iii) Every H-matrix $A \in \mathbb{IR}^{n\times n}$ is regular and satisfies

$$|A^{-1}| \leq \langle A\rangle^{-1}, \tag{15}$$

with equality, e.g., if A is an M-matrix.

Proof (i) If A is an H-matrix and $B \subseteq A$ then, by (7), we have $\langle B\rangle u \geq \langle A\rangle u > 0$ for some $u > 0$ and B is an H-matrix; together with the specialization $B = \tilde{A}$ this implies (i).

(ii) If $\check{A}$ is an H-matrix and (13) holds then $\langle\check{A}\rangle$ is an M-matrix, so that by Proposition 3.6.3(iii) the matrix $\langle\check{A}\rangle - \mathrm{rad}(A)$ is an M-matrix. Hence $(\langle\check{A}\rangle - \mathrm{rad}(A))u > 0$ for some $u > 0$, and by (2) we have $\langle A\rangle u > 0$. Hence A is an H-matrix. Conversely, if A is an H-matrix then $\check{A}$ is an H-matrix. Moreover, $\langle A\rangle$ is an M-matrix, and $\langle A\rangle_{ii} > 0$ for all i by Proposition 3.6.3(i). This implies $0 \notin A_{ii}$ for all i, so that (2) holds with equality. Thus, $(\langle\check{A}\rangle - \mathrm{rad}(A))u = \langle A\rangle u > 0$ for some $u > 0$, and we get $\mathrm{rad}(A)u < \langle\check{A}\rangle u$, and $\langle\check{A}\rangle^{-1}\mathrm{rad}(A)u < u$ since $\langle\check{A}\rangle^{-1} \geq 0$ by Proposition 3.6.3(i). This implies (13) and (14).

(iii) Suppose that $\tilde{A}\tilde{x} = 0$ for some $\tilde{A} \in A$, $\tilde{x} \in \mathbb{R}^n$. By (6) we have $0 = |\tilde{A}\tilde{x}| \geq \langle\tilde{A}\rangle|\tilde{x}|$, and multiplication with $\langle\tilde{A}\rangle^{-1} \geq 0$ gives $0 \geq |\tilde{x}|$, and therefore $\tilde{x} = 0$. Thus, every $\tilde{A} \in A$ is regular, so that A is regular. Now $\langle\tilde{A}\rangle \geq \langle A\rangle$ by (7), and $I = |\tilde{A}\tilde{A}^{-1}| \geq \langle\tilde{A}\rangle|\tilde{A}^{-1}| \geq \langle A\rangle|\tilde{A}^{-1}|$, and multiplication with $\langle A\rangle^{-1} \geq 0$ gives $|\tilde{A}^{-1}| \leq \langle A\rangle^{-1}$. This implies (15). If A is an M-matrix then $\langle A\rangle = \underline{A}$ so that $\langle A\rangle^{-1} = \underline{A}^{-1} = |\underline{A}^{-1}|$, and (15) is attained with equality. □

Depending on A the quality of the bound (15) can be good or bad.

3.7.6 Example For $\alpha > 1$, the matrices

$$A_1 = \begin{pmatrix} \alpha & 1 \\ -1 & \alpha \end{pmatrix} \quad \text{and} \quad A_2 = \begin{pmatrix} \alpha & 1 \\ 1 & \alpha \end{pmatrix}$$

are strictly diagonally dominant. Hence they are H-matrices. Their common comparison matrix is

$$B := \langle A_1\rangle = \langle A_2\rangle = \begin{pmatrix} \alpha & -1 \\ -1 & \alpha \end{pmatrix}.$$

Since

$$A_1^{-1} = \frac{1}{\alpha^2+1}\begin{pmatrix} \alpha & -1 \\ 1 & \alpha \end{pmatrix}, \quad A_2^{-1} = \frac{1}{\alpha^2-1}\begin{pmatrix} \alpha & -1 \\ -1 & \alpha \end{pmatrix},$$

$$B^{-1} = \frac{1}{\alpha^2-1}\begin{pmatrix} \alpha & 1 \\ 1 & \alpha \end{pmatrix},$$

$|A_1^{-1}|$ is overestimated by a factor $(\alpha^2+1)/(\alpha^2-1)$, which becomes arbitrarily large as $\alpha \to 1$, whereas $|A_2^{-1}|$ is estimated exactly. (As an explanation we note that A_1 remains regular for $\alpha \to 1$, whereas A_2 and $\langle A_1 \rangle = \langle A_2 \rangle$ become singular.)

Theorem 3.7.5 implies an easily computable bound for the hull inverse.

3.7.7 Theorem Let $A \in \mathbb{IR}^{n\times n}$ be an H-matrix and suppose that $u, v \in \mathbb{R}^n$ satisfy

$$u > 0, \quad \langle A \rangle u \geq v > 0. \tag{16}$$

Then

$$|A^{\mathrm{H}}b| \leq \langle A \rangle^{-1}|b| \leq \|b\|_v u \quad \text{for all } b \in \mathbb{IR}^n, \tag{17}$$

$$|A^{-1}| \leq uw^{\mathrm{T}}, \tag{18}$$

where w is the vector with components $w_i = 1/v_i$ $(i = 1, \ldots, n)$.

Proof We first note that $\langle A \rangle^{-1} \geq 0$ so that (16) implies $\langle A \rangle^{-1} v \leq u$. Since $\langle A \rangle^{-1} \geq |A^{-1}|$ and $|b| \leq \|b\|_v v$, we get

$$\begin{aligned} |A^{\mathrm{H}}b| &\leq |A^{\mathrm{H}}|\,|b| = |A^{-1}|\,|b| \leq \langle A \rangle^{-1}|b| \\ &\leq \|b\|_v \langle A \rangle^{-1} v \leq \|b\|_v u. \end{aligned}$$

Hence (17) holds, and since $|e^{(i)}| \leq w_i v$ we have

$$|A^{-1}|e^{(i)} \leq w_i|A^{-1}|v \leq w_i u \quad (i = 1, \ldots, n).$$

This implies (18). □

The theorem implies the rather crude enclosure

$$A^{\mathrm{H}}b \subseteq [-1, 1]\|b\|_v u. \tag{19}$$

As an example of a better enclosure of $A^{\mathrm{H}}b$, further discussed in Theorem 4.2.6, we prove the following.

3.7.8 Theorem For every H-matrix $A \in \mathbb{IR}^{n\times n}$, the mapping $A^{\mathrm{K}}: \mathbb{IR}^n \to \mathbb{IR}^n$ defined by

$$A^{\mathrm{K}}b := b + \langle A \rangle^{-1}|A - I|\,|b|[-1, 1] \quad \text{for } b \in \mathbb{IR}^n \tag{20}$$

is a sublinear mapping satisfying

$$A^{\mathrm{H}}b \subseteq A^{\mathrm{K}}b \quad \text{for all } b \in \mathbb{IR}^n. \tag{21}$$

We call A^{K} the *Krawczyk inverse* of A.

Proof If $\tilde{A}\tilde{x} = \tilde{b}$ with $\tilde{A} \in A$, $\tilde{b} \in b$ then $\tilde{b} - \tilde{x} = \tilde{A}^{-1}(\tilde{A} - I)\tilde{b} \in A^{-1}(A - I)b$ so that $|\tilde{b} - \tilde{x}| \leq |A^{-1}|\,|A - I|\,|b| \leq \langle A \rangle^{-1}|A - I|\,|b|$. Therefore $\tilde{x} \in b + \langle A \rangle^{-1}|A - I|\,|b|[-1, 1] = A^{\mathrm{K}}b$. Hence $\Sigma(A, b) \subseteq A^{\mathrm{K}}b$, and (21) follows. Sublinearity is straightforward. □

Remarks to Chapter 3

To 3.1 The first systematic treatment of interval vectors and matrices was given by Apostolatos & Kulisch (1968). See also O. Mayer (1970a).

To 3.2 The Perron–Frobenius theory (Perron, 1907; Frobenius, 1912) is described in many books on advanced linear algebra; see, e.g., Berman & Plemmons (1979) and Gantmacher (1959).

To 3.3 The distance for interval vectors and matrices was thoroughly discussed by O. Mayer (1970a). The contraction mapping theorem for vector-valued distances is due to Schröder (1956).

To 3.4 Solution sets of linear interval equations have first been considered by Oettli (1965); further structural results have been obtained by Beeck (1971, 1972, 1973), who also obtained Theorem 3.4.3. The equivalent formulation of Corollary 3.4.4 was obtained earlier by Oettli & Prager (1964); see also Oettli, Prager & Wilkinson (1965).

The method of Example 3.4.7 for the computation of $A^{\mathrm{H}}b$ when $n = 2$ is due to Apostolatos & Kulisch (1968).

The computation of $A^{\mathrm{H}}b$ must be sharply distinguished from the solution of the equation $Ax = b$ interpreted in interval arithmetic. The latter often has no solution even when A is square. For some results see Nickel (1982) and Ratschek & Sauer (1982).

To 3.5 Sublinear mappings were introduced in Neumaier (1984a) and further studied in Neumaier (1985a, 1987a).

To 3.6 M-matrices were introduced by Ostrowski (1937) and have become a large field of study; for many other properties and further equivalent definitions, see, e.g., Berman & Plemmons (1979). Interval M-matrices have first been considered by Barth & Nuding (1974). Proposition 3.6.6 is in Kuttler (1971).

Linear interval equations with an inverse positive coefficient matrix were considered by Beeck (1975).

To 3.7 H-matrices and the comparison matrix $\langle A \rangle$ were also introduced by Ostrowski (1937), who proved the important inequality (15) for real H-matrices. The generalization to interval matrices and the properties of the comparison matrix can be found in Neumaier (1984a).

4

The solution of square linear systems of equations

4.1 Preconditioning

In this section we introduce the class of strongly regular matrices. We show that for linear interval equations with a strongly regular coefficient matrix the solution set can be enclosed by the solution set of a related, preconditioned system whose coefficient matrix is an H-matrix. We also discuss an optimality result for preconditioned systems.

To motivate our approach let $\tilde{x}$ be a solution of the system

$$\tilde{A}\tilde{x} = \tilde{b} \quad (\tilde{A} \in A \text{ regular}, \tilde{b} \in b). \tag{1}$$

If $\mathrm{rad}(A)$ is sufficiently small then $\tilde{A} \approx \check{A}$ so that $\tilde{x} = \tilde{A}^{-1}\tilde{b} \approx \check{A}^{-1}\tilde{b}$. To analyze the error made we multiply (1) by $\check{A}^{-1}$, getting $\check{A}^{-1}\tilde{A}\tilde{x} = \check{A}^{-1}\tilde{b}$. Thus $\tilde{x}$ is the solution of a system with coefficient matrix $\check{A}^{-1}\tilde{A} \approx I$ and right-hand side $\check{A}^{-1}\tilde{b}$. Since $\check{A}^{-1}\tilde{A} \in \check{A}^{-1}A$ and $\check{A}^{-1}\tilde{b} \in \check{A}^{-1}b$ we conclude that $\tilde{x} \in (\check{A}^{-1}A)^{\mathrm{H}}(\check{A}^{-1}b)$ if $\check{A}^{-1}A$ is regular. But for small $\mathrm{rad}(A)$ we have $\check{A}^{-1}A \approx I$, and matrices close to the identity matrix are H-matrices and hence regular. The premultiplication with $\check{A}^{-1}$ therefore transforms the original system (1) into a new system

$$\tilde{A}'\tilde{x} = \tilde{b}' \quad (\tilde{A}' \in \check{A}^{-1}A, \tilde{b}' \in \check{A}^{-1}b) \tag{1'}$$

with a more tractable coefficient matrix. We refer to this transformation as *preconditioning with the midpoint inverse*, and call $\check{A}^{-1}$ the *preconditioning matrix* of the transformation.

In order to free ourselves from the vague terms 'small' and 'close to' we shall say that $A \in \mathbb{IR}^{n\times n}$ is *strongly regular* if $\check{A}^{-1}A$ is regular. We ask for precise conditions under which an interval matrix is strongly regular.

4.1.1 Proposition Let $A \in \mathbb{IR}^{n\times n}$ and suppose that $\check{A}$ is regular. Then the following conditions are equivalent:

(i) A is strongly regular;

(ii) A^{T} is strongly regular;
(iii) $\check{A}^{-1}A$ is regular;
(iv) $\rho(|\check{A}^{-1}|\mathrm{rad}(A)) < 1$;
(v) $\|I - \check{A}^{-1}A\|_u < 1$ for some $u > 0$;
(vi) $\check{A}^{-1}A$ is an H-matrix.

Proof By definition, (i) and (iii) are equivalent. Since $|(\check{A}^{\mathrm{T}})^{-1}|\mathrm{rad}(A^{\mathrm{T}})$, its transpose, $\mathrm{rad}(A)|\check{A}^{-1}|$, and $|\check{A}^{-1}|\mathrm{rad}(A)$ have the same spectral radius (by equation (3.2.9)), it is sufficient to prove the equivalence of (iii)–(vi).

(iii) ⇒ (vi): The matrix $B_0 := \check{A}^{-1}A$ satisfies $\check{B}_0 = \check{A}^{-1}\check{A} = I$ and $\mathrm{rad}(B_0) = |\check{A}^{-1}|\mathrm{rad}(A)$. If the spectral radius of $\mathrm{rad}(B)$ is at least one, then, by relation (3.2.6), there is a nonzero vector $u \geq 0$ with $\mathrm{rad}(B_0)u \geq u$. Now Corollary 3.4.5, applied with $\tilde{x} = u$, shows that B_0 is not regular. Hence, if B_0 is regular we must have $\rho(\mathrm{rad}(B_0)) < 1$.

(iv) ⇒ (v): Since $\rho(\mathrm{rad}(B_0)) < 1$, we have $\mathrm{rad}(B_0)u < u$ for some $u > 0$ by relation (3.2.5), so that $|I - \check{A}^{-1}A|u = |\check{B}_0 - B_0|u = \mathrm{rad}(B_0)u < u$, giving (v).

Since (v) ⇒ (vi) by Proposition 3.7.2, and (vi) ⇒ (iii) by Theorem 3.7.5(iii), the proof is complete. □

The motivating argument preceding the proposition does not really depend on $\check{A}^{-1}$ as preconditioning matrix; instead we can multiply A by any other *preconditioning matrix* $C \in \mathbb{R}^{n\times n}$ in place of $\check{A}^{-1}$ and ask whether CA is regular. This corresponds to a linear transformation of the right-hand side of (1). Furthermore, corresponding to a linear transformation of the variables, we can also postmultiply A by a matrix C'. We can even combine both transformations. The resulting matrix CAC' has to be regular in order to be useful, and we shall require that it is an H-matrix since this is a condition which is easy to check. (Note that $(CA)C' = C(AC')$ since both C and C' are thin, hence the notation CAC' is unambiguous.) Surprisingly, this freedom in the choice of C and C' does not extend the class of matrices A which can be proved to be regular in this way.

4.1.2 Theorem Let $A \in \mathbb{IR}^{n\times n}$ and $C, C' \in \mathbb{R}^{n\times n}$, and suppose that $B = CAC'$ is an H-matrix. Then A is strongly regular, and for all $b \in \mathbb{IR}^n$ the following inclusions are valid:

$$\Sigma(A, b) \subseteq \{C'\tilde{z} \mid \tilde{z} \in \Sigma(B, Cb\}, \tag{2}$$

$$A^{\mathrm{H}}b \subseteq C'(B^{\mathrm{H}}(Cb)). \tag{3}$$

Proof We have $\check{B} = C\check{A}C'$ and $\mathrm{rad}(B) = \mathrm{rad}(CAC') = |C|\mathrm{rad}(AC') = |C|\mathrm{rad}(A)|C'|$. Since B is an H-matrix, B and hence $\check{B}$ are regular; therefore,

C, $\check{A}$ and C' are regular, too. Now $|\check{A}^{-1}| = |C'(C\check{A}C')^{-1}C| = |C'\check{B}^{-1}C| \leq |C'|\,|\check{B}^{-1}|\,|C|$ so that

$$|\check{A}^{-1}| \leq |C'|\langle\check{B}\rangle^{-1}|C| \tag{4}$$

by Theorem 3.7.5. Now Proposition 3.2.4 and Theorem 3.7.5(ii) imply $\rho(|\check{A}^{-1}|\operatorname{rad}(A)) \leq \rho(|C'|\langle\check{B}\rangle^{-1}|C|\operatorname{rad}(A)) = \rho(\langle\check{B}\rangle^{-1}|C|\operatorname{rad}(A)|C'|) = \rho(\langle\check{B}\rangle^{-1}\operatorname{rad}(B)) < 1$; hence A is strongly regular.

To prove (2) and (3), suppose that $\tilde{x} \in \Sigma(A, b)$, so that $\tilde{A}\tilde{x} = \tilde{b}$ for suitable $\tilde{A} \in A$, $\tilde{b} \in \mathrm{b}$. Since $C\tilde{A}C' \in CAC' = B$ and $C\tilde{b} \in Cb$ we have $\tilde{x} = \tilde{A}^{-1}\tilde{b} = C'(C\tilde{A}C')^{-1}C\tilde{b} = C'\tilde{z}$ with $\tilde{z} = (C\tilde{A}C')^{-1}C\tilde{b} \in \Sigma(B, Cb)$. Hence (2) holds, and (3) follows since $A^{\mathrm{H}}b = \square\Sigma(A, B) \subseteq C'(\square\Sigma(B, Cb)) = C'B^{\mathrm{H}}(Cb))$. □

4.1.3 Corollary

(i) If $A \in \mathbb{IR}^{n\times n}$ is strongly regular and $B \subseteq A$ then B is strongly regular; in particular, every strongly regular matrix is regular.

(ii) Every H-matrix (and hence every M-matrix) is strongly regular.

Proof If A is strongly regular and $B \subseteq A$, the matrix $\check{A}^{-1}B$ is contained in $\check{A}^{-1}A$, and by Theorem 3.7.5 it is an H-matrix. By applying the above theorem with B in place of A and $C = \check{A}^{-1}$, $C' = I$ we find that B is also strongly regular. In particular, every $\tilde{A} \in A$ is strongly regular, and hence regular (since thin), so that A is regular, too. This proves (i), and (ii) follows by taking $C = C' = I$ in the theorem. □

4.1.4 Example We consider the linear system with coefficients given by

$$A = \begin{pmatrix} 2 & [-1,0] \\ [-1,0] & 2 \end{pmatrix}, \quad b = \begin{pmatrix} 1.2 \\ -1.2 \end{pmatrix}$$

already treated in Example 3.4.2. Putting $C = \check{A}^{-1}$, we have

$$C = \frac{2}{15}\begin{pmatrix} 4 & 1 \\ 1 & 4 \end{pmatrix}, \quad CA = \frac{2}{15}\begin{pmatrix} [\ \ 7,8] & [-2,2] \\ [-2,2] & [\ \ 7,8] \end{pmatrix}, \quad Cb = \begin{pmatrix} 0.48 \\ -0.48 \end{pmatrix}.$$

Since CA is strongly diagonally dominant it is an H-matrix and A is strongly regular; this also follows from the corollary since in fact A is an M-matrix.

The solution set of the preconditioned system can be computed with the help of Theorem 3.4.3 and is shown in Fig. 4.1, together with the (shaded) solution set of the original system.

Its hull is

$$(CA)^{\mathrm{H}}(Cb) = \begin{pmatrix} [\ \ 0.3\ ,\ \ \ 0.72] \\ [-0.72, -0.3\ ] \end{pmatrix},$$

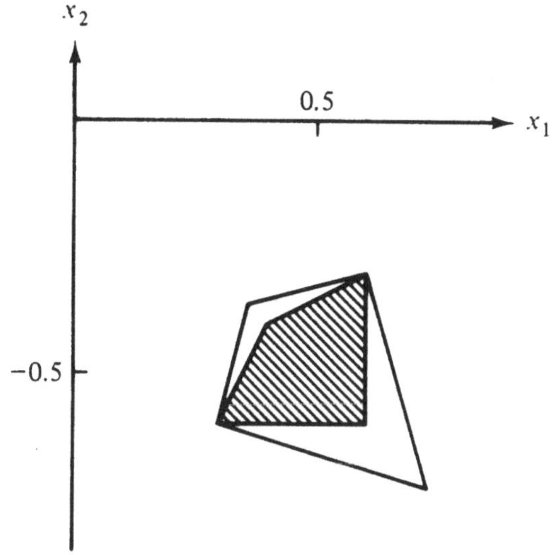

Fig. 4.1. Solution set (hatched) and preconditioned solution set.

with slightly wider intervals than

$$A^{\mathrm{H}}b = \begin{pmatrix} [\ \ 0.3, & 0.6] \\ [-0.6, & -0.3] \end{pmatrix}.$$

Thus, preconditioning may slightly increase the solution set and its hull. We shall show in the next section that for small rad(A) the overestimation of $A^{\mathrm{H}}b$ by $(CA)^{\mathrm{H}}(Cb)$ remains small (Corollary 4.2.5 below).

In a special case we can even guarantee that the hull of the solution set remains the same.

4.1.5 Theorem Let $A \in \mathbb{IR}^{n\times n}$ and $C \in \mathbb{R}^{n\times n}$. If CA is regular then A is regular and

$$A^{\mathrm{H}}b \subseteq (CA)^{\mathrm{H}}(Cb). \tag{5}$$

Equality holds if $C \geq 0$ and CA is an M-matrix.

Proof Relation (5) holds since $\tilde{A}^{-1}\tilde{b} = (C\tilde{A})^{-1}(C\tilde{b}) \in (CA)^{\mathrm{H}}(Cb)$ for all $\tilde{A} \in A$, $\tilde{b} \in$ b. Now assume that $C \geq 0$ and $B := CA$ is an M-matrix. Then $A^{-1} = A^{\mathrm{H}}I \subseteq (CA)^{\mathrm{H}}(CI) = B^{\mathrm{H}}C \subseteq B^{-1}C$, and since $B^{-1} \geq 0$ we find that $A^{-1} \geq 0$. We now use the notation of the proof of Theorem 3.6.5. In analogy with $A(\tilde{x})$, we define the matrix $B(\tilde{x}) \in \mathbb{R}^{n\times n}$ with components

$$B(\tilde{x})_{ik} = \overline{B}_{ik} \text{ if } \tilde{x}_k \geq 0 \text{ and } B(\tilde{x})_{ik} = \underline{B}_{ik} \text{ otherwise.}$$

If $\tilde{x}_k \geq 0$ then $B(\tilde{x})_{ik} = \overline{B}_{ik} = \sup(\Sigma\, C_{ij}A_{jk}) = \Sigma\, C_{ij}\overline{A}_{jk} = \Sigma\, C_{ij}A(\tilde{x})_{jk} =$

$(CA(\tilde{x}))_{ik}$, and if $\tilde{x}_k < 0$ then, by a similar argument, $(B(\tilde{x}))_{ik} = (CA(\tilde{x}))_{ik}$. Hence $B(\tilde{x}) = CA(\tilde{x})$, and using inequality (3.6.6) we get

$$\begin{aligned}\tilde{x} = \inf((CA)^{\mathrm{H}}(Cb)) &\Leftrightarrow B(\tilde{x})\tilde{x} = \inf(Cb)\\ &\Leftrightarrow CA(\tilde{x})\tilde{x} = C\underline{b}\\ &\Leftrightarrow A(\tilde{x})\tilde{x} = \underline{b}\\ &\Leftrightarrow \tilde{x} = \inf(A^{\mathrm{H}}b).\end{aligned}$$

Therefore, $\inf((CA)^{\mathrm{H}}(Cb)) = \inf(A^{\mathrm{H}}b)$ and $\sup((CA)^{\mathrm{H}}(Cb)) = -\inf((CA)^{\mathrm{H}}(-Cb)) = -\inf(A^{\mathrm{H}}(-b)) = \sup(A^{\mathrm{H}}b)$. This implies equality in (5). □

4.1.6 Corollary

(i) If $A^{-1} \geq 0$ then A is strongly regular and (5) holds with equality for $C = \bar{A}^{-1}$.
(ii) If A is an M-matrix and $\bar{A} \leq \tilde{A} \leq \operatorname{diag}(\bar{A})$ then (5) holds with equality for $C = \tilde{A}^{-1}$.

Proof (i) Clearly $C \geq 0$. Since $A \leq \bar{A}$ this implies $CA \leq \bar{A}^{-1}\bar{A} = I$, hence the off-diagonal elements of CA are ≤ 0. To show that CA is an M-matrix, pick $v > 0$ and define $u := \underline{A}^{-1}v$. Then $u > 0$ and $(CA)u \geq C\underline{A}u = Cv > 0$. Now apply the theorem.

(ii) Choose $u > 0$ such that $Au > 0$. Then $\tilde{A}u \geq Au > 0$ so that $\tilde{A}$ is an M-matrix. Now $A^{-1} \geq 0$ and $\tilde{A}^{-1} \geq 0$, so that the argument of (i) applies with $\tilde{A}$ in place of $\bar{A}$. □

4.1.7 Proposition Let $A \in \mathbb{IR}^{n\times n}$ be regular.

(i) If $\check{A}^{-1} \geq 0$ then A is strongly regular.
(ii) If $\check{A}$ is an M-matrix then A is an H-matrix.

Proof (i) Suppose that A is not strongly regular. Then $1 \leq \rho(|\check{A}^{-1}|\operatorname{rad}(A)) = \rho(\operatorname{rad}(A)|\check{A}^{-1}|)$ by equation (3.2.9) so that relation (3.2.7) implies the existence of a vector $u \geq 0$ such that $\operatorname{rad}(A)|\check{A}^{-1}|u \geq u \neq 0$. In particular, $\tilde{x} := |\check{A}^{-1}|u = \check{A}^{-1}u$ is nonzero and nonnegative, and $|\check{A}\tilde{x}| = |u| = u \leq \operatorname{rad}(A)|\check{A}^{-1}|u = \operatorname{rad}(A)\tilde{x} = \operatorname{rad}(A)|\tilde{x}|$. By Corollary 3.4.5, A cannot be regular. Hence (i) holds.

(ii) If $\check{A}$ is an M-matrix then $\check{A}^{-1} \geq 0$, and by (i) A is strongly regular. Since $\langle\check{A}\rangle^{-1} = \check{A}^{-1} = |\check{A}^{-1}|$, Proposition 4.1.1 implies $\rho(\langle\check{A}\rangle^{-1}\operatorname{rad}(A)) = \rho(|\check{A}^{-1}|\operatorname{rad}(A)) < 1$, so that A is an H-matrix by Theorem 3.7.5. □

The preceding results show that H-matrices, inverse positive matrices and regular matrices with inverse positive midpoints are strongly regular. Moreover, since

$$\rho(|\check{A}^{-1}|\mathrm{rad}(A)) \le \| \, |\check{A}^{-1}|\mathrm{rad}(A)\|_\infty \le \|\check{A}^{-1}\|_\infty \|\mathrm{rad}(A)\|_\infty,$$

all regular interval matrices with fixed midpoints and sufficiently small radii are strongly regular. But, unfortunately, there are also regular matrices which are not strongly regular.

4.1.8 Example The matrices

$$A = \begin{pmatrix} [0,2] & 1 \\ -1 & [0,2] \end{pmatrix} \quad \text{and} \quad B = \begin{bmatrix} 3 & [0,2] & [0,2] \\ [0,2] & 3 & [0,2] \\ [0,2] & [0,2] & 3 \end{bmatrix}$$

are regular. Indeed, $\det(\tilde{A}) \in A_{11}A_{22} - A_{12}A_{21} = [1,5] > 0$ for all $\tilde{A} \in A$, and B was shown to be regular in Example 3.4.6. However, neither A nor B is strongly regular. Indeed, with the all-one vector e we have

$$\check{A}^{-1} = \frac{1}{2}\begin{pmatrix} 1 & -1 \\ 1 & 1 \end{pmatrix}, \qquad |\check{A}^{-1}|\mathrm{rad}(A)e = |\check{A}^{-1}|e = e,$$

$$\check{B}^{-1} = \begin{bmatrix} 0.4 & -0.1 & -0.1 \\ -0.1 & 0.4 & -0.1 \\ -0.1 & -0.1 & 0.4 \end{bmatrix}, \quad |\check{B}^{-1}|\mathrm{rad}(B)e = 2|\check{B}^{-1}|e = 1.2e,$$

so that $\rho(|\check{A}^{-1}|\mathrm{rad}(A)) = 1$, $\rho(|\check{B}^{-1}|\mathrm{rad}(B)) = 1.2$. By Proposition 4.1.1, A and B are not strongly regular.

The next result generalizes the bounds on $A^{\mathrm{H}}b$ deduced for H-matrices in Theorem 3.7.7 to similar bounds for strongly regular matrices.

4.1.9 Proposition Let $A \in \mathbb{IR}^{n\times n}$ and $C, C' \in \mathbb{R}^{n\times n}$.

(i) If CAC' is an H-matrix then, for all $b \in \mathbb{IR}^n$, we have

$$|A^{\mathrm{H}}b| \le |C'|\langle CAC'\rangle^{-1}|Cb|. \tag{6}$$

(ii) If $\langle CAC'\rangle u \ge v > 0$ for some $u \ge 0$ then

$$|A^{\mathrm{H}}b| \le \|Cb\|_v \cdot |C'|u, \tag{7}$$

$$A^{\mathrm{H}}b \subseteq \|Cb\|_v \cdot |C'|[-u,u]. \tag{8}$$

Proof Put $B = CAC'$. If B is an H-matrix then by (3) and Theorem 3.7.7 we have

$$|A^{\mathrm{H}}b| \le |C'(B^{\mathrm{H}}(Cb))| \le |C'|\,|B^{\mathrm{H}}(Cb)| \le |C'|\langle B\rangle^{-1}|Cb|$$

which gives (6). Moreover, the assumption in (ii) implies that $|Cb| \le \|Cb\|_v v \le \langle B\rangle \|Cb\|_v u$ so that $|A^H b| \le |C'|\,\|Cb\|_v u$, which gives (7) and (8). □

We now consider the choice of C and C' which minimizes the upper bound (6). Note that under the assumption (i) of Proposition 4.1.9, A is strongly regular by Theorem 4.1.2.

4.1.10 Theorem Let $A \in \mathbb{IR}^{n\times n}$ be strongly regular. Then, for all $b \in \mathbb{IR}^n$ and all $C, C' \in \mathbb{R}^{n\times n}$ such that CAC' is an H-matrix, we have

$$|A^H b| \le \langle \check{A}^{-1}A\rangle^{-1}|\check{A}^{-1}b| \le |C'|\langle CAC'\rangle^{-1}|Cb|; \tag{9}$$

i.e. the bound in (6) becomes minimal for the choice $C = \check{A}^{-1}$ and $C' = I$.

Proof Put $B = CAC'$ and $B_0 := \check{A}^{-1}A$. Since B_0 is an H-matrix, $\langle B_0\rangle = \langle \check{B}_0\rangle - \text{rad}(B_0) = I - |\check{A}^{-1}|\text{rad}(A) \ge I - |C'|\langle\check{B}\rangle^{-1}|C|\text{rad}(A)$ by (4), so that

$$\begin{aligned}\langle B_0\rangle|C'| &\ge |C'| - |C'|\langle\check{B}\rangle^{-1}|C|\text{rad}(A)|C'|\\ &= |C'| - |C'|\langle\check{B}\rangle^{-1}\text{rad}(B)\\ &= |C'|\langle\check{B}\rangle^{-1}(\langle\check{B}\rangle - \text{rad}(B))\\ &= |C'|\langle\check{B}\rangle^{-1}\langle B\rangle\end{aligned}$$

since B is an H-matrix. Now (6), applied with $\check{A}$ in place of A, yields

$$\begin{aligned}|\check{A}^{-1}b| &\le |C'|\langle\check{B}\rangle^{-1}|Cb| = (|C'|\langle\check{B}\rangle^{-1}\langle B\rangle)\langle B\rangle^{-1}|Cb|\\ &\le \langle B_0\rangle|C'|\langle B\rangle^{-1}|Cb|\end{aligned}$$

since $\langle B\rangle^{-1} \ge 0$. Multiplication with $\langle B_0\rangle^{-1} \ge 0$ gives $\langle B_0\rangle^{-1}|\check{A}^{-1}b| \le |C'|\langle B\rangle^{-1}|Cb|$. This is the right-hand inequality of (9). Since the left-hand inequality is a special case of (6) with $C = \check{A}^{-1}$, $C' = I$, the assertion now follows. □

Because of this optimality result, and of further natural properties to be proved later (e.g. Theorem 4.1.12), the midpoint inverse plays a distinguished role as a preconditioning matrix. However, in most practical calculations this midpoint inverse is not exactly known. Hence, it is important that one can use the enclosure (5) with a preconditioning matrix $C \approx \check{A}^{-1}$ calculated with finite precision arithmetic by a standard inversion routine. Because of the results of a rounding error analysis for inversion methods it is advisable to use the transpose of the approximate inverse of $\check{A}^T$ computed by Gauss elimination for C, since then $\|I - C\check{A}\|$ and hence $\|I - CA\|$ will generally be smaller than for the computed inverse of $\check{A}$ itself.

Note, however, that in special cases other preconditioning matrices may yield better results since $\check{A}^{-1}$ only optimizes the upper bound (6). In particular, a different choice is more profitable in the cases mentioned in Corollary 4.1.6.

Preconditioning also yields simple enclosures for the inverse of a strongly regular interval matrix.

4.1.11 Theorem Let $A \in \mathbb{IR}^{n\times n}$ and $0 < u \in \mathbb{R}^n$. Then for any matrix $C \in \mathbb{R}^{n\times n}$ satisfying $\|CA - I\|_u \le \beta < 1$ and any nonnegative vector $w \in \mathbb{R}^n$ with $|C| \le uw^{\mathrm{T}}$, the following enclosure holds:

$$A^{-1} \subseteq C + \frac{\beta}{1-\beta}[-uw^{\mathrm{T}}, uw^{\mathrm{T}}]. \qquad (10)$$

Proof The assumptions imply $\langle CA\rangle u \ge (I - |CA - I|)u \ge (1-\beta)u$; hence CA is an H-matrix and A is strongly regular. Now any $\tilde{A} \in A$ satisfies

$$|\tilde{A}^{-1} - C| = |-(C\tilde{A})^{-1}(C\tilde{A} - I)C| \le \langle CA\rangle^{-1}|CA - I|\,|C|$$
$$\le \langle CA\rangle^{-1} \cdot \beta uw^{\mathrm{T}} \le \frac{\beta}{1-\beta} uw^{\mathrm{T}}.$$

This implies (10). □

We end this section by noting some monotonicity properties of the expression

$$\mathrm{r}(A) := |\check{A}^{-1}|\mathrm{rad}(A) \qquad (11)$$

for strongly regular matrices $A \in \mathbb{IR}^{n\times n}$.

4.1.12 Theorem (Krawczyk) Let $A \in \mathbb{IR}^{n\times n}$ be strongly regular and $B \subseteq A$. Then, for all $b \in \mathbb{IR}^n$ and all nonnegative $u \in \mathbb{R}^n$, the following relations hold:

(i) $|\check{A}^{-1}b| \le u - \mathrm{r}(A)u \Rightarrow |\check{B}^{-1}b| \le u - \mathrm{r}(B)u$;
(ii) $(I - \mathrm{r}(B))^{-1}|\check{B}^{-1}b| \le (I - \mathrm{r}(A))^{-1}|\check{A}^{-1}b|$;
(iii) $(I - \mathrm{r}(B))^{-1} \le (I - \mathrm{r}(A))^{-1}$.

Proof Put $E := \check{B}^{-1}(\check{A} - \check{B})$. Then

$$|E| \le |\check{B}^{-1}|\,|\check{A} - \check{B}| \le |\check{B}^{-1}|(\mathrm{rad}(A) - \mathrm{rad}(B))$$
$$= |(I + E)\check{A}^{-1}|\mathrm{rad}(A) - r(B) \le (I + |E|)\mathrm{r}(A) - \mathrm{r}(B)$$

so that

$$(I + |E|)(I - \mathrm{r}(A)) \le I - \mathrm{r}(B). \qquad (12)$$

Now (i) holds since $|\check{A}^{-1}b| \le u - \mathrm{r}(A)u$ implies

$$|\check{B}^{-1}b| \le |\check{B}^{-1}\check{A}|\,|\check{A}^{-1}b| \le (I + |E|)(I - \mathrm{r}(A))u \le (I - \mathrm{r}(B))u.$$

Since A is strongly regular, $(I - \mathrm{r}(A))^{-1} \geq 0$; hence (ii) follows from (i) by taking $u = (I - \mathrm{r}(A))^{-1}|\check{A}^{-1}b|$. Finally, (12) implies

$$I \leq I + |E| \leq (I - \mathrm{r}(B))(I - \mathrm{r}(A))^{-1}.$$

Since B is also strongly regular by Corollary 4.1.3, multiplication with $(I - \mathrm{r}(B))^{-1} \geq 0$ proves (iii). □

Remark It is not known whether the implication $B \subseteq A \Rightarrow \mathrm{r}(B) \leq \mathrm{r}(A)$ holds whenever A is strongly regular.

4.2 Krawczyk's method and quadratic approximation

In this section we discuss Krawczyk's iteration method for the solution of linear interval equations. In particular, we show that this method provides both computable bounds for the overestimation of an enclosure of $A^{\mathrm{H}}b$ and enclosures which have the quadratic approximation property.

In Proposition 4.1.9 we found a first enclosure for the hull $A^{\mathrm{H}}b$ of the solution set of a square system of linear equations whose coefficient matrix is strongly regular. For convenience we repeat a special case ($C' = I$) of formula (4.1.8), namely

$$A^{\mathrm{H}}b \subseteq \|Cb\|_v[-u, u] \quad \text{if } u > 0, \langle CA \rangle u \geq v > 0. \tag{1}$$

Note that u, v can be obtained as indicated in the remark after Proposition 3.7.3. Generally, the enclosure (1) is rather crude. However, if the right-hand side b is of order $\mathrm{O}(\varepsilon)$ then the enclosure (1) and therefore the overestimation in (1) is also of order $\mathrm{O}(\varepsilon)$. This suggests that we might get a better enclosure from (1) if we first find an approximation $\tilde{x}$ to a particular solution and then add to $\tilde{x}$ an enclosure for the residual correction $A^{\mathrm{H}}(b - A\tilde{x})$ obtained via (1).

4.2.1 Lemma Let $A \in \mathbb{IR}^{n\times n}$ be regular and $b \in \mathbb{IR}^n$, $\tilde{x} \in \mathbb{R}^n$. Then

$$A^{\mathrm{H}}b \subseteq \tilde{x} + A^{\mathrm{H}}(b - A\tilde{x}). \tag{2}$$

Proof If $\tilde{A} \in A$, $\tilde{b} \in b$ then $\tilde{A}^{-1}\tilde{b} = \tilde{x} + \tilde{A}^{-1}(\tilde{b} - \tilde{A}\tilde{x}) \in \tilde{x} + A^{\mathrm{H}}(b - A\tilde{x})$, and (2) follows. □

Thus, if $b - A\tilde{x} = \mathrm{O}(\varepsilon)$ and $A^{\mathrm{H}}(b - A\tilde{x})$ is enclosed as in (1) then (2) yields the bound

$$A^{\mathrm{H}}b \subseteq \tilde{x} + \|C(b - A\tilde{x})\|_v[-u, u] \quad \text{if } u > 0, \langle CA \rangle u \geq v > 0, \tag{3}$$

with a radius of order $\mathrm{O}(\varepsilon)$ and hence an overestimation of order $\mathrm{O}(\varepsilon)$.

Therefore, (3) is an enclosure for $A^{\mathrm{H}}b$ with linear approximation order. Note that

$$|b - A\tilde{x}| = \mathrm{rad}(b) + \mathrm{rad}(A)|\tilde{x}| + |\check{b} - \check{A}\tilde{x}|$$

so that $b - A\tilde{x} = \mathrm{O}(\varepsilon)$ if $\mathrm{rad}(b)$, $\mathrm{rad}(A)$ and $\check{b} - \check{A}\tilde{x}$ are of order $\mathrm{O}(\varepsilon)$. In particular, a good approximation of $\check{A}^{-1}\check{b}$ (or of $C\check{b}$) is a good choice for $\tilde{x}$.

As we have already seen in the context of function evaluation, a linear approximation order does not necessarily imply a small overestimation. Therefore it is desirable to dispose of methods with a quadratic approximation order. We can achieve this by improving our initial enclosure (3) iteratively. To improve an enclosure x of $A^{\mathrm{H}}b$ we may use the relation

$$\tilde{A}^{-1}\tilde{b} = C\tilde{b} - (C\tilde{A} - I)(\tilde{A}^{-1}\tilde{b}) \in Cb - (CA - I)x$$

which holds for all $\tilde{A} \in A$ and $\tilde{b} \in b$, so that

$$A^{\mathrm{H}}b \subseteq x \Rightarrow A^{\mathrm{H}}b \subseteq (Cb - (CA - I)x) \cap x. \tag{4}$$

This leads us to consider the *Krawczyk iteration*

$$z^0 := x, \quad z^{l+1} := (Cb - (CA - I)z^l) \cap z^l \quad (l = 0, 1, 2, \ldots). \tag{5}$$

Clearly, the z^l form a nested sequence of interval vectors, and by construction

$$A^{\mathrm{H}}b \subseteq x \Rightarrow A^{\mathrm{H}}b \subseteq z^l \quad \text{for all } l \geq 0. \tag{6}$$

One can also use the Krawczyk iteration in a residual form by applying (4) to an enclosure d of $A^{\mathrm{H}}(b - A\tilde{x})$; this gives the *residual Krawczyk iteration*

$$d^0 := d, \quad d^{l+1} := (C(b - A\tilde{x}) - (CA - I)d^l) \cap d^l,$$

and corresponding enclosures $y^l := \tilde{x} + d^l$ of $A^{\mathrm{H}}b$ if $A^{\mathrm{H}}b \subseteq d$. However, since

$$\tilde{x} + C(b - A\tilde{x}) = \tilde{x} + Cb - C(A\tilde{x}) = Cb - (CA)\tilde{x} + \tilde{x} = Cb - (CA - I)\tilde{x}$$

by Propositions 3.1.2 and 3.1.4, we have

$$\begin{aligned} \tilde{x} + C(b - A\tilde{x}) - (CA - I)d^l &= Cb - (CA - I)\tilde{x} - (CA - I)d^l \\ &\supseteq Cb - (CA - I)(\tilde{x} + d^l) \\ &= Cb - (CA - I)y^l. \end{aligned}$$

Therefore

$$y^{l+1} = \tilde{x} + d^{l+1} \supseteq (Cb - (CA - I)y^l) \cap y^l,$$

and a comparison with (5) shows inductively that

$$z^0 = y^0 \Rightarrow z^l \subseteq y^l \quad \text{for all } l \geq 0.$$

Thus, the residual Krawczyk iteration never gives better results than the simple version (4), and it will not be further discussed. (But in the presence of rounding errors there is an exception to this rule when A and b are thin and the residual $b - A\tilde{x}$ can be enclosed with higher precision; then a final residual step often yields an optimal enclosure of $A^{-1}b$.)

In order to show that the Krawczyk iteration can produce enclosures with a quadratic approximation order we need a way to bound the overestimation of an enclosure x of $A^H b$. Note that the following result not only bounds the distance $q(A^H b, x)$ but also the overestimation in the radii, since by formula (3.3.14) we have

$$0 \le \operatorname{rad}(x) - \operatorname{rad}(A^H b) \le q(A^H b, x) \quad \text{if } A^H b \subseteq x. \tag{7}$$

4.2.2 Proposition Let $A \in \mathbb{IR}^{n \times n}$ be regular and $b \in \mathbb{IR}^n$. Then for every enclosure $x \in \mathbb{IR}^n$ of $A^H b$ and every $\tilde{x} \in \mathbb{R}^n$ and $C \in \mathbb{R}^{n \times n}$ we have

$$C(b - A\tilde{x}) \subseteq A^H b - \tilde{x} + (CA - I)(x - \tilde{x}), \tag{8}$$

$$q(x, A^H b) \le q(C(b - A\tilde{x}), x - \tilde{x} + (CA - I)(x - \tilde{x})). \tag{9}$$

Proof By equations (3.1.5) and (3.1.10), every $\tilde{r} \in r := b - A\tilde{x}$ has the form $\tilde{r} = \tilde{b} - \tilde{A}\tilde{x}$ with $\tilde{A} \in A$, $\tilde{b} \in B$. Hence $C\tilde{r} = C\tilde{A}(\tilde{A}^{-1}\tilde{b} - \tilde{x}) \in (CA)$ $(A^H b - \tilde{x})$, and, since $Cr = \square\{C\tilde{r} \mid \tilde{r} \in r\}$, this implies $Cr \subseteq (CA)$ $(A^H b - \tilde{x}) \subseteq A^H b - \tilde{x} + (CA - I)(A^H b - \tilde{x}) \subseteq A^H b - \tilde{x} + (CA - I)$ $(x - \tilde{x})$, giving (8). Adding x to both sides, we get $x + Cr \subseteq$ $A^H b + (x - \tilde{x}) + (CA - I)(x - \tilde{x})$. Hence, if we denote the right-hand side of (9) by u then formula (3.3.13) implies $(x - \tilde{x}) +$ $(CA - I)(x - \tilde{x}) \subseteq Cr + [-u, u]$. Thus $x + Cr \subseteq A^H b + Cr + [-u, u]$, and, since $A^H b \subseteq x$, equation (3.3.18) and Proposition 3.3.1 show that $q(x, A^H b) = q(x + Cr, A^H b + Cr) \le u$, which is (9). $\qquad\square$

Note that the bound in (9) is computable from the given data; thus, (9) can be used to bound *a posteriori* the amount of overestimation of an enclosure x of $A^H b$ found by an arbitrary method. In particular, applied to Krawczyk's iteration we obtain the following:

4.2.3 Theorem Let $A \in \mathbb{IR}^{n \times n}$ be regular, $b \in \mathbb{IR}^n$ and $C \in \mathbb{R}^{n \times n}$. Then, for every enclosure $x \in \mathbb{IR}^n$ of $A^H b$, the Krawczyk iterates z^l defined by (5) satisfy

$$A^H b \subseteq z^{l+1} \subseteq A^H b + (CA - I)(z^l - z^l), \tag{10}$$

$$q(z^{l+1}, A^H b) \le 2|CA - I|\operatorname{rad}(z^l). \tag{11}$$

Proof Put $s = (CA - I)(z^l - \tilde{x})$. By definition of z^{l+1}, subdistributivity, and (8), we have

$$z^{l+1} \subseteq Cb - (CA - I)z^l \subseteq Cb - (CA - I)\check{x} - (CA - I)(z^l - \check{x})$$
$$\overset{(*)}{=} C(b - A\check{x}) + \check{x} - s \subseteq A^{\mathrm{H}}b - \check{x} + s + \check{x} - s = A^{\mathrm{H}}b + s - s.$$

Here we used the distributive and the associative law in (∗) which hold in this special case since C and $\check{x}$ are thin (Propositions 3.1.2 and 3.1.4). In particular, if we take $\check{x} = \check{z}^l$ then $s = |CA - I|\mathrm{rad}(z^l)[-1, 1]$; therefore $-s = s$ and $s - s = 2s = (CA - I)(z^l - z^l)$. Since $A^{\mathrm{H}}b \subseteq z^{l+1}$ by (6), the assertion follows from Proposition 3.3.1. □

We now show that, for problems with sufficiently small radius, Theorem 4.2.3 implies that the Krawczyk iterates have the quadratic approximation property. To this end we consider the class of problems with fixed midpoint $\check{A}$ and suppose that $\mathrm{rad}(A)$, $\mathrm{rad}(b)$, $C\check{A} - I$ and $\check{b} - \check{A}\check{x}$ are all of order $\mathrm{O}(\varepsilon)$, with a small number ε (this requires that C and $\check{x}$ are sufficiently good approximations of $\check{A}^{-1}$ and $\check{A}^{-1}\check{b}$). Then $|CA - I| = |C\check{A} - I| + |C|\mathrm{rad}(A)$ is also of order $\mathrm{O}(\varepsilon)$ since $|C| \le (|C\check{A} - I| + I)|\check{A}^{-1}|$ is bounded. In particular, CA is an H-matrix, and the enclosure (3) can be used as an initial enclosure for the iteration,

$$z^0 = x := \check{x} + \|C(b - A\check{x})\|_v[-u, u]. \tag{12}$$

Since $z^l \subseteq z^0$ we have $\mathrm{rad}(z^l) \le \mathrm{rad}(z^0) = \|C(b - A\check{x}\|_v u \le \|C\|_v\|b - A\check{x}\|_v u$, and (11) implies

$$q(A^{\mathrm{H}}b, z^{l+1}) \le 2\|C\|_v\|b - A\check{x}\|_v|CA - I|u.$$

Apart from two factors of order $\mathrm{O}(\varepsilon)$, this bound contains bounded factors only; therefore $q(A^{\mathrm{H}}b, z^{l+1}) = \mathrm{O}(\varepsilon^2)$ for all $l \ge 0$. Hence, with the initial enclosure (12), the Krawczyk iterates z^l with $l \ge 1$ enclose the hull $A^{\mathrm{H}}b$ of the solution set $\Sigma(A, b)$ with quadratic approximation order. If the weaker initial enclosure

$$z^0 = x := \|Cb\|_v[-u, u] \tag{12a}$$

from (1) is used in place of (12), we get from (10) and (2) the enclosure

$$z^1 \subseteq A^{\mathrm{H}}b + (CA - I)(z^0 - z^0) \subseteq \check{x} + A^{\mathrm{H}}(b - A\check{x}) + (CA - I)(z^0 - z^0)$$

from which we deduce that $\mathrm{rad}(z^1) = \mathrm{O}(\varepsilon)$, and hence $\mathrm{rad}(z^l) = \mathrm{O}(\varepsilon)$ for $l \ge 1$. Now (11) shows as before that, with the initial enclosure (12a), the Krawczyk iterates z^l with $l \ge 2$ again enclose $A^{\mathrm{H}}b$ with quadratic approximation order.

We conclude that for both initial enclosures (12) or (12a), Krawczyk's iteration is a reliable method for obtaining realistic enclosures of the solution set of systems of linear equations whose coefficients are narrow intervals. However, note that Gauss–Seidel iteration, discussed in the next section, is still superior to Krawczyk's method.

For CA close to I we now show that in the limit the Krawczyk iteration produces enclosures of $A^{\mathrm{H}}b$ which have a bounded radius overestimation factor independent of the right-hand side. Hence, after sufficiently many iterations, reliable enclosures are also obtained when b contains wide intervals.

4.2.4 Theorem Let $A \in \mathbb{IR}^{n\times n}$ and $b \in \mathbb{IR}^n$. If $C \in \mathbb{R}^{n\times n}$ is chosen such that

$$\|CA - I\|_u \leq \beta < 1$$

for some $u > 0$ then, for every initial enclosure x of $A^{\mathrm{H}}b$, the limit $z = \lim_{l\to\infty} z^l$ of Krawczyk's iteration (5) satisfies

$$\|\mathrm{rad}(A^{\mathrm{H}}b)\|_u \leq \|\mathrm{rad}(z)\|_u \leq \frac{1+\beta}{1-\beta}\|\mathrm{rad}(A^{\mathrm{H}}b)\|_u. \tag{13}$$

Proof Write $E := CA - I$, so that $\|E\|_u \leq \beta < 1$. We first show that

$$z \subseteq Cb - Ez \subseteq A^{\mathrm{H}}b + E(A^{\mathrm{H}}b - z). \tag{14}$$

Since $A^{\mathrm{H}}b \subseteq z^l$ and the z^l form a nested sequence of interval vectors, the limit $z = \lim z^l$ exists, and by taking limits in (5) we find $z \subseteq Cb - Ez$. To show the other half of (14) we fix an index i and pick

$$\zeta \in (Cb - Ez)_i = \sum_k C_{ik}b_k - \sum_k E_{ik}z_k.$$

Since the latter expression contains every interval variable only once, Theorem 1.4.3 implies that $\zeta = \Sigma\, C_{ik}\tilde{b}_k - \Sigma\, \tilde{E}_{ik}\tilde{z}_k$ for suitable $\tilde{b}_k \in b_k$, $\tilde{E}_{ik} \in E_{ik}$ and $\tilde{z}_k \in z_k$. Since

$$E_{ik} = \sum_j C_{ij}A_{jk} - \delta_{ik},$$

we can similarly find $\tilde{A}_{jk} \in A_{jk}$ such that $\tilde{E}_{ik} = \Sigma\, C_{ij}\tilde{A}_{jk} - \delta_{ik}$ for $k = 1, \ldots, n$. Therefore, ζ is the ith coordinate of a vector of the form $\tilde{x} = C\tilde{b} - (C\tilde{A} - I)\tilde{z}$ with $\tilde{b} \in b$, $\tilde{A} \in A$ and $\tilde{z} \in z$. Now

$$\tilde{x} = \tilde{A}^{-1}\tilde{b} + (C\tilde{A} - I)(\tilde{A}^{-1}\tilde{b} - \tilde{z}) \in A^{\mathrm{H}}b + (CA - I)(A^{\mathrm{H}}b - z)$$

so that ζ is an element of the ith component of the right-hand side of (14). Since i and $\zeta \in (Cb - Ez)_i$ were arbitrary, this implies the right-hand enclosure of (14).

Now, since $A^{\mathrm{H}}b \subseteq z$, formulae (3.3.12), (3.3.18) and (3.3.16) yield

$$\begin{aligned} q(z, A^{\mathrm{H}}b) &\leq q(A^{\mathrm{H}}b, A^{\mathrm{H}}b + E(A^{\mathrm{H}}b - z)) = |E(A^{\mathrm{H}}b - z)| \\ &\leq |E|\,|A^{\mathrm{H}}b - z| \leq |E|(q(z, A^{\mathrm{H}}b) + 2\cdot\mathrm{rad}(A^{\mathrm{H}}b)). \end{aligned}$$

Taking norms we find

$$\|q(z, A^{\mathrm{H}}b)\|_u \le \frac{2\beta}{1-\beta}\|\mathrm{rad}(A^{\mathrm{H}}b)\|_u,$$

and (13) follows by applying relation (3.3.14). □

4.2.5 Corollary Let $A \in \mathbb{IR}^{n\times n}$ and $b \in \mathbb{IR}^n$. If $C \in \mathbb{R}^{n\times n}$ is chosen such that $\|CA - I\|_u \le \beta < 1$ for some $u > 0$ then

$$\|\mathrm{rad}(A^{\mathrm{H}}b)\|_u \le \|\mathrm{rad}((CA)^{\mathrm{H}}(Cb))\|_u \le \frac{1+\beta}{1-\beta}\|\mathrm{rad}(A^{\mathrm{H}}b)\|_u.$$

Proof Since replacing A, b and C by CA, Cb and I, respectively, does not change the Krawczyk iteration (5), (6) also implies

$$(CA)^{\mathrm{H}}(Cb) \subseteq x \Rightarrow (CA)^{\mathrm{H}}(Cb) \subseteq z^l \quad \text{for all } l \ge 0. \tag{6'}$$

In particular, if $x = (CA)^H(Cb)$ then $x \subseteq z^l \subseteq z^0 = x$ and therefore $z^l = x$ for all $l \ge 0$. Thus, the theorem applies with $z = (CA)^{\mathrm{H}}(Cb)$. □

The preceding results show that preconditioning is an effective method yielding reliable enclosures when $\|CA - I\|_u \le \beta \le \frac{1}{2}$ since then the u-norm of the radius of $A^{\mathrm{H}}b$ is overestimated by a factor of at most $(1+\beta)/(1-\beta) \le 3$. However, when β approaches unity, the quality of the enclosure may deteriorate quickly. In this case, one must use (9) and (11) to assess whether the enclosure obtained is sufficiently tight.

In the special case where $C = \check{A}^{-1}$, the limit of Krawczyk's iteration is related to the enclosure by the preconditioned Krawczyk inverse

$$A^{\mathrm{H}}b \subseteq (CA)^{\mathrm{K}}(Cb) = Cb + \langle CA\rangle^{-1}|CA - I|\,|Cb|[-1, 1] \tag{15}$$

obtained (for general C such that CA is an H-matrix) by applying Theorem 3.7.8 to the preconditioned system. Note that the matrices $\langle CA\rangle$ and $|CA - I|$ commute when $C = \check{A}^{-1}$, so that in this case $(CA)^{\mathrm{K}}(Cb)$ can be considered as the first iterate z^1 of Krawczyk's iteration (5) applied to the simple enclosure $z^0 = \langle CA\rangle^{-1}|Cb|[-1, 1]$ of $A^{\mathrm{H}}b$.

4.2.6 Theorem Let $A \in \mathbb{IR}^{n\times n}$ be strongly regular, $b \in \mathbb{IR}^n$ and $C = \check{A}^{-1}$. Then the Krawczyk iteration (5) satisfies

$$\lim_{l\to\infty} z^l \subseteq (CA)^{\mathrm{K}}(Cb) \tag{16}$$

for all initial enclosures x of $A^{\mathrm{H}}b$. Equality holds in (16) iff $(CA)^{\mathrm{K}}(Cb) \subseteq x$. In particular, if $\|I - CA\|_u \le \beta < 1$ then

$$\|\mathrm{rad}(A^{\mathrm{H}}b)\|_u \le \|\mathrm{rad}(CA)^{\mathrm{K}}(Cb)\|_u \le \frac{1+\beta}{1-\beta}\|\mathrm{rad}(A^{\mathrm{H}}b)\|_u. \tag{17}$$

Proof Put $z := \lim z^l$, $z^* := (CA)^K(Cb)$ and $E := CA - I$. Since $\text{mid}(CA) = I$ we have $\langle CA\rangle = I - |E|$ (so that in particular $\langle CA\rangle$ and $|E|$ commute), $\check{E} = 0$, and by assumption, the spectral radius of $|E| = \text{rad}(E) = |\check{A}^{-1}|\text{rad}(A)$ is less than one. We construct nonnegative vectors $w^l \in \mathbb{R}^n$ such that

$$z^l \subseteq Cb + w^l[-1, 1]. \tag{18}$$

This can certainly be satisfied for $l = 0$ with a suitable $w^0 > 0$. If (18) holds for some l then $z^{l+1} \subseteq Cb - Ez^l \subseteq Cb + |E|\,|z^l|[-1, 1] \subseteq Cb + |E|(|Cb| + w^l)[-1, 1]$, so that (18) holds for $l + 1$ in place of l with $w^{l+1} = |E|(|Cb| + w^l)$. Since $\rho(|E|) < 1$, the sequence w^l $(l = 0, 1, 2, \ldots)$ converges to a vector w with $w = |E|(|Cb| + w)$, giving $w = \langle CA\rangle^{-1}|E|\,|Cb|$. Taking limits in (18) we get $z \subseteq Cb + w[-1, 1] = z^*$. Hence (16) holds.

To characterize the equality case we first note that $|z^*| = |Cb| + \langle CA\rangle^{-1}|E|\,|Cb| = \langle CA\rangle^{-1}(\langle CA\rangle + |E|)|Cb| = \langle CA\rangle^{-1}|Cb|$, so that $|E|\,|z^*| = |E|\langle CA\rangle^{-1}|Cb| = w$. Since $\check{E} = 0$, this implies $Ez^* = |E|\,|z^*|[-1, 1] = w[-1, 1]$, so that $z^* = Cb + w[-1, 1] = Cb - Ez^*$. In particular, if $z^* \subseteq z^l$ then $z^* = (Cb - Ez^*) \cap z^* \subseteq (Cb - Ez^l) \cap z^l = z^{l+1}$, and a trivial induction step yields $z^* \subseteq z$ when $z^* \subseteq x$. Since $z \subseteq x$ and $z \subseteq z^*$ this shows that $z = z^*$ precisely when $z^* \subseteq x$. In particular, (17) follows by applying Theorem 4.2.4. □

Thus, the remarks on the quality of Krawczyk's iteration also apply to the enclosure (15), at least when $C = \check{A}^{-1}$. When $C \neq \check{A}^{-1}$ and $\rho(|CA - I|) < 1$ then (16) must be replaced by the weaker relation

$$\lim_{l\to\infty} z^l \subseteq Cb + (I - |CA - I|)^{-1}|CA - I|\,|Cb|[-1, 1]; \tag{16'}$$

this follows by a similar argument as used above. Since $\langle CA\rangle \geq I - |CA - I|$ we have $\langle CA\rangle^{-1} \leq (I - |CA - I|)^{-1}$ so that the bound (16′) is weaker than $(CA)^K(Cb)$.

The following simple example demonstrates both the excellent performance of preconditioning methods for problems with narrow intervals and the possibility of dramatic overestimation when the coefficients of A are so wide that $\|CA - I\|$ is close to one.

4.2.7 Example We consider a linear system with coefficients given by

$$A = \begin{pmatrix} p & p \\ -p & p \end{pmatrix}, \quad b = \begin{pmatrix} q \\ q \end{pmatrix},$$

where $p = [1 - \varepsilon, 1 + \varepsilon]$, $q = [1 - \delta, 1 + \delta]$ and $0 \leq \delta \leq 1$, $0 \leq \varepsilon$. By computing the determinant of $\tilde{A} \in A$ we find that A is regular iff $0 \notin p$, i.e. iff $\varepsilon < 1$. In this case, the method of Example 3.4.7 gives

$$x^* := A^{\mathrm{H}}b = \begin{pmatrix}0\\1\end{pmatrix} + \frac{\delta+\varepsilon}{1-\varepsilon}\begin{pmatrix}[-1,1]\\ [-(1-\varepsilon)/(1+\varepsilon),1]\end{pmatrix}.$$

We have

$$C := \check{A}^{-1} = \frac{1}{2}\begin{pmatrix}1 & -1\\ 1 & 1\end{pmatrix}, \quad CA = \begin{pmatrix}p & p-1\\ p-1 & p\end{pmatrix}, \quad Cb = \begin{pmatrix}q-1\\ q\end{pmatrix},$$

and since

$$|\check{A}^{-1}|\operatorname{rad}(A) = |CA - I| = \begin{pmatrix}\varepsilon & \varepsilon\\ \varepsilon & \varepsilon\end{pmatrix}$$

has spectral radius 2ε we see that A is strongly regular iff $\varepsilon < \frac{1}{2}$. In particular, preconditioning does not work for the whole range of ε where A is regular. If $\varepsilon < \frac{1}{2}$, the explicit enclosure (15) yields

$$A^{\mathrm{H}}b \subseteq x' := (CA)^{\mathrm{K}}(Cb) = \begin{pmatrix}0\\1\end{pmatrix} + \frac{\delta+\varepsilon}{1-2\varepsilon}\begin{pmatrix}[-1,1]\\ [-1,1]\end{pmatrix}.$$

The overestimation factors of the radii, $(1-\varepsilon)/(1-2\varepsilon)$ and $(1-\varepsilon^2)/(1-2\varepsilon)$, are excellent for small ε, but become arbitrarily bad for $\varepsilon \to \frac{1}{2}$. The same analysis also holds for the enclosure (12) with the natural values

$$\tilde{x} = \check{A}^{-1}\check{b} = \begin{pmatrix}0\\1\end{pmatrix}, \quad u = \begin{pmatrix}1\\1\end{pmatrix}, \quad v = (1-2\varepsilon)\begin{pmatrix}1\\1\end{pmatrix},$$

since it gives the same vector x' as (15).

Krawczyk's iteration starting with the weaker enclosure (12a) gives

$$z^0 = x = \frac{\delta+1}{1-2\varepsilon}\begin{pmatrix}[-1,1]\\ [-1,1]\end{pmatrix}, \quad z^l = \begin{pmatrix}0\\1\end{pmatrix} + \gamma_l\begin{pmatrix}[-1,1]\\ [-1,1]\end{pmatrix} \quad \text{for } l > 0,$$

where

$$\gamma_l = \frac{\delta+\varepsilon+2^{l-1}\varepsilon^l}{1-2\varepsilon}.$$

In particular, we see that $z^l \to (CA)^{\mathrm{K}}(Cb)$ for $l \to \infty$, and for small $\operatorname{rad}(A)$ (i.e. small ε) all iterates z^l $(l \geq 2)$ have the quadratic approximation property.

4.3 Interval Gauss–Seidel iteration

In this section we discuss another method for the iterative improvement of an initial enclosure x of $A^{\mathrm{H}}b$, namely an interval version of the Gauss–Seidel iteration. Among the many interesting properties of this iteration we mention that

(i) for M-matrices A, Gauss–Seidel iteration converges to the hull $A^{\mathrm{H}}b$ of the solution set (cf. Theorem 4.4.8);
(ii) applied to a preconditioned system, Gauss–Seidel iteration always yields tighter intervals than the Krawczyk iteration (cf. Theorem 4.3.5).

Since Gauss–Seidel iteration is also useful when we want to find improved enclosures for the set of those solutions of a linear interval system which lie in a specified initial box x, we shall not assume that the initial box x contains $A^{\mathrm{H}}b$. Thus, for arbitrary $x \in \mathbb{IR}^n$, we are interested in good enclosures for the *truncated solution set*

$$\Sigma(A, b) \cap x = \{\tilde{x} \in x \quad | \quad \tilde{A}\tilde{x} = \tilde{b} \text{ for some } \tilde{A} \in A, \tilde{b} \in b\}.$$

Note that the truncated solution set is bounded even when A is singular; and, indeed, Gauss–Seidel iteration can be applied to singular systems, although with limited success only (cf. Example 4.3.3(i) and Proposition 4.3.7).

Gauss–Seidel iteration is based on writing the system $\tilde{A}\tilde{x} = \tilde{b}$ explicitly in components as

$$\sum_{k=1}^{n} \tilde{A}_{ik}\tilde{x}_k = \tilde{b}_i \quad (i = 1, \ldots, n), \tag{1}$$

and, assuming that $\tilde{A}_{ii} \neq 0$, solving the ith equation for the ith variable. This gives

$$\begin{aligned} \tilde{x}_i &= \Big(\tilde{b}_i - \sum_{k \neq i} \tilde{A}_{ik}\tilde{x}_k\Big)\Big/\tilde{A}_{ii} \\ &\subseteq \Big(b_i - \sum_{k \neq i} A_{ik}x_k\Big)\Big/A_{ii} =: x_i' \end{aligned} \tag{2}$$

if $0 \notin A_{ii}$ and an interval vector x containing $\tilde{x}$ is known.

If $0 \notin A_{ii}$ for all i, we can apply (2) for $i = 1, \ldots, n$ and obtain another enclosure x' for $\tilde{x}$. Since this works for all $\tilde{x} \in x$ with $\tilde{A}\tilde{x} = \tilde{b}$, $\tilde{A} \in A$ and $\tilde{b} \in b$, we have

$$\Sigma(A, b) \cap x \subseteq x' \cap x.$$

However, we have not yet made the best use of the available information. For, in the ith step, an enclosure for $\tilde{x}_i$ and improved enclosures for $\tilde{x}_1, \ldots, \tilde{x}_{i-1}$ are already available; we can use them to obtain an improved enclosure of the right-hand side of (2) to get an improved enclosure for $\tilde{x}_i$. Continuing in this way for $i = 2, \ldots, n$ we see that $\tilde{x}$ is contained in the vector y determined by

$$y_i := \left(b_i - \sum_{k<i} A_{ik} y_k - \sum_{k>i} A_{ik} x_k\right) \Big/ A_{ii} \cap x_i \quad (i = 1, \ldots, n). \tag{3}$$

The rule (3) breaks down when $0 \in A_{ii}$ for some i. However, even in this case it is often possible to reduce the size of x_i.

To obtain a concise notation we shall treat the scalar case $n = 1$ first. In this case the coefficient matrix $A = a$, the right-hand side b and the initial box x are intervals, and the hull of the truncated solution set can be determined by a simple calculation.

4.3.1 Proposition For intervals $a, b, x \in \mathbb{IR}$, define

$$\Gamma(a, b, x) := \Box\{\tilde{x} \in x \mid \tilde{a}\tilde{x} = \tilde{b} \text{ for some } \tilde{a} \in a, \tilde{b} \in b\}. \tag{4}$$

Then

$$\Gamma(a, b, x) = \begin{cases} b/a \cap x & \text{if } 0 \notin a, \\ \Box(x \setminus]\underline{b}/\underline{a}, \underline{b}/\overline{a}[) & \text{if } b > 0 \in a, \\ \Box(x \setminus]\overline{b}/\overline{a}, \overline{b}/\underline{a}[) & \text{if } b < 0 \in a, \\ x & \text{if } 0 \in a, b. \end{cases} \tag{5}$$

Proof If $0 \notin a$ then $\Sigma(a, b) = b/a$. If $0 \in a$ and $0 \in b$ then $\Sigma(a, b) = \mathbb{R}$. And if $0 \in a$ and $0 \notin b$ then either $b > 0$ and $\Sigma(a, b) = \mathbb{R} \setminus (\underline{b}/\underline{a}, \underline{b}/\overline{a})$ or $b < 0$ and $\Sigma(a, b) = \mathbb{R} \setminus]\underline{b}/\underline{a}, \overline{b}/\underline{a}[$. Now take the intersection with x. □

For an implementation of an outward rounding $\Gamma(a, b, x)$ on a computer we note that (for $c =]\overline{b}/\overline{a}, \underline{b}/\overline{a}[$ or $c =]\overline{b}/\overline{a}, \overline{b}/\underline{a}[$) we have

$$\Box(x \backslash c) \subseteq \Box(x \setminus]\Delta \underline{c}, \nabla \overline{c}[) \tag{6}$$

but not necessarily $\Box(x \backslash c) \subseteq \Box(x \backslash \Diamond c)$, so that rounding in (5) must be performed according to (6). Moreover, one has to account for the possibility that $\Gamma(a, b, x)$ may be empty. With a proper implementation the evaluation of $\Gamma(a, b, x)$ can be achieved using two real divisions, but only one when $0 \in a$ and either $x \geq 0$ or $x \leq 0$.

We mention the following properties of $\Gamma(a, b, x)$:

4.3.2 Proposition Let $a, a', b, b', x, x' \in \mathbb{IR}$. Then

$$\Sigma(a, b) \cap x \subseteq \Gamma(a, b, x) \subseteq x, \tag{7}$$

$$\emptyset \neq x' = \Gamma(a, b, x) \subseteq \operatorname{int}(x) \Rightarrow 0 \notin a \text{ and } x' = b/a. \tag{8}$$

$$x \subseteq \Gamma(a, b, x) \text{ and } 0 \notin a \Rightarrow x \subseteq b/a, \tag{9}$$

$$\Gamma(1, b, x) = b \cap x, \tag{10}$$

$$\Gamma(a,b,x)=\emptyset \Leftrightarrow ax\cap b=\emptyset, \tag{11}$$

$$a\subseteq a', b\subseteq b', x\subseteq x' \Rightarrow \Gamma(a,b,x)\subseteq\Gamma(a',b',x'), \tag{12}$$

$$a\subseteq a', b\subseteq b', x=\Gamma(a',b',x') \Rightarrow \Gamma(a,b,x)=\Gamma(a,b,x'), \tag{13}$$

$$a=a'+a'' \Rightarrow \Gamma(a,b,x)\subseteq\Gamma(a',b-a''x,x). \tag{14}$$

Proof Formulae (7), (10) and (12) directly follow from (4). Formulae (8) and (9) follow from (5), and (11) holds since $ax=\{\tilde{a}\tilde{x} \mid \tilde{a}\in a, \tilde{x}\in x\}$. Now suppose that the hypothesis of (13) holds. If $z:=\Gamma(a,b,x')\neq\emptyset$ then we have $\tilde{a}\underline{z}\in b\subseteq b'$ for some $\tilde{a}\in a\subseteq a'$, hence $\underline{z}\in\Gamma(a',b',x')=x$, and therefore $\underline{z}\in\Gamma(a,b,x)$. By the same argument, $\bar{z}\in\Gamma(a,b,x)$ so that $z\subseteq\Gamma(a,b,x)$, and by (12) we must have equality. Hence (13) holds. Finally, suppose that $a=a'+a''$. Then every $\tilde{a}\in a$ can be written as $\tilde{a}=\tilde{a}'+\tilde{a}''$ with $\tilde{a}'\in a'$, $\tilde{a}''\in a''$. If now $\tilde{x}\in x$, $\tilde{a}\tilde{x}=\tilde{b}\in b$ then $\tilde{a}'\tilde{x}=(\tilde{a}-\tilde{a}'')\tilde{x}=\tilde{b}-\tilde{a}''\tilde{x}\in b-a''x$; therefore $\Gamma(a,b,x)\subseteq\Gamma(a',b-a''x,x)$. This proves (14). □

We now return to the general case, and consider the equation

$$\tilde{A}_{ii}\tilde{x}_i=\tilde{b}_i-\sum_{k\neq i}\tilde{A}_{ik}\tilde{x}_k. \tag{2a}$$

With an enclosure x for $\tilde{x}$ and improved enclosures $y_1,\dots,y_{i-1}$ for $\tilde{x}_1,\dots,\tilde{x}_{i-1}$, (2a) implies improved enclosures

$$y_i:=\Gamma\Big(A_{ii}, b_i-\sum_{k<i}A_{ik}y_k-\sum_{k>i}A_{ik}x_k, x_i\Big) \quad (i=1,\dots,n) \tag{15}$$

for $\tilde{x}_i$. Since it may happen that some y_i becomes empty, we shall adopt the conventions that the value of an arithmetical expression involving an empty set has as value the empty set, and that $\Gamma(a,b,x)=\emptyset$ if $b=\emptyset$ or $x=\emptyset$. In extension of the notation for the scalar case we shall write $\Gamma(A,b,x)$ for the vector y defined by (15), and call $\Gamma(A,b,x)$ the *Gauss–Seidel operator* applied to A, b and x. Note that (15) and (3) define the same vector when no diagonal element of A contains zero. Before considering the properties of the Gauss–Seidel operator we give some illustrative numerical examples.

4.3.3 Examples (i) Let

$$A=\begin{pmatrix}[0,1] & 0.1\\ 0.1 & [1,2]\end{pmatrix},\quad b=\begin{pmatrix}1\\2\end{pmatrix},\quad x=\begin{pmatrix}[-1,1]\\ [-1,1]\end{pmatrix}.$$

Since

$$A\begin{pmatrix}15\\1\end{pmatrix}=\begin{pmatrix}[-0.1,14.9]\\ [-0.5,\ 0.5]\end{pmatrix}\ni 0,$$

A is singular. Nevertheless, (15) gives

$$y_1 = \Gamma(A_{11}, b_1 - A_{12}x_2, x_1) = \Gamma([0,1],[0.9,1.1],[-1,1]) = [0.9,1],$$
$$y_2 = \Gamma(A_{22}, b_2 - A_{21}y_1, x_2) = \Gamma([1,2],[1.9,1.91],[-1,1]) = [0.95,1],$$

Thus,

$$\Sigma(A,b) \cap x \subseteq \Gamma(A,b,x) = \begin{pmatrix}[0.90,1]\\ [0.95,1]\end{pmatrix},$$

In this case, the enclosure is optimal since

$$\begin{pmatrix}1.0 & 0.10\\ 0.1 & 1.91\end{pmatrix}\begin{pmatrix}0.9\\ 1.0\end{pmatrix} = \begin{pmatrix}1\\ 2\end{pmatrix} = \begin{pmatrix}0.905 & 0.1\\ 0.100 & 2.0\end{pmatrix}\begin{pmatrix}1.00\\ 0.95\end{pmatrix}.$$

(ii) Let

$$A = \begin{pmatrix}1 & 0\\ 2 & 1\end{pmatrix}, \quad b = \begin{pmatrix}0\\ 2\end{pmatrix}, \quad x = \begin{pmatrix}[-1,1]\\ [-1,1]\end{pmatrix}.$$

Then (15) gives

$$y_1 = \Gamma(1,0,[-1,1]) = 0, \quad y_2 = \Gamma(1,2,[-1,1]) = \emptyset,$$

so that $\Gamma(A,b,x) = \emptyset$ and therefore $\Sigma(A,b) \cap x = \emptyset$. Since

$$Ax \cap b = \begin{pmatrix}[-1,1]\\ [-3,3]\end{pmatrix} \cap b = \begin{pmatrix}0\\ 2\end{pmatrix} \neq \emptyset$$

this shows that (11) does not extend to the multidimensional case (however, cf. (18) below).

(iii) Let

$$A = \begin{bmatrix}1 & 0 & 0\\ 1 & 1 & 0\\ 1 & 1 & 1\end{bmatrix}, \quad b = \begin{bmatrix}[-1,1]\\ [-1,1]\\ 0\end{bmatrix}, \quad x = \begin{bmatrix}[-1,1]\\ [-2,2]\\ [\ 2,3]\end{bmatrix}.$$

Then

$$A^{-1} = \begin{bmatrix}1 & 0 & 0\\ -1 & 1 & 0\\ 0 & -1 & 1\end{bmatrix}, \quad \Sigma(A,b) \subseteq A^{-1}b = \begin{bmatrix}[-1,1]\\ [-2,2]\\ [-1,1]\end{bmatrix}$$

so that $\Sigma(A,b) \cap x = \emptyset$. However,

$$y_1 = \Gamma(1,[-1,1],[-1,1]) = [-1,1],$$
$$y_2 = \Gamma(1,[-2,2],[-2,2]) = [-2,2],$$
$$y_3 = \Gamma(1,[-3,3],[2,3]) = [2,3],$$

so that $\Gamma(A,b,x) = x$ without improvement.

We now extend some of the properties mentioned in Proposition 4.3.2 to the multidimensional case.

4.3.4 Proposition Let $A \in \mathbb{IR}^{n\times n}$, $b, x \in \mathbb{IR}^n$. Then

$$\Sigma(A,b)\cap x \subseteq \Gamma(A,b,x)\subseteq x, \tag{16}$$

$$\Gamma(I,b,x) = b\cap x, \tag{17}$$

$$\Gamma(A,b,x) = \emptyset \quad \text{if } Ax\cap b = \emptyset, \tag{18}$$

$$A'\subseteq A, b'\subseteq b, x'\subseteq x \Rightarrow \Gamma(A',b',x')\subseteq \Gamma(A,b,x), \tag{19}$$

$$A = A' + A'' \Rightarrow \Gamma(A,b,x)\subseteq \Gamma(A', b - A''x, x). \tag{20}$$

Proof Let $y = \Gamma(A,b,x)$, so that (15) holds. Formula (16) follows by the argument leading to the derivation of (15), and (17) holds since $y_i = \Gamma(1, b_i, x_i) = b_i \cap x_i$ when $A = I$.

To prove (18) suppose that $Ax\cap b = \emptyset$. Then

$$\left(\sum_k A_{ik}x_k\right)\cap b_i = \emptyset \tag{21}$$

for some index i, and since in (21) every interval variable occurs only once, the equation

$$\sum_k \tilde{A}_{ik}\tilde{x}_k = \tilde{b}_i$$

has no solution with $\tilde{A}_{ik}\in A_{ik}$, $\tilde{x}_k \in x_k$ $(k = 1,\ldots,n)$ and $\tilde{b}_i \in b_i$. Since $y_k \subseteq x_k$ for $k < i$, this implies that

$$\tilde{A}_{ii}\tilde{x}_i = \tilde{b}_i - \sum_{k<i}\tilde{A}_{ik}\tilde{y}_k - \sum_{k>i}\tilde{A}_{ik}\tilde{x}_k$$

has no solution with $\tilde{A}_{ik}\in A_{ik}$ $(k = 1,\ldots,n)$, $\tilde{y}_k \in y_k$ $(k<i)$, $\tilde{x}_k\in x_k$ $(k \geq i)$ and $\tilde{b}_i \in b$. Therefore,

$$A_{ii}x_i \cap \left(b_i - \sum_{k<i} A_{ik}y_k - \sum_{k>i} A_{ik}x_k\right) = \emptyset, \tag{22}$$

since in (22), too, every interval variable occurs only once. By (11), this implies $y_i = \emptyset$ so that $\Gamma(A,b,x) = \emptyset$, and (18) holds.

Formula (19) follows inductively from (12) and (15). Finally, to prove (20), suppose that $A = A' + A''$. By induction we show that y is contained in $y' := \Gamma(A', b - A''x, x)$. Assume that $y_k \subseteq y'_k$ for all $k < i$; this certainly holds for $i = 1$. Then

$$A_{ik}x_k = (A'_{ik} + A''_{ik})x_k \subseteq A'_{ik}x_k + A''_{ik}x_k,$$

and since $y_k \subseteq x_k \cap y'_k$ for $k < i$, we also have

$$A_{ik}y_k \subseteq A'_{ik}y_k + A''_{ik}y_k \subseteq A'_{ik}y'_k + A''_{ik}x_k.$$

Therefore,

$$\begin{aligned} a_i &:= b_i - \sum_{k<i} A_{ik}y_k - \sum_{k>i} A_{ik}x_k \\ &\subseteq b_i - \sum_{k<i} (A'_{ik}y'_k + A''_{ik}x_k) - \sum_{k>i} (A'_{ik}x_k + A''_{ik}x_k), \end{aligned}$$

so that

$$a_i - A''_{ii}x_i \subseteq (b - A''x)_i - \sum_{k<i} A'_{ik}y'_k - \sum_{k>i} A'_{ik}x_k =: a'_i.$$

By (14), (15) and (12) we thus have

$$\begin{aligned} y_i = \Gamma(A_{ii}, a_i, x_i) &\subseteq \Gamma(A'_{ii}, a_i - A''_{ii}x_i, x_i) \\ &\subseteq \Gamma(A'_{ii}, a'_i, x_i) = y'_i. \end{aligned}$$

Therefore $y_i \subseteq y'_i$, which completes the induction and proves (20). □

If $y = \Gamma(A, b, x)$ is strictly contained in x we may hope to get a further improved enclosure of $A^{\mathrm{H}}b \cap x$ by repeating the process. This leads us to consider the iteration

$$x^0 := x, \quad x^{l+1} := \Gamma(A, b, x^l) \quad (l = 0, 1, 2, \ldots) \tag{23}$$

which we call the *interval Gauss–Seidel iteration*. Quite often the iteration (23) is applied in conjunction with preconditioning, resulting in the *preconditioned Gauss–Seidel iteration*

$$x^0 := x, \quad x^{l+1} := \Gamma(CA, Cb, x^l) \quad (l = 0, 1, 2, \ldots) \tag{24}$$

for a suitable preconditioning matrix $C \in \mathbb{R}^{n\times n}$. By (16), we have

$$\Sigma(A, b) \cap x \subseteq \Sigma(CA, Cb) \cap x \subseteq x^l \quad \text{for all } l \geq 0. \tag{25}$$

We now compare the preconditioned Gauss–Seidel iteration with the Krawczyk iteration

$$z^0 = x, \quad z^{l+1} := (Cb - (CA - I)z^l) \cap z^l \quad (l = 0, 1, 2, \ldots). \tag{26}$$

The Krawczyk iteration is in fact a special case of a more general iteration family defined by

$$z^0 = x, \quad z^{l+1} := \Gamma(A_0, Cb - Ez^l, z^l) \quad (l = 0, 1, 2, \ldots) \tag{27}$$

in terms of a *splitting* $CA = A_0 + E$ of the preconditioned matrix. By (17), the iteration (27) reduces to (26) when $A_0 = I$, $E = CA - I$. For $A_0 = A$, $E = 0$, (27) reduces to (24).

It is a remarkable fact that, in contrast to the case of noninterval linear systems, splittings do *not* improve the performance of the interval Gauss–Seidel iteration. In particular, no advantage can be gained by using so-called overrelaxation methods.

4.3.5 Theorem Let $A \in \mathbb{IR}^{n\times n}$, $b, x \in \mathbb{IR}^n$ and $C \in \mathbb{R}^{n\times n}$. If $CA = A_0 + E$ then the iterates defined by (27) contain the preconditioned Gauss–Seidel iterates defined by (24):

$$x^l \subseteq z^l \quad \text{for all } l \geq 0.$$

In particular, this holds for the Krawczyk iterates z^l defined by (26).

Proof Suppose that $x^l \subseteq z^l$ for some $l \geq 0$. Then $x^{l+1} = \Gamma(CA, Cb, x^l) \subseteq \Gamma(CA, Cb, z^l) \subseteq \Gamma(A_0, Cb - Ez^l, z^l) = z^{l+1}$ by (19) and (20). Since $x^0 \subseteq z^0$ by definition, we have $x^l \subseteq z^l$ for all $l \geq 0$. In particular, this applies with $A_0 = I$, $E = CA - I$ to the Krawczyk iteration (26). □

4.3.6 Example The improved qualities of Gauss–Seidel iteration when compared with Krawczyk's iteration is already visible in the scalar case $n = 1$. Take $A = [1/2, 3/2]$, $b = 1$, so that $A^{\mathrm{H}}b = b/A = [2/3, 2]$. By construction, Gauss–Seidel iteration yields for $n = 1$ the optimal solution in one step from every initial enclosure of $A^{\mathrm{H}}b$. On the other hand, the Krawczyk iteration preconditioned with $C = \check{A}^{-1} = 1$ (where $|CA - I| = 1/2 < 1$) proceeds according to $z^{l+1} = (1 - [-1/2, 1/2]z^l) \cap z^l$. Hence starting with $z^0 = [-2, 2]$ we get $z^1 = [0, 2] = z^2$, and we are stuck with a larger interval. Preconditioned with $C = \bar{A}^{-1} = 2/3$ (where $|CA - I| = 2/3 < 1$), the Krawczyk iteration proceeds according to $z^{l+1} = (2/3 - [-2/3, 0]z^l) \cap z^l$, and leads from $z^0 = [-2, 2]$ to $z^1 = [-2/3, 2]$, $z^2 = [2/9, 2]$, $z^3 = [2/3, 2] = A^{\mathrm{H}}b$ with more work than for Gauss–Seidel iteration.

Example 4.3.3(iii) showed that not every initial box x is improved by Gauss–Seidel iteration. But could we hope that at least every large initial box x (with sufficiently large $\|x\|_\infty$, say) is improved? As the next result shows this is not necessarily the case.

4.3.7 Proposition Let $A \in \mathbb{IR}^{n\times n}$. If A is not an H-matrix then there are vectors $x \in \mathbb{IR}^n$ of arbitrary norm $\|x\|_\infty$ such that $\Gamma(A, 0, x) = x$.

Proof By proposition 3.7.3 there is a nonzero vector $u \geq 0$ with $\langle A\rangle u \leq 0$, so that $\Sigma_{k\neq i} |A_{ik}|u_k \geq \langle A_{ii}\rangle u_i$ $(i = 1, \ldots, n)$. Now put $b = 0$, $x = [-a, a]u$, $y = \Gamma(A, b, x)$ and suppose that $y_k = x_k$ for $k < i$. Then

$$b_i - \sum_{k<i} A_{ik} y_k - \sum_{k>i} A_{ik} x_k = -\sum_{k\neq i} A_{ik} x_k = [-\alpha, \alpha] \sum_{k\neq i} |A_{ik}| u_k$$
$$\supseteq [-\alpha, \alpha]\langle A_{ii}\rangle u_i = \langle A_{ii}\rangle x_i.$$

Thus, (15) and (19) imply $x_i \supseteq y_i \supseteq \Gamma(A_{ii}, \langle A_{ii}\rangle x_i, x_i) = x_i$ since $\check{x}_i = 0$. Hence $x_i = y_i$, and by induction $x = y$. □

We can even show that when *all* lower and upper bounds of x are improved by Gauss–Seidel iteration then A must be an H-matrix. The proof is based on the following lemma.

4.3.8 Lemma For all $l \geq 0$, the components of the Gauss–Seidel iteration (23) satisfy

$$x_i^{l+1} = \Gamma\left(A_{ii}, b_i - \sum_{k<i} A_{ik} x_k^{l+1} - \sum_{k>i} A_{ik} x_k^l, x_i\right) \quad \text{for } i = 1, \ldots, n. \tag{28}$$

Proof Write $r_i^l := b_i - \Sigma_{k<i} A_{ik} x_k^{l+1} - \Sigma_{k>i} A_{ik} x_k^l$. Since (28) trivially holds for $l = 0$, we assume that it holds when l is replaced by $l-1$, so that $x_i^l = \Gamma(A_{ii}, r_i^{l-1}, x_i)$. Since $x^{l+1} \subseteq x^l \subseteq x^{l-1}$ we have $r_i^l \subseteq r_i^{l-1}$ and hence $x_i^{l+1} = \Gamma(A_{ii}, r_i^l, x_i^l) = \Gamma(A_{ii}, r_i^l, x_i)$ by (13). By induction, (28) holds generally. □

4.3.9 Theorem If the Gauss–Seidel iteration (23) satisfies $\emptyset \neq x^j \subseteq \text{int}(x)$ for some $j \geq 1$ then A is an H-matrix.

Proof From (28), our assumption and (8) we find that $0 \notin A_{ii}$ for $i = 1, \ldots, n$ and

$$x_i^{l+1} = \left(b_i - \sum_{k<i} A_{ik} x_k^{l+1} - \sum_{k>i} A_{ik} x_k^l\right) \Big/ A_{ii} \tag{29}$$

for $l = j-1$. If (29) holds for some $l \geq j-1$ then, with r_i^l as in the previous proof, $r_i^{l+1}/A_{ii} \subseteq r_i^l/A_{ii} = x_i^{l+1} \subseteq x_i^j \subseteq \text{int}(x_i)$. Thus, by (28), we have $x_i^{l+2} = \Gamma(A_{ii}, r_i^{l+1}, x_i) = r_i^{l+1}/A_{ii}$, i.e. (29) holds with $l+1$ in place of l. Hence (29) is valid for all $l \geq j-1$.

Now choose $\tilde{A} \in A$ such that $\langle\tilde{A}\rangle = \langle A\rangle$. Let $y^l := x^l$ for $l \leq j$, and, for $l \geq j$,

$$y_i^{l+1} := \left(b_i - \sum_{k<i} \tilde{A}_{ik} y_k^{l+1} - \sum_{k>i} \tilde{A}_{ik} y_k^l\right) \Big/ \tilde{A}_{ii} \quad (i = 1, \ldots, n). \tag{30}$$

By induction we find from (30) and (29) that $y^l \subseteq x^l$ and $y^{l+1} \subseteq y^l$ for all l; in particular, $y^l \subseteq \text{int}(x)$ for $l \geq j$. As a nested sequence of interval vectors, the y^l $(l \geq j)$ converge to some $y \in \mathbb{IR}^n$. Clearly, $y \subseteq \text{int}(x)$ and from (30) we get

$$y_i = \left(b_i - \sum_{k \neq i} \tilde{A}_{ik} y_k\right) \Big/ \tilde{A}_{ii} \quad (i = 1, \ldots, n). \tag{31}$$

Now suppose that A is not an H-matrix. Then Proposition 3.7.3 implies the existence of a nonzero vector $u \in \mathbb{R}^n$ satisfying $\langle A \rangle u \leq 0 \leq u$. Since $y \subseteq \text{int}(x)$ there is a real number $\alpha > 0$ such that $z := y + [-\alpha, \alpha]u \subseteq x = x^0$. Now

$$\begin{aligned} \tilde{A}_{ii} z_i &= \tilde{A}_{ii}(y_i + [-\alpha, \alpha]u_i) = \tilde{A}_{ii} y_i + [-\alpha, \alpha]|\tilde{A}_{ii}|u_i \\ &\subseteq b_i - \sum_{k \neq i} \tilde{A}_{ik} y_k + [-\alpha, \alpha] \sum_{k \neq i} |\tilde{A}_{ik}| u_k \\ &= b_i - \sum_{k \neq i} \tilde{A}_{ik}(y_k + [-\alpha, \alpha]u_k) \\ &= b_i - \sum_{k \neq i} \tilde{A}_{ik} z_k, \end{aligned}$$

so that

$$z_i \subseteq \left(b_i - \sum_{k \neq i} \tilde{A}_{ik} z_k\right) \Big/ \tilde{A}_{ii} \quad (i = 1, \ldots, n). \tag{32}$$

This implies $z \subseteq \Gamma(\tilde{A}, \tilde{b}, z) \subseteq \Gamma(A, b, z)$, and an induction argument using (19) yields $z \subseteq x^l$ for all $l \geq 0$. In particular $z \subseteq y^l$ for $l \leq j$. Another induction argument using (31) and (32) yields $z \subseteq y^l$ for all $l \geq 0$. But then $z \subseteq y$, so that $\alpha u = \text{rad}(z) - \text{rad}(y) \leq 0$, contradicting $\alpha > 0$ and $0 \neq u \geq 0$. Hence A must be an H-matrix. □

4.3.10 Corollary If A is not an H-matrix and all Gauss–Seidel iterates (23) are nonempty then some component bound $\underline{x}_i^l$ or $\overline{x}_i^l$ remains fixed throughout the iteration. □

On the other hand, if A is an H-matrix, then *all* sufficiently large initial boxes are improved by Gauss–Seidel iteration. This is a consequence of the following result which still holds for arbitrary $A \in \mathbb{IR}^n$.

4.3.11 Proposition Let $A \in \mathbb{IR}^n$ and $b, x \in \mathbb{IR}^n$. Then

$$\Gamma(A, b, x) = x \Rightarrow \langle A \rangle |x| \leq |b|. \tag{33}$$

Proof We have

$$(\langle A \rangle |x|)_i = \langle A_{ii} \rangle |x_i| - \sum_{k \neq i} |A_{ik}|\,|x_k|.$$

If $0 \in A_{ii}$ then this expression is ≤ 0 and hence $\leq |b_i|$. And if $0 \notin A_{ii}$ then $x_i \subseteq (b_i - \Sigma_{k \neq i} A_{ik} x_k)/A_{ii}$. Therefore

$$|x_i| \leq \left(|b_i| + \sum_{k \neq i} |A_{ik}|\,|x_k|\right) \Big/ \langle A_{ii} \rangle,$$

which again yields $(\langle A \rangle |x|)_i \leq |b|_i$. Hence this holds for all i, giving (33). □

4.3.12 Corollary Let $A \in \mathbb{IR}^n$ be an H-matrix, and let $b, x \in \mathbb{IR}^n$. Then

$$\Gamma(A, b, x) = x \Rightarrow |x| \leq \langle A \rangle^{-1} |b|.$$

In particular, the limit of the Gauss–Seidel iteration (23) satisfies

$$\left|\lim_{l \to \infty} x^l\right| \leq \langle A \rangle^{-1} |b|.$$

Proof Multiply (33) by $\langle A \rangle^{-1} \geq 0$. □

We note some further bounds which will be useful later.

4.3.13 Proposition Let $A \in \mathbb{IR}^{n \times n}$ and define the $n \times n$ matrix $r^*(A) \in \mathbb{R}^{n \times n}$ by

$$r^*(A)_{ik} := \begin{cases} 0 & \text{if } i = k, \\ \max(0, \operatorname{rad}(A_{ik}) - |\check{A}_{ik}|) & \text{if } i \neq k. \end{cases} \tag{34}$$

Then

$$0 \in x \subseteq \Gamma(A, b, x) \tag{35}$$

implies

$$\langle A \rangle \operatorname{rad}(x) \leq \operatorname{rad}(b) + r^*(A)|\check{x}| \tag{36}$$

and

$$\operatorname{rad}(x) \leq \operatorname{rad}(b) + |I - A|\,|x|. \tag{37}$$

Proof Suppose first that $0 \notin A_{ii}$. Then (35) implies

$$0 \in x_i \subseteq \left(b_i - \sum_{k \neq i} A_{ik} x_k\right) \Big/ A_{ii}.$$

Using Proposition 1.6.8(iv) we get

$$\operatorname{rad}(x_i) \leq \operatorname{rad}\left(\left(b_i - \sum_{k \neq i} A_{ik} x_k\right) \Big/ \langle A_{ii} \rangle\right),$$

so that

$$\begin{aligned}\langle A_{ii}\rangle \mathrm{rad}(x_i) &\le \mathrm{rad}(b_i) + \sum_{k\ne i} \mathrm{rad}(x_k A_{ik})\\ &\le \mathrm{rad}(b_i) + \sum_{k\ne i} (\mathrm{rad}(x_k)|A_{ik}| + |\check{x}_k| r^*(A)_{ik})\end{aligned}$$

by Proposition 1.6.8(ii). The same formula holds if $0 \in A_{ii}$ since then $\langle A_{ii}\rangle = 0$. Hence, without restriction,

$$\begin{aligned}(\langle A\rangle \mathrm{rad}(x))_i &= \langle A_{ii}\rangle \mathrm{rad}(x_i) - \sum_{k\ne i} |A_{ik}| \mathrm{rad}(x_k)\\ &\le \mathrm{rad}(b_i) + \sum_{k\ne i} r^*(A)_{ik}\, |\check{x}_k|,\end{aligned}$$

which implies (36). Since $I - \langle A\rangle \le |I - A|$ and $r^*(A) \le |I - A|$, we also have

$$\begin{aligned}\mathrm{rad}(x) &\le \mathrm{rad}(b) + r^*(A)|\check{x}| + (I - \langle A\rangle)\mathrm{rad}(x)\\ &\le \mathrm{rad}(b) + |I - A|(|\check{x}| + \mathrm{rad}(x)),\end{aligned}$$

which gives (37). □

4.3.14 Corollary Let A be an H-matrix. Then

$$\check{x} = 0, \quad x \subseteq \Gamma(A, b, x) \Rightarrow |x| \le \langle A\rangle^{-1}\, \mathrm{rad}(b).$$

Proof Insert $\check{x} = 0$ into (36), multiply by $\langle A\rangle^{-1} \ge 0$, and observe that here $|x| = \mathrm{rad}(x)$. □

4.3.15 Corollary If $A \in \mathbb{IR}^{n\times n}$ satisfies $\|I - A\| \le \beta \le 1$ (for some scaled maximum norm) then

$$0 \in x \subseteq \Gamma(A, b, x) \Rightarrow \|\mathrm{rad}(x)\| - \beta\|x\| \le \|\mathrm{rad}(b)\|.$$

Proof Take norms in (37). □

For the further investigation of Gauss–Seidel iteration for H-matrices we need some facts about linear fixed point equations. They are treated in the next section, presented for the sake of clarity in the more abstract setting of sublinear mappings.

4.4 Linear fixed point equations

In this section we discuss fixed point equations defined by sublinear mappings and a related iteration method. The results are applied to derive properties of the fixed point inverse of an H-matrix. The fixed point inverse is closely related to the Gauss–Seidel iteration, and provides important insight into its behavior.

4.4.1 Theorem Let $S, T: \mathbb{IR}^n \to \mathbb{IR}^n$ be sublinear mappings satisfying $\rho(|S|\,|T|) < 1$. Then, for every $b \in \mathbb{IR}^n$, the following statements hold.

(i) The equation

$$z = S(b - Tz) \tag{1}$$

has a unique solution $z \in \mathbb{IR}^n$.

(ii) For all starting vectors $z^0 \in \mathbb{IR}^n$, the iteration

$$z^{l+1} := S(b - Tz^l) \quad (l = 0, 1, \ldots) \tag{2}$$

converges to the solution of (1), and

$$\|q(z^{l+1}, z)\| \le \beta \|q(z^l, z)\| \tag{3}$$

holds for any scaled maximum norm satisfying

$$\|\,|S|\,|T|\,\| \le \beta < 1. \tag{4}$$

(iii) For $l \ge k \ge 0$, we have

$$z^1 \subseteq z^0 \Rightarrow z \subseteq z^l \subseteq z^k$$
$$z^1 \supseteq z^0 \Rightarrow z \supseteq z^l \supseteq z^k.$$

Proof Since $\rho(|S|\,|T|) < 1$ there exist β and a scaled maximum norm such that (4) holds. For fixed $b \in \mathbb{IR}^n$, the mapping $\Phi: \mathbb{IR}^n \to \mathbb{IR}^n$ defined by

$$\Phi z := S(b - Tz) \quad \text{for } z \in \mathbb{IR}^n$$

satisfies $q(\Phi y, \Phi z) = q(S(b - Ty), S(b - Tz)) \le |S| q(b - Ty, b - Tz) = |S| q(Ty, Tz) \le |S|\,|T| q(y, z)$ by formulae (3.5.10) and (3.3.19), so that $\|q(\Phi y, \Phi z)\| \le \beta \|q(y, z)\|$ by (4). Now Schröder's fixed point theorem 3.3.3 shows that Φ has a unique fixed point $z \in \mathbb{IR}^n$, and, for arbitrary $z^0 \in \mathbb{IR}^n$, the iteration $z^{l+1} := \Phi z^l$, i.e. (2), converges to z with speed determined by (3). This proves (i) and (ii). For the proof of (iii) we first note that Φ is inclusion isotone. Hence if $z^l \subseteq z^{l-1}$ then $z^{l+1} = \Phi z^l \subseteq \Phi z^{l-1} = z^l$. So if $z^1 \subseteq z^0$ then $z^l \subseteq z^k$ for all $l \ge k$, and by taking the limit $l \to \infty$ we find $z \subseteq z^k$. Similarly if $z^1 \supseteq z^0$ then $z^{l+1} \supseteq z^l$ for all l and $z \supseteq z^k$. This implies (iii). □

4.4.2 Theorem Let $S, T: \mathbb{IR}^n \to \mathbb{IR}^n$ be sublinear mappings satisfying $\rho(|S|\,|T|) < 1$. Then:

(i) The mapping $P: \mathbb{IR}^n \to \mathbb{IR}^n$ which associates with each $b \in \mathbb{IR}^n$ the unique solution $Pb := z$ of (1) is sublinear, with absolute value

$$|P| \le (I - |S|\,|T|)^{-1}|S|, \tag{5}$$

and, for all $b, x \in \mathbb{IR}^n$, we have

$$S(b - Tx) = x \Leftrightarrow Pb = x, \tag{6}$$

$$S(b - Tx) \subseteq x \Rightarrow Pb \subseteq x, \tag{7}$$

$$S(b - Tx) \supseteq x \Rightarrow Pb \supseteq x, \tag{8}$$

(ii) If S and T are normal then P is also normal, and (5) holds with equality.

(iii) If S and T are linear then P is also linear, and

$$\text{cor}(P) = (I + \text{cor}(S)\text{cor}(T))^{-1}\,\text{cor}(S). \tag{9}$$

Proof (i) The uniqueness of P and (6)–(8) are immediate consequences of Theorem 4.4.1. We verify the sublinearity axioms for P. If $x \subseteq y$ then $S(x - T(Py)) \subseteq S(y - T(Py)) = Py$ so that $Px \subseteq Py$ by (7); hence P is inclusion isotone. Homogeneity is immediate. To show subadditivity we note that $S(x + y - T(Px + Py)) \subseteq S(x + y - T(Px) - T(Py)) \subseteq S(x - T(Px)) + S(y - T(Py)) = Px + Py$, so that $P(x + y) \subseteq Px + Py$ by (7). Therefore P is sublinear. Moreover, if S, T are linear then we have equality throughout this argument so that P is also linear. To show (5) we observe that formula (3.5.11) implies $|Px| = |S(x - T(Px))| \le |S|\,|x - T(Px)| \le |S|(|x| + |T(Px)|) \le |S|(|x| + |T|\,|Px|)$, so that $(I - |S|\,|T|)|Px| \le |S|\,|x|$. Multiplication with the nonnegative matrix $(I - |S|\,|T|)^{-1}$ gives $|Px| \le B|x|$, where $B = (I - |S|\,|T|)^{-1}|S|$. Hence $|P| = |P[-I, I]| \le B|[-I, I]| \le B$, which is (5). This completes the proof of (i).

(ii) If S, T are normal then $\text{rad}(Px) = \text{rad}(S(x - T(Px))) \ge |S|\text{rad}(x - T(Px)) = |S|(\text{rad}(x) + \text{rad}(T(Px))) \ge |S|(\text{rad}(x) + |T|\text{rad}(Px))$. Therefore $(I - |S|\,|T|)\text{rad}(Px) \ge |S|\text{rad}(x)$, giving $\text{rad}(Px) \ge B \cdot \text{rad}(x)$. Hence $|P| = |P[-I, I]| = \text{rad}(P[-I, I]) \ge B \cdot \text{rad}([-I, I]) = B$, and we must have $|P| = B$. Thus $\text{rad}(Px) \ge |P|\text{rad}(x)$, which shows that P is normal.

(iii) Let S, T be linear. It has already been shown that then P is linear, and it remains to verify (9). But, for linear P and thin b, we have $Pb = \text{cor}(P)b$, so that, by (6), $\text{cor}(P)b = S(b - T(\text{cor}(P)b)) = S(b - \text{cor}(T)\text{cor}(P)b) = \text{cor}(S)(b - \text{cor}(T)\text{cor}(P)b)$, and, since this involves thin objects only, we get $\text{cor}(P)b = (I - \text{cor}(S)\text{cor}(T))^{-1}\text{cor}(S)b$ for all b, which implies (9). □

We shall refer to P as the *fixed point mapping* of the iteration (1). For later use we record an easily verifiable existence and enclosure test for the fixed point mapping.

4.4.3 Proposition Let $S, T: \mathbb{IR}^n \to \mathbb{IR}^n$ be normal sublinear mappings, and let $b, x \in \mathbb{IR}^n$. Then

$$\operatorname{rad}(S(b - Tx)) < \operatorname{rad}(x) \Rightarrow \rho(|S|\,|T|) < 1, \tag{10}$$

and the fixed point mapping P of the iteration (1) exists. In particular, this holds if

$$S(b - Tx) \subseteq \operatorname{int}(x); \tag{11}$$

in this case we have the enclosure

$$Pb \subseteq \operatorname{int}(x).$$

Proof Since S, T are normal, the hypothesis of (10) implies $\operatorname{rad}(x) > \operatorname{rad}(S(b - Tx)) \geq |S|\operatorname{rad}(b - Tx) = |S|(\operatorname{rad}(b) + \operatorname{rad}(Tx)) \geq |S|\operatorname{rad}(Tx) \geq |S|\,|T|\operatorname{rad}(x) \geq 0$. Hence the vector $u := \operatorname{rad}(x)$ is positive and satisfies $|S|\,|T|u < u$. This yields $\rho(|S|\,|T|) < 1$, so that P exists by Theorem 4.4.2. Clearly, (11) implies the hypothesis of (10), and by (7) we then have $Pb \subseteq x$ and hence $Pb = S(b - T(Pb)) \subseteq S(b - Tx) \subseteq \operatorname{int}(x)$. □

Now we apply the preceding abstract results to a concrete situation related to the enclosure (4.3.2), and then deduce further properties of Gauss–Seidel iteration.

4.4.4 Theorem Let $A \in \mathbb{IR}^{n\times n}$ be an H-matrix. Then:

(i) For all $b \in \mathbb{IR}^n$, the system of equations

$$z_i = \left(b_i - \sum_{k\neq i} A_{ik}z_k\right)\Big/ A_{ii} \quad (i = 1, \ldots, n) \tag{12}$$

has a unique solution $z \in \mathbb{IR}^n$.

(ii) The mapping $A^{\mathrm{F}}: \mathbb{IR}^n \to \mathbb{IR}^n$ which associates with each $b \in \mathbb{IR}^n$ the solution $A^{\mathrm{F}}b = z$ of (12) is sublinear and normal, with absolute value

$$|A^{\mathrm{F}}| = \langle A\rangle^{-1}. \tag{13}$$

(iii) If $B \subseteq A$ then $B^{\mathrm{F}}b \subseteq A^{\mathrm{F}}b$ for all $b \in \mathbb{IR}^n$; in particular,

$$A^{\mathrm{H}}b \subseteq A^{\mathrm{F}}b \quad \text{for all } b \in \mathbb{IR}^n. \tag{14}$$

(iv) If A is thin then A^{F} is linear and $\operatorname{cor}(A^{\mathrm{F}}) = A^{-1}$.

Proof Denote by D the diagonal matrix with diagonal entries $D_{ii} = A_{ii}$ $(i = 1, \ldots, n)$, and by E the matrix obtained from A by replacing the diagonal entries by zeros. Then

$$A = D + E \quad \text{with } D_{ik} = 0 \text{ for } i \neq k \text{ and } E_{ik} = 0 \text{ for } i = k. \tag{15}$$

Equation (15) is called the *Jacobi splitting* of A. Since A is an H-matrix, $0 \notin A_{ii} = D_{ii}$; hence D is regular, and D^{-1} is the diagonal matrix with diagonal entries $(D^{-1})_{ii} = (A_{ii})^{-1}$. Since $|a^{-1}| = \langle a\rangle^{-1}$ for $0 \notin a \in \mathbb{IR}$, we have

$$|D^{-1}| = \langle D\rangle^{-1}, \quad \langle A\rangle = \langle D\rangle - |E|, \tag{16}$$

and since $b/a = ba^{-1}$, we can write the system (12) in the equivalent form

$$z = D^{-1}(b - Ez). \tag{17}$$

Thus we have the special case of (1) with $S = (D^{-1})^{\mathrm{M}}$ and $T = E^{\mathrm{M}}$. Since $|S| = |D^{-1}| = \langle D\rangle^{-1}$ and $|T| = |E|$, the condition $\rho(|S|\,|T|) < 1$ is equivalent with $\rho(\langle D\rangle^{-1}|E|) < 1$, and by Proposition 3.6.3(iii), this holds if $\langle D\rangle - |E| = \langle A\rangle$ is an M-matrix, i.e. if A is an H-matrix. Hence part (i) follows from Theorem 4.4.1(i). Since both S and T are normal, (ii) follows from Theorem 4.4.2 by noting that $|A^{\mathrm{F}}| = (I - |S|\,|T|)^{-1}|S| = (I - \langle D\rangle^{-1}|E|)^{-1}\langle D\rangle^{-1} = (\langle D\rangle - |E|)^{-1} = \langle A\rangle^{-1}$.

To prove (iii), let $D' + E'$ be the Jacobi splitting of $B \subseteq A$, and put $x := B^{\mathrm{F}}b$. Then $x = D'^{-1}(b - E'x) \subseteq D^{-1}(b - Ex)$, so that $x \subseteq A^{\mathrm{F}}b$ by (8). In particular, if $\tilde{A} \in A$, $\tilde{b} \in b$ then $\tilde{z} := \tilde{A}^{-1}\tilde{b}$ satisfies the equations (12) with $\tilde{z}, \tilde{b}, \tilde{A}$ in place of z, b, A, respectively. Hence $\tilde{A}^{-1}\tilde{b} = \tilde{A}^{\mathrm{F}}\tilde{b} \subseteq A^{\mathrm{F}}b$ which implies (14). Finally, if A is thin then S and T are linear so that A^{F} is linear by Theorem 4.4.2(iii), and $\mathrm{cor}(A^{\mathrm{F}}) = (I + \mathrm{cor}(S)\mathrm{cor}(T))^{-1}\mathrm{cor}(S) = (I + D^{-1}E)^{-1}D^{-1} = (D + E)^{-1} = A^{-1}$. □

We call the mapping A^{F} the *fixed point inverse* of A. The fixed point inverse is closely related to Gauss–Seidel iteration.

4.4.5 Theorem Let $A \in \mathbb{IR}^{n\times n}$ and $b, x \in \mathbb{IR}^n$.

(i) If A is an H-matrix then

$$\Gamma(A, b, x) = x \Rightarrow x \subseteq A^{\mathrm{F}}b; \tag{18}$$

$$A^{\mathrm{F}}b \subseteq x \Rightarrow A^{\mathrm{F}}b \subseteq \Gamma(A, b, x); \tag{19}$$

$$x = A^{\mathrm{F}}b \Rightarrow x = \Gamma(A, b, x). \tag{20}$$

(ii) If $\emptyset \neq \Gamma(A, b, x) \subseteq \mathrm{int}(x)$ then A is an H-matrix, and $A^{\mathrm{F}}b \subseteq \mathrm{int}(x)$.

Proof We have $\Gamma(A, b, x) = \Gamma(D + E, b, x) \subseteq \Gamma(D, b - Ex, x) \subseteq D^{-1}(b - Ex)$. Hence if $\Gamma(A, b, x) = x$ then $x \subseteq D^{-1}(b - Ex)$, so that $x \subseteq A^{\mathrm{F}}b$ by (8). This proves (18). If $z := A^{\mathrm{F}}b \subseteq x$ then $z_i = (b_i - \Sigma_{k\neq i} A_{ik}z_k)/A_{ii} \cap z_i$, and an induction argument shows $z \subseteq \Gamma(A, b, x)$, giving (19). In particular, for $x = z$ we get (20) since $\Gamma(A, b, x) \subseteq x$. Finally, (ii) is a special case of Theorem 4.3.9. □

Note that (18) implies that, for H-matrices A, Gauss–Seidel iteration improves every initial box $x \not\subseteq A^F b$, in sharp contrast to the behavior for general A (Proposition 4.3.7). In particular, we have the following:

4.4.6 Corollary If A is an H-matrix then the Gauss–Seidel iteration $x^0 := x$, $x^{l+1} := \Gamma(A, b, x^l)$ $(l = 0, 1, 2, \ldots)$ satisfies

$$\Sigma(A, b) \cap x \subseteq \lim_{l\to\infty} x^l \subseteq A^F b \cap x. \qquad \square$$

4.4.7 Corollary Let $A \in \mathbb{IR}^{n\times n}$, $C \in \mathbb{R}^{n\times n}$ and $b, x \in \mathbb{IR}^n$.

(i) If A is an H-matrix then $Cb + (I - CA)x \subseteq x$ implies $A^F b \subseteq x$.
(ii) If $Cb + (I - CA)x \subseteq \mathrm{int}(x)$ then A is an H-matrix and $A^F b \subseteq \mathrm{int}(x)$.

Proof Noting that $Cb + (I - CA)x$ is the first iterate z^1 of the Krawczyk iteration (4.3.26), the assertions follow from Theorems 4.3.5 and 4.4.5. $\square$

In an important special case (14) holds with equality.

4.4.8 Theorem (Barth & Nuding) If $A \in \mathbb{IR}^{n\times n}$ is an M-matrix then

$$A^F b = A^H b \quad \text{for all } b \in \mathbb{IR}^n.$$

Proof Let $z := A^F b$. Since $-A_{ik} \geq 0$ for $i \neq k$ we have

$$z_i = \left(b_i - \sum_{k\neq i} A_{ik} z_k\right) \Big/ A_{ii}$$

$$= \left[\underline{b}_i - \sum_{k\neq i} \underset{\sim}{A}_{ik} \underline{z}_k,\ \overline{b}_i - \sum_{k\neq i} \tilde{A}_{ik} \overline{z}_k\right] \Big/ A_{ii}$$

$$= \left[\left(\underline{b}_i - \sum_{k\neq i} \underset{\sim}{A}_{ik} \underline{z}_k\right) \Big/ \underset{\sim}{A}_{ii},\ \left(\overline{b}_i - \sum_{k\neq i} \tilde{A}_{ik} \overline{z}_k\right) \Big/ \tilde{A}_{ii}\right]$$

for suitable numbers $\underset{\sim}{A}_{ik}, \tilde{A}_{ik} \in A$. Comparison of the lower bounds gives $\sum_k \underset{\sim}{A}_{ik} \underline{z}_k = \underline{b}_i$, hence $\underset{\sim}{A}\underline{z} = \underline{b}$, $\underline{z} = \underset{\sim}{A}^{-1} \underline{b} \in A^H b$. Similarly, we find for the upper bound $\overline{z} = \tilde{A}^{-1} \overline{b} \in A^H b$. Hence $z \subseteq A^H b$, and since $A^H b \subseteq A^F b = z$, the assertion follows. $\square$

Remarks (i) By Corollary 4.4.6 and the above theorem, Gauss–Seidel iteration satisfies

$$\square(\Sigma(A, b) \cap x) \subseteq \lim_{l\to\infty} x^l \subseteq A^H b \cap x$$

when A is an M-matrix. It is an open problem whether we must always have equality on the left-hand side; this is certainly the case when $\Sigma(A, b) \subseteq x$.

(ii) In combination with Theorem 4.4.5 and Corollary 4.1.6(i) this theorem provides a way for the computation of $A^{\mathrm{H}}b$ for any inverse positive interval matrix A.

When A is not an M-matrix (and cannot be transformed to one by diagonal preconditioning) then we usually have $A^{\mathrm{F}}b \neq A^{\mathrm{H}}b$. For example, if $\check{b} = 0$ then we have $A^{\mathrm{H}}b \subseteq |A^{-1}|b$, but since A^{F} is normal, we have $A^{\mathrm{F}}b = |A^{\mathrm{F}}|b = \langle A\rangle^{-1}b$. Hence, as shown by Example 3.7.6, excessive overestimation of the hull $A^{\mathrm{H}}b$ of the solution set by $A^{\mathrm{F}}b$ is possible. However, in conjunction with preconditioning, $(CA)^{\mathrm{F}}(Cb)$ cannot overestimate $A^{\mathrm{H}}b$ more than the limit of Krawczyk's iteration, and hence has the quadratic approximation property. More specifically, we prove the following analog of Theorem 4.2.4:

4.4.9 Theorem Let $A \in \mathbb{IR}^{n\times n}$, $C \in \mathbb{R}^{n\times n}$ and suppose that CA is an H-matrix. Then

$$x := A^{\mathrm{H}}b \subseteq (CA)^{\mathrm{F}}(Cb) \subseteq \check{x} + \langle CA\rangle^{-1}|CA|(x - \check{x}). \tag{21}$$

In particular, if CA is strictly diagonally dominant, i.e.

$$\max_i \sum_{k\neq i} |CA|_{ik}/\langle CA\rangle_{ii} \leq \beta < 1$$

then

$$\|\mathrm{rad}(A^{\mathrm{H}}b)\|_\infty \leq \|\mathrm{rad}(CA)^{\mathrm{F}}(Cb)\|_\infty \leq \frac{1+\beta}{1-\beta}\|\mathrm{rad}(A^{\mathrm{H}}b)\|_\infty. \tag{22}$$

Proof Write $B := CA$, $a := Cb$, $x := A^{\mathrm{H}}b$ and $z := (CA)^{\mathrm{F}}(Cb)$. Then

$$B_{ik} = \sum_j C_{ij}A_{jk}, \quad a_i = \sum_k C_{ik}b_k,$$

$$z_i = \left(a_i - \sum_{k\neq i} B_{ik}z_k\right)\Big/ B_{ii}.$$

For fixed i, each interval variable A_{jk}, B_k, z_k occurs only once; hence these expressions are equal to the range. In particular, for each $\zeta \in z_i$ we can find $\tilde{z} \in z$, $\tilde{b} \in b$ and $\tilde{A} \in A$ such that

$$\zeta = \left(\tilde{a}_i - \sum_{k\neq i} \tilde{B}_{ik}\tilde{z}_k\right)\Big/ \tilde{B}_{ii},$$

where $\tilde{B} = C\tilde{A}$ and $\tilde{a} = C\tilde{b}$. Now $\tilde{x} := \tilde{A}^{-1}\tilde{b} \in A^{\mathrm{H}}b = x$, and $\tilde{a} = C\tilde{b} = C\tilde{A}\tilde{x} = \tilde{B}\tilde{x}$. Hence

$$\tilde{a}_i = \sum_k \tilde{B}_{ik}\tilde{x}_k,$$

$$\zeta = \Big(\sum_k \tilde{B}_{ik}\tilde{x}_k - \sum_{k\neq i} \tilde{B}_{ik}\tilde{z}_k\Big)\Big/\tilde{B}_{ii}$$

$$= \tilde{x}_i - \Big(\sum_{k\neq i} \tilde{B}_{ik}(\tilde{z}_k - \tilde{x}_k)\Big)\Big/\tilde{B}_{ii}$$

$$\in x_i - \Big(\sum_{k\neq i} B_{ik}(z_k - x_k)\Big)\Big/B_{ii}.$$

Since $\zeta \in z_i$ was arbitrary, we get

$$z_i \subseteq x_i - \Big(\sum_{k\neq i} B_{ik}(z_k - x_k)\Big)\Big/B_{ii}.$$

This holds for all i, and implies

$$|z_i - \check{x}_i| \leq \operatorname{rad}(x_i) + \left|\Big(\sum_{k\neq i} B_{ik}(z_k - x_k)\Big)\Big/B_{ii}\right|,$$

hence

$$|z_i - \check{x}_i| \leq \operatorname{rad}(x_i) + \langle B_{ii}\rangle^{-1}\sum_{k\neq i} |B_{ik}|(|z_k - \check{x}_k| + \operatorname{rad}(x_k)). \tag{23}$$

Since $\langle B_{ii}\rangle \leq |B_{ii}|$ we deduce

$$(\langle B\rangle|z - \check{x}|)_i = \langle B_{ii}\rangle|z_i - \check{x}_i| - \sum_{k\neq i} |B_{ik}|\,|z_k - \check{x}_k|$$

$$\leq \sum_k |B_{ik}|\operatorname{rad}(x_k) = (|B|\operatorname{rad}(x))_i,$$

hence $\langle B\rangle|z - \check{x}| \leq |B|\operatorname{rad}(x)$. Now multiplication by $\langle B\rangle^{-1} \geq 0$ gives (21). To prove (22) we put $\alpha_0 = \|\operatorname{rad}(x)\|_\infty$ and $\alpha = \|z - \check{x}\|_\infty$. Then, for some index i,

$$\alpha = |z_i - \check{x}_i| \leq \alpha_0 + \langle B_{ii}\rangle^{-1}\sum_{k\neq i} |B_{ik}|(\alpha + \alpha_0)$$

by (23). Hence $\alpha \leq \alpha_o + \beta(\alpha + \alpha_0)$. This implies

$$\alpha \leq \frac{1+\beta}{1-\beta}\alpha_0$$

and thus leads to (22). □

After preconditioning a strongly regular matrix with the midpoint inverse we always obtain an H-matrix whose midpoint is the identity matrix. Our next result shows that in this case the fixed point inverse can be obtained without iteration.

4.4.10 Theorem Let $A \in \mathbb{IR}^{n\times n}$ be an H-matrix and suppose that $\check{A}$ is diagonal. Then

$$|A^{\mathrm{F}}b| = \langle A\rangle^{-1}|b|, \tag{24}$$

and $z := A^{\mathrm{F}}b$ can be expressed in terms of $u := \langle A\rangle^{-1}|b|$ as

$$z_i = \frac{b_i + (\langle A_{ii}\rangle u_i - |b_i|)[-1, 1]}{A_{ii}} \quad (i = 1, \ldots, n). \tag{25}$$

Proof We write the fixed point equations (12) as

$$z_i = y_i/A_{ii}, \quad \text{where } y_i = b_i - \sum_{k\neq i} A_{ik}z_k.$$

Since $\check{A}_{ik} = 0$ for $i \neq k$ we have $\mathrm{rad}(y_i) = \mathrm{rad}(b_i) + \Sigma_{k\neq i}|A_{ik}|\,|z_k|$ and $\check{y}_i = \check{b}_i$. Since $|z_i| = |y_i|/\langle A_{ii}\rangle$ this implies

$$\langle A_{ii}\rangle|z_i| = |y_i| = |b_i| + \sum_{k\neq i}|A_{ik}|\,|z_k|. \tag{26}$$

In particular, we have $(\langle A\rangle|z|)_i = |b_i|$, hence $\langle A\rangle|z| = |b|$, so that $|z| = \langle A\rangle^{-1}|b| = u$, and (24) holds. Since $A_{ik}z_k = |A_{ik}|\,|z_k|[-1, 1]$ for $i \neq k$ we get from (26) the further relation $y_i = b_i + (\langle A_{ii}\rangle u_i - |b_i|)[-1, 1]$. This implies (25). □

Note that the assertion of the theorem is equivalent to the statement that, under the given hypothesis, Gauss–Seidel iteration converges in one step when started with $z^0 = [-u, u]$. In practice, rounding errors will lead to a preconditioned matrix whose midpoint is only approximately diagonal. We suggest the following finite algorithm for computing a good approximation to $(CA)^{\mathrm{F}}(Cb)$.

Algorithm (for the enclosure of $A^{\mathrm{H}}b$)

Step 1 Find an approximate inverse $C \approx \check{A}^{-1}$ and compute $A' = CA$, $b' = Cb$.

Step 2 Find an approximate solution $\tilde{u} > 0$ of $\langle A'\rangle u = |b'| + (\varepsilon, \ldots, \varepsilon)^{\mathrm{T}}$ for some small $\varepsilon > 0$ and a number $\alpha > 0$ such that $\langle A'\rangle\tilde{u} \geq \alpha|b'|$. (If this is not possible we conclude that either A was not strongly regular or the precision of the calculation was not high enough.)

Step 3 Perform a few (one or two) steps of preconditioned Gauss–Seidel iteration, starting with $z^0 = \alpha^{-1}\tilde{u}[-1, 1]$.

Each iterate in Step 3 is an enclosure of $A^{\mathrm{H}}b$.

Note that $|(CA)^{\mathrm{F}}(Cb)| \leq \langle A' \rangle^{-1}|b'| \leq \alpha^{-1}\tilde{u}$, so that z^0 is an enclosure for $(CA)^{\mathrm{F}}(Cb)$. Hence the same holds for all iterates z^l. For the choice $C = \check{A}^{-1}$, $\tilde{u} = u$, $\alpha = 1$ and exact arithmetic, the theorem implies that $z^1 = (CA)^{\mathrm{F}}(Cb)$. Hence in general we may expect that z^1 is already a good approximation to $(CA)^{\mathrm{F}}(Cb)$.

We end this section with some regularity results for the fixed point inverse which have important applications to nonlinear systems (cf. Section 5.2).

4.4.11 Theorem Let $A \in \mathbb{IR}^{n\times n}$ be strongly regular and let $C \in \mathbb{R}^{n\times n}$ be such that $\|CA - I\| \leq \beta < \frac{1}{2}$ in some scaled maximum norm. Then the fixed point inverse $(CA)^{\mathrm{F}}$ is regular, and

$$0 \in (CA)^{\mathrm{F}}(Cb) \Rightarrow \|\mathrm{rad}(CA)^{\mathrm{F}}(Cb)\| \leq \frac{\|\mathrm{rad}(Cb)\|}{1 - 2\beta}. \tag{27}$$

Proof We treat the case $C = I$ first. Suppose $x := A^{\mathrm{F}}b \ni 0$. Then $0 \in x = \Gamma(A, b, x)$ and $\|x\| \leq \|\check{x}\| + \|\mathrm{rad}(x)\| \leq 2\|\mathrm{rad}(x)\|$. Hence Corollary 4.3.15 yields $(1 - 2\beta)\|\mathrm{rad}(x)\| \leq \|\mathrm{rad}(b)\|$ so that (27) holds. In particular, if $b \in \mathbb{R}^n$ and $0 \in A^{\mathrm{F}}b$ then $\mathrm{rad}(b) = 0$ so that $\mathrm{rad}(A^{\mathrm{F}}b) = 0$ and hence $A^{\mathrm{F}}b = 0$. Hence A^{F} is regular. The general case follows by substituting CA for A and Cb for b. □

4.4.12 Theorem (Thiel) Let $A \in \mathbb{IR}^{n\times n}$ and suppose that the matrix $\langle A \rangle^*$ defined by

$$\langle A \rangle^*_{ik} := \begin{cases} \langle A_{ii} \rangle & \text{for } i = k, \\ -\max(|A_{ik}|, 2 \cdot \mathrm{rad}(A_{ik})) & \text{otherwise,} \end{cases}$$

is an M-matrix. Then the fixed point inverse A^{F} is regular, and

$$0 \in A^{\mathrm{F}}b \Rightarrow \mathrm{rad}(A^{\mathrm{F}}b) \leq (\langle A \rangle^*)^{-1}\,\mathrm{rad}(b). \tag{28}$$

Proof Define $r^*(A)$ as in equation (4.3.34). Since $|A_{ik}| = |\check{A}_{ik}| + \mathrm{rad}(A_{ik})$ we have $\langle A \rangle^* = \langle A \rangle - r^*(A)$. Now suppose that $x := A^{\mathrm{F}}b \ni 0$. Then $0 \in x = \Gamma(A, b, x)$ and $|\check{x}| \leq \mathrm{rad}(x)$. Hence Proposition 4.3.13 implies $\langle A \rangle^*\mathrm{rad}(x) \subseteq \mathrm{rad}(b)$. Since $(\langle A \rangle^*)^{-1} \geq 0$, this implies (28). In particular, if $b \in \mathbb{R}^n$ and $0 \in A^{\mathrm{F}}b$ then $\mathrm{rad}(b) = 0$ so that $\mathrm{rad}(A^{\mathrm{F}}b) = 0$ and hence $A^{\mathrm{F}}b = 0$. Therefore A^{F} is regular. □

4.4.13 Corollary The hull inverse of an M-matrix A is regular, and

$$0 \in A^{\mathrm{H}}b \Rightarrow \mathrm{rad}(A^{\mathrm{H}}b) \leq \underline{A}^{-1}\,\mathrm{rad}(b). \tag{29}$$

Proof Apply the previous theorem, noting that, for M-matrices, $A^{\mathrm{H}} = A^{\mathrm{F}}$ and $\langle A\rangle^* = \underline{A}$. □

4.5 Interval Gauss elimination

In this section we discuss the interval version of Gauss elimination. We show that Gauss elimination can be performed on H-matrices (although considerable overestimation is possible), and that it yields the hull when applied to an M-matrix and a nonnegative right-hand side. Preconditioned Gauss elimination turns out to be even more effective than Gauss–Seidel iteration; in particular, it has the quadratic approximation property.

It is assumed that the reader is familiar with standard Gauss elimination; however, we shall begin with a quick review to motivate the interval approach.

Let $A \in \mathbb{R}^{n\times n}$ and $b \in \mathbb{R}^n$. To eliminate the first variable in the system

$$\begin{bmatrix} A_{11} & A_{12} & \cdots & A_{1n} \\ A_{21} & A_{22} & \cdots & A_{2n} \\ \vdots & \vdots & & \vdots \\ A_{n1} & A_{n2} & \cdots & A_{nn} \end{bmatrix} \begin{bmatrix} x_1 \\ x_2 \\ \vdots \\ x_n \end{bmatrix} = \begin{bmatrix} b_1 \\ b_2 \\ \vdots \\ b_n \end{bmatrix}, \tag{1}$$

we subtract a suitable multiple of the first row from the other rows such that the subdiagonal entries of the first column become zero. This requires that $A_{11} \neq 0$; then the multiplication factor for the ith row is $L_{i1} = A_{i1}/A_{11}$ $(i > 1)$. After subtraction we have the reduced system

$$\begin{bmatrix} A_{11} & A_{12} & \cdots & A_{1n} \\ 0 & A_{22}^{(1)} & \cdots & A_{2n}^{(1)} \\ \vdots & \vdots & & \vdots \\ 0 & A_{n2}^{(1)} & \cdots & A_{nn}^{(1)} \end{bmatrix} \begin{bmatrix} x_1 \\ x_2 \\ \vdots \\ x_n \end{bmatrix} = \begin{bmatrix} b_1 \\ b_2^{(1)} \\ \vdots \\ b_n^{(1)} \end{bmatrix},$$

where

$$A_{ik}^{(1)} = A_{ik} - L_{i1}A_{1k} = A_{ik} - A_{i1}A_{11}^{-1}A_{1k}$$
$$b_i^{(1)} = b_i - L_{i1}b_1 = b_i - A_{i1}A_{11}^{-1}b_1$$

for $i, k > 1$. Therefore the last $n-1$ components of x, i.e. the vector $x^{(1)} := (x_2, \ldots, x_n)^{\mathrm{T}}$ must be determined from the smaller system $A^{(1)}x^{(1)} = b^{(1)}$, and the first variable is obtained from $x_2, \ldots, x_n$ as

$$x_1 = \left(b_1 - \sum_{k>1} A_{1k}x_k\right)\Big/A_{11}.$$

As long as the corresponding diagonal elements $A_{jj}^{(j-1)}$ remain nonzero we may eliminate further variables in the same way by

$$\left.\begin{aligned} L_{ij} &= A_{ij}^{(j-1)}/A_{jj}^{(j-1)}, \\ A_{ik}^{(j)} &= A_{ik}^{(j-1)} - L_{ij}A_{jk}^{(j-1)}, \quad (i, k > j), \\ b_i^{(j)} &= b_i^{(j-1)} - L_{ij}b_j^{(j-1)}. \end{aligned}\right\} \tag{2}$$

The variable x_j can then be obtained from $x_{j+1}, \ldots, x_n$ by

$$x_j = \left(b_j^{(j-1)} - \sum_{k>j} A_{jk}^{(j-1)} x_k\right) \Big/ A_{jj}^{(j-1)}, \tag{3}$$

and the first step is subsumed under (2) and (3) by putting $A^{(0)} = A$ and $b^{(0)} = b$.

As usual, the multiplication factors L_{ij} $(i > j)$ and the coefficients $R_{jk} := A_{jk}^{(j-1)}$ $(j \le k)$ required for the back substitution (3) are combined to triangular matrices

$$L = \begin{bmatrix} 1 & 0 & \cdots & & 0 \\ L_{21} & 1 & \cdots & & 0 \\ \cdot & L_{32} & \cdot & & \cdot \\ \cdot & \cdot & \cdot & \cdot & \cdot \\ \cdot & \cdot & & \cdot & \cdot \\ L_{n1} & L_{n2} & \cdots & L_{nn-1} & 1 \end{bmatrix}, \quad R = \begin{bmatrix} R_{11} & R_{12} & \cdots & & R_{1n} \\ 0 & R_{21} & \cdots & & R_{2n} \\ \cdot & 0 & \cdot & & \cdot \\ \cdot & \cdot & \cdot & \cdot & \cdot \\ \cdot & \cdot & \cdot & \cdot & \cdot \\ 0 & 0 & \cdots & 0 & R_{nn} \end{bmatrix}$$

From (2) one easily obtains the relations

$$\begin{aligned} A_{ik}^{(l)} &= A_{ik} - \sum_{j \le l} L_{ij}R_{jk}, \\ & \qquad\qquad\qquad\qquad (i, k > j), \\ b_i^{(l)} &= b_i - \sum_{j \le l} L_{ij}y_i, \end{aligned}$$

where $y_j = b_j^{(j-1)}$; in particular, we have

$$L_{ik} = \left(A_{ik} - \sum_{j>k} L_{ij}R_{jk}\right) \Big/ R_{kk} \quad \text{for } i > k, \tag{4}$$

$$R_{ik} = A_{ik} - \sum_{j<i} L_{ij}R_{jk} \quad \text{for } i \le k, \tag{5}$$

$$y_i = b_i - \sum_{j<i} L_{ij}y_j \quad (i = 1, \ldots, n), \tag{6}$$

$$x_i = \left(y_i - \sum_{k>i} R_{ik}x_k\right) \Big/ R_{ii} \quad (i = n, n-1, \ldots, 1). \tag{7}$$

In compact form, this can be written as

$$A = LR, \quad Ly = b, \quad Rx = y; \tag{8}$$

i.e. Gauss elimination effectively consists in the construction of a factorization of A into the product of two triangular matrices, thereby reducing the solution of $Ax = b$ to the two triangular systems $Ly = b$ and $Rx = y$. The latter can easily be solved by forward substitution (6) and backward substitution (7).

Now we consider the interval case. Suppose that $A \in \mathbb{IR}^{n \times n}$ and $b \in \mathbb{IR}^n$. Then (2) and (3) can still be computed as long as the intervals $A_{jj}^{(j-1)} = R_{jj}$ do not contain zero. If this holds, the relations (4), (5), (6) and (7) remain valid, and we say that A has the *triangular decomposition* (L, R). Note, however, that, in general, (8) is not valid; indeed, we only have

$$A \subseteq LR, \quad Ly \supseteq b, \quad Rx \supseteq y. \tag{9}$$

The vector x defined by the *forward substitution* (6) followed by the *backward substitution* (7) is denoted by $A^{\mathrm{G}}b$, and the mapping $A^{\mathrm{G}}: \mathbb{IR}^n \to \mathbb{IR}^n$ defined in this way is called the *Gauss inverse* of A. The name Gauss inverse is justified since $A^{\mathrm{G}}b$ always encloses the solution set $\Sigma(A, b)$:

4.5.1 Theorem Let $A \in \mathbb{IR}^{n \times n}$ have the triangular decomposition (L, R). Then:

(i) If $A' \subseteq A$ then A' has a triangular decomposition (L', R') with $L' \subseteq L$, $R' \subseteq R$, and

$$A' \subseteq A, b' \subseteq b \Rightarrow (A')^{\mathrm{G}}b' \subseteq A^{\mathrm{G}}b.$$

(ii) A is regular, and $A^{\mathrm{H}}b \subseteq A^{\mathrm{G}}b$.

(iii) The Gauss inverse A^{G} is a sublinear and normal mapping, and

$$A^{\mathrm{G}}b = R^{\mathrm{F}}(L^{\mathrm{F}}b) \qquad \text{for all } b \in \mathbb{IR}^n,$$
$$|A^{\mathrm{G}}| = \langle R\rangle^{-1}\langle L\rangle^{-1}.$$

(iv) If A is thin then A^{G} is linear, and

$$\operatorname{cor}(A^{\mathrm{G}}) = A^{-1}.$$

Proof (i) follows since the defining equations (2)–(7) are inclusion isotone. In the special case where $A' = \tilde{A}$ and $b' = \tilde{b}$ are thin we get $\tilde{A}^{-1}\tilde{b} = \tilde{A}^{\mathrm{G}}\tilde{b} \in A^{\mathrm{G}}b$ for all $\tilde{A} \in A$, $\tilde{b} \in b$. Therefore (ii) holds.

(iii) Equation (7) is the fixed point equation for $y = L^{\mathrm{F}}b$, and (8) is the fixed point equation for $x = R^{\mathrm{F}}y$. Hence $A^{\mathrm{G}}b := x = R^{\mathrm{F}}y = R^{\mathrm{F}}(L^{\mathrm{F}}b)$ for all $b \in \mathbb{IR}^n$, i.e. $A^{\mathrm{G}} = R^{\mathrm{F}}L^{\mathrm{F}}$. Hence, by Proposition 3.5.7, A^{G} is sublinear and normal, and $|A^{\mathrm{G}}| = |R^{\mathrm{F}}|\,|L^{\mathrm{F}}| = \langle R\rangle^{-1}\langle L\rangle^{-1}$.

(iv) If A is thin then L and R are thin; therefore R^{F}, L^{F} and $A^{\mathrm{G}} = R^{\mathrm{F}}L^{\mathrm{F}}$ are linear. Moreover, $\operatorname{cor}(A^{\mathrm{G}}) = A^{\mathrm{G}}I = A^{-1}$ since, for thin A and b, we just have standard Gauss elimination. □

For the analysis of Gauss elimination it is useful to have a compact form of the first elimination step. We write $\Sigma(A)$ for the matrix $A^{(1)}$ obtained after the first elimination step, and call it the *Schur complement* (of the (1, 1)-entry) of $A \in \mathbb{IR}^{n\times n}$. If we partition A as

$$A = \begin{pmatrix} \alpha & a^{\mathrm{T}} \\ a' & A' \end{pmatrix} \quad (\alpha \in \mathbb{IR}, a, a' \in \mathbb{IR}^{n-1}, A' \in \mathbb{IR}^{(n-1)\times(n-1)}) \tag{10}$$

then (2) requires $0 \notin \alpha$, and implies that the Schur complement can be written as

$$\Sigma(A) = A' - a'\alpha^{-1}a^{\mathrm{T}}. \tag{11}$$

Note that the matrix $a'\alpha^{-1}a^{\mathrm{T}}$ has the (i, k)-entry $(a_i'\alpha^{-1})a_k = a_i'(\alpha^{-1}a_k)$ so that no brackets are necessary in (11). We now have

4.5.2 Proposition Let $A \in \mathbb{IR}^{n\times n}$ $(n > 1)$ be partitioned according to (10). Then A has a triangular decomposition (L, R) iff $0 \notin A_{11} = \alpha$ and $\Sigma(A)$ has a triangular decomposition (L', R'). In this case the triangular decompositions are related by

$$L = \begin{pmatrix} 1 & 0 \\ a'\alpha^{-1} & L' \end{pmatrix}, \quad R = \begin{pmatrix} \alpha & a^{\mathrm{T}} \\ 0 & R' \end{pmatrix}. \tag{12}$$

Moreover, for all $\beta \in \mathbb{IR}$ and $b' \in \mathbb{IR}^{n-1}$, we have

$$A^{\mathrm{G}}\begin{pmatrix} \beta \\ b' \end{pmatrix} = \begin{pmatrix} \xi \\ x' \end{pmatrix}, \tag{13}$$

where

$$x' = \Sigma(A)^{\mathrm{G}}(b - a'\alpha^{-1}\beta), \quad \xi = (\beta - a^{\mathrm{T}}x')/\alpha. \tag{14}$$

Proof Immediate from the above exposition. □

4.5.3 Corollary If $A \in \mathbb{IR}^{n\times n}$ is a regular lower or upper triangular matrix then $A^{\mathrm{F}} = A^{\mathrm{G}}$, and $x = A^{\mathrm{F}}b$ can be computed in finitely many steps by

$$x_i = \left(b_i - \sum_{k<i} A_{ik}x_k\right)\bigg/ A_{ii} \quad (i = 1, \ldots, n) \tag{15a}$$

if A is lower triangular, and by

$$x_i = \left(b_i - \sum_{k>i} A_{ik}x_k\right)\bigg/ A_{ii} \quad (i = n, n-1, \ldots, 1) \tag{15b}$$

if A is upper triangular.

Proof For triangular A, (15a,b) are the fixed point equations for $x = A^{\mathrm{F}}b$.

Now if A is lower triangular then (10) holds with $a = 0$, hence $\Sigma(A) = A'$, and, by the proposition and induction on n, it is easy to establish (15a) for $x = A^{G}b$. And if A is upper triangular then $(L, R) = (I, A)$ is its triangular decomposition; hence $A^{G}b = R^{F}(L^{F}b) = A^{F}(I^{F}b) = A^{F}b$. Hence both cases give $A^{F} = A^{G}$. □

4.5.4 Proposition (Alefeld, Reichmann) Let $A \in \mathbb{IR}^{2\times 2}$. Then A has a triangular decomposition iff A is regular and $0 \notin A_{11}$.

Proof

$$A = \begin{pmatrix} A_{11} & A_{12} \\ A_{21} & A_{22} \end{pmatrix}$$

has a triangular decomposition iff $0 \notin A_{11}$ and $0 \notin \Sigma(A) = A_{22} - A_{21}A_{11}^{-1}A_{12}$. In the equation for $\Sigma(A)$, every term occurs only once; hence $0 \notin \Sigma(A)$ iff $\tilde{A}_{22} - \tilde{A}_{21}\tilde{A}_{11}^{-1}\tilde{A}_{12} \neq 0$ for all $\tilde{A}_{ik} \in A_{ik}$ $(i, k = 1, 2)$, i.e. iff $\det(\tilde{A}) = \tilde{A}_{11}\tilde{A}_{22} - \tilde{A}_{21}\tilde{A}_{12} \neq 0$ for all $\tilde{A} \in A$, i.e. iff A is regular. □

We note that for regular matrices $A \in \mathbb{IR}^{2\times 2}$, either $0 \notin A_{11}$ or $0 \notin A_{21}$; hence the existence of a triangular decomposition can be enforced here by a suitable permutation of the rows. (However, dependence often produces some overestimation in $A^{G}b$, while the method of Example 3.4.7 gives the hull.) For $n > 2$, the effect of row (or column) permutations is highly matrix dependent.

4.5.5 Examples

(i) The matrix

$$A := \begin{bmatrix} [0,4] & -3 & -3 \\ -3 & [0,4] & -3 \\ -3 & -3 & [0,4] \end{bmatrix} \text{ has } \check{A}^{-1} = \frac{1}{20}\begin{bmatrix} 1 & -3 & -3 \\ -3 & 1 & -3 \\ -3 & -3 & 1 \end{bmatrix},$$

and since $|\check{A}^{-1}|\operatorname{rad}(A)e = 2|\check{A}^{-1}|e = 0.7e$, A is strongly regular. Interchanging the first two rows gives a matrix PA with

$$\Sigma(PA) = \begin{pmatrix} [-3, \frac{7}{3}] & [-7, 3] \\ [-7, -3] & [3, 7] \end{pmatrix} \ni \begin{pmatrix} 2 & -3 \\ -3 & 4.5 \end{pmatrix}.$$

Thus $\Sigma(PA)$ is singular and $(PA)^{G}$ does not exist. By symmetry, row and column interchanges cannot produce a matrix for which the triangular decomposition exists.

(ii) The M-matrix

$$A := \begin{bmatrix} [0.1, 0.9] & 0 & 0 \\ -1 & 3 & -1 \\ 0 & -1 & 2 \end{bmatrix}$$

has a triangular decomposition. However, standard column pivoting would lead to an interchange of the first two rows, giving a matrix PA with

$$\Sigma(PA) = \begin{pmatrix} [0.3, 2.7] & [-0.9, -0.1] \\ -1 & 2 \end{pmatrix} \ni \begin{pmatrix} 0.4 & -0.8 \\ -1 & 2 \end{pmatrix}.$$

Thus $\Sigma(PA)$ is singular and $(PA)^{\mathrm{G}}$ does not exist.

In view of the fact that Gauss elimination without pivoting works very well for M-matrices (Theorem 4.5.8 below), this example shows that pivoting cannot be recommended as a procedure for stabilizing interval Gauss elimination. Instead, we recommend elimination with *diagonal* pivoting (for the largest $\langle A_{ii}\rangle$), resorting to preconditioning (cf. Theorem 4.5.11) when this fails or gives unacceptably wide results.

We now show that every H-matrix has a triangular decomposition. We need the following auxiliary result.

4.5.6 Lemma Let $A \in \mathbb{IR}^{n\times n}$ $(n > 1)$. Then

(i) If $0 \notin A_{11}$ then $\langle\Sigma(A)\rangle \geq \Sigma(\langle A\rangle)$.
(ii) If $\langle A\rangle = \underline{A}$ then $\langle\Sigma(A)\rangle = \Sigma(\underline{A})$.
(iii) If A is an M-matrix then $\Sigma(A)$ is an M-matrix, and

$$\Sigma(A) = [\Sigma(\underline{A}), \Sigma(\bar{A})].$$

(iv) If A is an H-matrix then $\Sigma(A)$ is an H-matrix.

Proof (i) Using the partition (10) we have

$$\begin{aligned}\langle\Sigma(A)\rangle &= \langle A' - a'\alpha^{-1}a^{\mathrm{T}}\rangle \geq \langle A'\rangle - |a'\alpha^{-1}a^{\mathrm{T}}| \\ &= \langle A'\rangle - |a'|\langle\alpha\rangle^{-1}|a|^{\mathrm{T}} = \Sigma(\langle A\rangle).\end{aligned}$$

(ii) If $\langle A\rangle = \underline{A}$ then the preceding argument holds with equality.

(iii) If A is an M-matrix then $\alpha > 0$ and $a, a' \leq 0$ in (10) so that for some $\omega \in \mathbb{R}$, $u' \in \mathbb{R}^{n-1}$,

$$\begin{aligned}\Sigma(A) = A' - a'\alpha^{-1}a^{\mathrm{T}} &= [\underline{A}', \bar{A}'] - [\underline{a}', \bar{a}'][\underline{\alpha}, \bar{\alpha}]^{-1}[\underline{a}^{\mathrm{T}}, \bar{a}^{\mathrm{T}}] \\ &= [\underline{A}', \bar{A}'] - [\underline{a}', \bar{a}'][\underline{\alpha}^{-1}\underline{a}^{\mathrm{T}}, \bar{\alpha}^{-1}\bar{a}^{\mathrm{T}}] \\ &= [\underline{A}', \bar{A}'] - [\underline{a}'\underline{\alpha}^{-1}\underline{a}^{\mathrm{T}}, \bar{a}'\bar{\alpha}^{-1}\bar{a}^{\mathrm{T}}] \\ &= [\underline{A}' - \underline{a}'\underline{\alpha}^{-1}\underline{a}^{\mathrm{T}}, \bar{A}' - \bar{a}'\bar{\alpha}^{-1}\bar{a}^{\mathrm{T}}] = [\Sigma(\underline{A}), \Sigma(\bar{A})].\end{aligned}$$

Since A is an M-matrix, there is a vector $u \in \mathbb{R}^n$ such that for some $\omega \in \mathbb{R}$, $u' \in \mathbb{R}^{n-1}$,

$$0 < u = \begin{pmatrix} \omega \\ u' \end{pmatrix}, \quad 0 < \underline{A}u = \begin{pmatrix} \underline{\alpha}\omega + \underline{a}^{\mathrm{T}}u' \\ \underline{a}'\omega + \underline{A}'u' \end{pmatrix}.$$

Therefore

$$\Sigma(A)u' \geq \Sigma(\underline{A})u' = (\underline{A}' - \underline{a}'\underline{\alpha}^{-1}\underline{a}^{\mathrm{T}})u' = \underline{A}'u' - \underline{a}'\underline{\alpha}^{-1}\underline{a}^{\mathrm{T}}u'$$
$$= (\underline{a}'\omega + \underline{A}'u') - \underline{a}'\underline{\alpha}^{-1}(\underline{\alpha}\omega + \underline{a}^{\mathrm{T}}u') \geq 0,$$

and since $\Sigma(A)_{ik} \leq A'_{ik} \leq 0$ for $i \neq k$, we see that $\Sigma(A)$ is an M-matrix.

(iv) If A is an H-matrix then $\langle A\rangle$ is a thin M-matrix, and, by (ii), $\Sigma(\langle A\rangle)$ is an M-matrix. Hence there is a vector $u' > 0$ such that $\Sigma(\langle A\rangle)u' > 0$, and we get $\langle\Sigma(A)\rangle u' > 0$ by (i). Therefore $\Sigma(A)$ is an H-matrix. □

4.5.7 Theorem (Alefeld) Let $A \in \mathbb{IR}^{n\times n}$ be an H-matrix. Then A has a triangular decomposition (L, R), and we have

$$\langle A\rangle \leq \langle L\rangle\langle R\rangle, \tag{16}$$
$$|A^{\mathrm{G}}| \leq \langle A\rangle^{-1}. \tag{17}$$

If $\langle A\rangle = \underline{A}$ (and in particular if $\check{A}$ is diagonal) then (16) and (17) hold with equality.

Proof Since the case $n = 1$ is trivial, we may assume that the statement holds with $n-1$ in place of n. If $A \in \mathbb{IR}^{n\times n}$ is an H-matrix then $\Sigma(A)$ is an H-matrix, and the induction hypothesis implies that $\Sigma(A)$ has a triangular decomposition (L', R'). Since $0 \notin A_{11}$, Proposition 4.5.2 shows that A has a triangular decomposition (L, R). Moreover, using (12), we see that

$$\langle L\rangle\langle R\rangle = \begin{bmatrix} 1 & 0 \\ -|a'|\langle\alpha\rangle^{-1} & \langle L'\rangle \end{bmatrix} \begin{bmatrix} \langle\alpha\rangle & -|a|^{\mathrm{T}} \\ 0 & \langle R'\rangle \end{bmatrix}$$
$$= \begin{bmatrix} \langle\alpha\rangle & -|a|^{\mathrm{T}} \\ -|a'| & |a'|\langle\alpha\rangle^{-1}|a|^{\mathrm{T}} + \langle L'\rangle\langle R'\rangle \end{bmatrix}$$

By the induction hypothesis and Lemma 4.5.6,

$$\langle L'\rangle\langle R'\rangle \geq \langle\Sigma(A)\rangle \geq \Sigma(\langle A\rangle) = \langle A'\rangle - |a'|\langle\alpha\rangle^{-1}|a|^{\mathrm{T}}.$$

Since

$$\langle A\rangle = \begin{bmatrix} \langle\alpha\rangle & -|a|^{\mathrm{T}} \\ -|a'| & \langle A'\rangle \end{bmatrix}, \tag{18}$$

this implies that (16) holds. Multiplication with $\langle A\rangle^{-1}$ on the left and with $\langle R\rangle^{-1}$ and $\langle L\rangle^{-1}$ on the right gives $\langle A\rangle^{-1} \geq \langle R\rangle^{-1}\langle L\rangle^{-1} = |A^{\mathrm{G}}|$, hence (17). Finally, if $\langle A\rangle = \underline{A}$ then all arguments hold with equality. □

Remark Formula (17) implies the relation

$$A^{\mathrm{G}}b \subseteq A^{\mathrm{F}}b \quad \text{when } \check{b} = 0,$$

suggesting that A^{G} might generally be better than A^{F}. This is the case for special H-matrices A (see Theorem 4.5.11 below). But for M-matrices A we even have the reverse inequality $A^{\mathrm{F}}b \subseteq A^{\mathrm{G}}b$ since $A^{\mathrm{F}} = A^{\mathrm{H}}$; and strict inequality is possible (Example 4.5.9(i)). However, for M-matrices, $A^{\mathrm{G}}b$ cannot be too large.

4.5.8 Theorem (Barth & Nuding, Beeck) Let $A \in \mathbb{IR}^{n\times n}$ be an M-matrix. Then

$$A^{\mathrm{H}}b \subseteq A^{\mathrm{G}}b \subseteq A^{-1}b \quad \text{for all } b \in \mathbb{IR}^n, \tag{19}$$

with equality if $0 < b$, $0 \in b$ or $0 > b$.

Proof Since A is an M-matrix we have $\alpha > 0$ and $a, a' \leq 0$ in (10), so that

$$\begin{aligned} \underline{L}\underline{R} &= \begin{pmatrix} 1 & 0 \\ \underline{a}'\underline{\alpha}^{-1} & \underline{L}' \end{pmatrix}\begin{pmatrix} \underline{\alpha} & \underline{a}^{\mathrm{T}} \\ 0 & \underline{R}' \end{pmatrix} \\ &= \begin{pmatrix} \underline{\alpha} & \underline{a}^{\mathrm{T}} \\ \underline{a}' & \underline{a}'\underline{\alpha}^{-1}\underline{a}^{\mathrm{T}} + \underline{L}'\underline{R}' \end{pmatrix} = \underline{A}, \end{aligned}$$

and similarly $\bar{L}\bar{R} = \bar{A}$. Since $\Sigma(A)$ also is an M-matrix, a trivial induction argument shows that $L_{ik} \leq 0$ and $R_{ik} \leq 0$ for $i \neq k$. Since L and R are triangular, it follows that L and R are M-matrices. Hence $L^{-1} \geq 0$ and $R^{-1} \geq 0$, and we find

$$\begin{aligned} A^{-1} &= [\bar{A}^{-1}, \underline{A}^{-1}] = [\bar{R}^{-1}\bar{L}^{-1}, \underline{R}^{-1}\underline{L}^{-1}] \\ &= [\bar{R}^{-1}, \underline{R}^{-1}][\bar{L}^{-1}, \underline{L}^{-1}] = R^{-1}L^{-1} \end{aligned}$$

by formulae (3.6.6) and (3.1.15). Using Theorem 4.4.8 and (3.1.11) this gives

$$A^{\mathrm{G}}b = R^{\mathrm{F}}(L^{\mathrm{F}}b) = R^{\mathrm{H}}(L^{\mathrm{H}}b) \subseteq R^{-1}(L^{-1}b) \subseteq (R^{-1}L^{-1})b = A^{-1}b.$$

Thus (19) holds, and since $A^{-1}b = A^{\mathrm{H}}b$ for $0 < b$, $0 \in b$ or $0 > b$ we must have equality in (19) in these cases. □

4.5.9 Examples

(i) Consider

$$A = \begin{pmatrix} 2 & [-1, 0] \\ [-1, 0] & 2 \end{pmatrix}, \quad b = \begin{pmatrix} 1.2 \\ -1.2 \end{pmatrix}.$$

A is an M-matrix with triangular decomposition (L, R), where

$$L = \begin{pmatrix} 1 & 0 \\ [-0.5, 0] & 1 \end{pmatrix}, \quad R = \begin{pmatrix} 2 & [-1.0, 0] \\ 0 & [1.5, 2] \end{pmatrix}.$$

Straightforward computation gives

$$A^{\mathrm{G}}b = R^{\mathrm{F}}(L^{\mathrm{F}}b) = R^{\mathrm{F}}\begin{pmatrix} 1.2 \\ [-1.2, -0.6] \end{pmatrix} = \begin{pmatrix} [0.2, 0.6] \\ [-0.8, -0.3] \end{pmatrix},$$

and a comparison with Example 3.4.2 shows that

$$A^{\mathrm{H}}b \subsetneqq A^{\mathrm{G}}b \subsetneqq A^{-1}b.$$

(ii) Consider

$$A = \begin{pmatrix} 2 & -1 \\ [-6, -2] & 3.5 \end{pmatrix}, \quad b = \begin{pmatrix} -2 \\ 5 \end{pmatrix}.$$

A is an M-matrix with triangular decomposition (L, R), where

$$L = \begin{pmatrix} 1 & 0 \\ [-3, -1] & 1 \end{pmatrix}, \quad R = \begin{pmatrix} 2 & -1 \\ 0 & [0.5, 2.5] \end{pmatrix}.$$

Hence

$$A^{\mathrm{G}}b = R^{\mathrm{F}}(L^{\mathrm{F}}b) = R^{\mathrm{F}}\begin{pmatrix} -2 \\ [-1, 3] \end{pmatrix} = \begin{pmatrix} [-2, 2] \\ [-2, 6] \end{pmatrix}.$$

We note that $0 \in A^{\mathrm{G}}b$ although $b \neq 0$ is thin. Hence A^{G} need not be regular, and (19) implies that the matrix inverse A^{-1} of an M-matrix need not be regular either. This can also be seen directly from

$$A^{-1} = \begin{pmatrix} [0.7, 3.5] & [0.2, 1.0] \\ [0.4, 6.0] & [0.4, 2.0] \end{pmatrix} \ni \begin{pmatrix} 2 & 1 \\ 2 & 1 \end{pmatrix}.$$

Remarks (i) Theorem 4.5.8 does not generalize to inverse positive matrices. For example, the matrix

$$A = \begin{bmatrix} [4, 5] & [-3, -2] & 1 \\ [-3, -2] & 4 & [-3, -2] \\ 1 & [-3, -2] & [4, 5] \end{bmatrix}$$

is inverse positive but A^{G} does not even exist. (This case could be rescued by permuting A; and indeed it is unknown whether there are inverse positive matrices for which no matrix obtained by row and column permutations has a triangular decomposition.)

(ii) For H-matrices which are not scaled M-matrices, $A^{\mathrm{G}}b$ may be of much poorer quality. For example, if A is the lower triangular matrix with entries $A_{ik} = 1$ if $i \geq k$ and $b = ([-\varepsilon, \varepsilon], 0, 0, \ldots, 0)^{\mathrm{T}}$ then

$$A^{\mathrm{H}}b = A^{-1}b = ([-\varepsilon, \varepsilon], [-\varepsilon, \varepsilon], 0, \ldots, 0)^{\mathrm{T}},$$

whereas

$$A^{\mathrm{G}}b = A^{\mathrm{F}}b = ([-\varepsilon, \varepsilon], [-\varepsilon, \varepsilon], [-2\varepsilon, 2\varepsilon], \ldots, [-2^{n-2}\varepsilon, 2^{n-2}\varepsilon])$$

with exponential overestimation. This is due to the fact that any vector $u > 0$ with $\langle A\rangle u > 0$ must have entries of exponentially differing magnitudes.

When A is diagonally dominant, this does not happen, and indeed Theorem 4.5.11 below (applied with $C = I$) provides an overestimation bound in this case. We conclude:

> *Unless* A *is an* M-*matrix or diagonally dominant, Gauss elimination should be combined with preconditioning to yield useful results.*

After preconditioning a strongly regular matrix with the midpoint inverse we always obtain an H-matrix whose midpoint is the identity matrix. We show that for such a matrix the Gauss inverse gives enclosures for $A^{\mathrm{H}}b$ which are at least as good as the enclosures obtained from the fixed point inverse. The proof involves the following auxiliary result.

4.5.10 Lemma Let $a, a', a'', b, x \in \mathbb{IR}$ and suppose that $0 \notin a, a'$. Then

$$x \subseteq b/a, \quad a = a' + a'' \Rightarrow x \subseteq (b - a''x)/a'.$$

Proof The proof of the implication (4.3.14) applies with trivial changes. □

4.5.11 Theorem Let $A \in \mathbb{IR}^{n\times n}$ be an H-matrix, and suppose that $\check{A}$ is diagonal. Then

$$A^{\mathrm{G}}b \subseteq A^{\mathrm{F}}b, \tag{20}$$

$$|A^{\mathrm{G}}b| = |A^{\mathrm{F}}b| = \langle A\rangle^{-1}|b|. \tag{21}$$

Proof This is trivial for $n = 1$; hence assume that the assertion holds with $n - 1$ in place of n. Put $x := A^{\mathrm{G}}b$ and $z := A^{\mathrm{F}}b$. We use Proposition 4.5.2. If the matrix A' is indexed by $\{2, \ldots, n\}$ the partition (10) holds with

$$\alpha = A_{11}, \quad \beta = b_1, \quad a_i' = A_{i1}, \quad a_k = A_{1k}, \quad A_{ik} = A_{ik}' \quad (i, k > 1);$$

in particular, $\check{a}_i' = \check{a}_k = 0$ for $i, k > 1$. Let

$$A^* = \Sigma(A), \quad b^* = b - a'\alpha^{-1}\beta, \quad z^* := (A^*)^{\mathrm{F}}b^*.$$

Then

$$A_{ik}^* = A_{ik} - A_{i1}A_{11}^{-1}A_{1k},$$

$$b_i^* = b_i - A_{i1}A_{11}^{-1}b_1.$$

Since $\check{A}^*$ is diagonal, the induction hypothesis applies and gives

$$x' = (A^*)^{\mathrm{G}}b^* \subseteq z^*, \quad |x'| = |z^*|.$$

By Theorem 4.4.10 we have

$$z_i = y_i/A_{ii}, \quad \text{where } y_i = b_i + (\langle A_{ii}\rangle|z_i| - |b_i|)[-1, 1],$$

$$z_i^* = y_i^*/A_{ii}^*, \quad \text{where } y_i^* = b_i^* + (\langle A_{ii}^*\rangle|z_i^*| - |b_i^*|)[-1, 1],$$

and

$$|z| = \langle A\rangle^{-1}|b|, \quad |z^*| = \langle A^*\rangle^{-1}|b^*|.$$

Now $\underline{A}$ is an M-matrix since it is an H-matrix and $\underline{A}_{ik} \le \check{A}_{ik} = 0$ for $i \neq k$. Hence $\langle A^*\rangle = \Sigma(\langle A\rangle)$ by Lemma 4.5.6. Since $|b^*| = |b| + |a'|\langle\alpha\rangle^{-1}|\beta|$, Proposition 4.5.2 and (18) imply

$$|z| = \begin{pmatrix} \zeta \\ |z^*| \end{pmatrix}$$

with $|z_1| = \zeta = (|\beta| + |a|^{\mathrm{T}}|z^*|)/\langle\alpha\rangle = (|\beta| + |a|^{\mathrm{T}}|x'|)/\langle\alpha\rangle = |(\beta - a^{\mathrm{T}}x')/\alpha| = |x_1|$ and $|z_i| = |z_i^*| = |x_i'| = |x_i|$ for $i > 1$. Hence $|x| = |z|$, so that (21) holds.

Next, Lemma 4.5.10 applied to $x_i' \subseteq z_i^* = y_i^*/A_{ii}^*$ gives

$$x_i' \subseteq y_i'/A_{ii}, \quad \text{where } y_i' = y_i^* + A_{i1}A_{11}^{-1}A_{1i}x_i'.$$

Now $\check{y}_i' = \check{y}_i^* = \check{b}_i^* = \check{b}_i = \check{y}_i$ and, since $\check{A}_{i1} = 0$ and $|x_i'| = |z_i^*| = |z_i|$, we have

$$\begin{aligned} |y_i'| &= |y_i^*| + |A_{i1}A_{11}^{-1}A_{1i}|\,|x_i'| \\ &= \langle A_{ii}^*\rangle|z_i^*| + |A_{i1}A_{11}^{-1}A_{1i}|\,|z_i^*| \\ &= \langle A_{ii}\rangle|z_i| = |y_i|. \end{aligned}$$

Hence y_i' has the same midpoint and absolute value as y_i. Therefore $y_i' = y_i$, giving $x_i' \subseteq y_i/A_{ii} = z_i$ for $i > 1$. Finally,

$$x_1 = \xi = \Big(b_1 - \sum_{k\neq 1} A_{1k}x_k'\Big)\Big/A_{11} \subseteq \Big(b_1 - \sum_{k\neq 1} A_{1k}z_k\Big)\Big/A_{11} = z_1.$$

Hence $x \subseteq z$, so that (20) holds. By induction, (20) and (21) now hold generally. □

Note that (20) and (21) imply that one endpoint of each component of $A^{\mathrm{G}}b$ agrees with that of $A^{\mathrm{F}}b$, whereas the other endpoint may be sharper. This allows us to deduce the quadratic approximation property of the preconditioned Gauss inverse and its regularity when $\|CA - I\| < \frac{1}{2}$.

4.5.12 Theorem Let $A \in \mathbb{IR}^{n\times n}$ and $C \in \mathbb{R}^{n\times n}$. If CA is strictly diagonally dominant, i.e.

$$\max_i \Big(\sum_{k\neq i} |CA|_{ik}/\langle CA\rangle_{ii}\Big) \le \beta < 1 \tag{22}$$

then

$$\|\mathrm{rad}\, A^{\mathrm{H}}b\|_\infty \le \|\mathrm{rad}(CA)^{\mathrm{G}}(Cb)\|_\infty \le \frac{1+\beta}{1-\beta}\|\mathrm{rad}\, A^{\mathrm{H}}b\|_\infty. \tag{23}$$

In particular, (23) holds if $C = \check{A}^{-1}$ and

$$\| |\check{A}^{-1}|\operatorname{rad}(A)\|_\infty \leq \beta < 1. \tag{24}$$

Proof Define $B \in \mathbb{IR}^{n\times n}$ by $B_{ii} := (CA)_{ii}$ and $B_{ik} := [-1, 1](CA)_{ik}$ for $i \neq k$. Then $\check{B}$ is strictly diagonally dominant and $\langle B\rangle = \langle CA\rangle$; in particular (22) implies that B is an H-matrix. Since $CA \subseteq B$, Theorem 4.5.1(i) and Theorem 4.5.11 imply

$$A^{\mathrm{H}}b \subseteq (CA)^{\mathrm{G}}(Cb) \subseteq B^{\mathrm{G}}(Cb) \subseteq B^{\mathrm{F}}(Cb).$$

Now (23) follows from Theorem 4.4.9. Finally, (24) implies $\Sigma_k |CA - I|_{ik} \leq \beta$ and hence (22) and (23). □

4.5.13 Theorem Let $A \in \mathbb{IR}^{n\times n}$ be strongly regular and let $C \in \mathbb{R}^{n\times n}$ be such that $\|CA - I\| \leq \beta < \frac{1}{2}$ in some scaled maximum norm. Then the preconditioned Gauss inverse $(CA)^{\mathrm{G}}$ is regular, and

$$0 \in (CA)^{\mathrm{G}}(Cb) \Rightarrow \|\operatorname{rad}(CA)^{\mathrm{G}}(Cb)\| \leq \frac{\|\operatorname{rad}(Cb)\|}{1 - 2\beta}.$$

Proof This follows from Theorem 4.4.11, using the argument of the previous proof. □

While preconditioned Gauss elimination gives good results when $\operatorname{rad}(A)$ is small, preconditioning matrices with wide intervals may lead to a singular $\check{A}^{-1}A$ even when A^{G} exists. An example of this situation is the matrix

$$A = \begin{pmatrix} 0.5 & [0, 1] \\ [-1, 0] & 0.5 \end{pmatrix}, \quad \text{where} \quad \check{A}^{-1}A = \begin{pmatrix} [-1, 1] & [-1, 1] \\ [-1, 1] & [\ \ 1, 3] \end{pmatrix} \ni \begin{pmatrix} 1 & 1 \\ 1 & 1 \end{pmatrix}.$$

This leads to the question whether the class of matrices $A \in \mathbb{IR}^{n\times n}$ for which A^{G} exists can be characterized by easily recognizable properties. This seems to be a very difficult problem.

Finally we prove a perturbation theorem for Gauss elimination. By continuity of Gauss elimination as a function of the matrix coefficients it is clear that the existence of A^{G} implies the existence of B^{G} for all $B \in \mathbb{IR}^{n\times n}$ sufficiently close to A. We shall prove that, in this sense, B is sufficiently close to A if $\rho(|A^{\mathrm{G}}|q(A, B)) < 1$. The proof is based on the following result.

4.5.14 Lemma Let (L, R) be a triangular decomposition of $A \in \mathbb{IR}^{n\times n}$, and let $A' \supseteq A$ be such that for suitable $u \geq 0$ and $v > 0$ we have

$$q(A, A')u \leq \langle L\rangle\langle R\rangle u - v.$$

Then A' has a triangular decomposition (L', R'), and

$$\langle L'\rangle\langle R'\rangle u \geq v.$$

Proof This is trivial for $n = 1$; hence we proceed by induction on n and assume that the statement is true for some dimension n. Let $A_0 \in \mathbb{IR}^{(n+1)\times(n+1)}$ have the triangular decomposition (L_0, R_0). Then

$$A_0 = \begin{pmatrix} \alpha & a^{\mathrm{T}} \\ b & B \end{pmatrix}, \quad L_0 = \begin{pmatrix} 1 & 0 \\ b\alpha^{-1} & L \end{pmatrix}, \quad R_0 = \begin{pmatrix} \alpha & a^{\mathrm{T}} \\ 0 & R \end{pmatrix},$$

where (L, R) is the triangular decomposition of the Schur complement

$$A := B - b\alpha^{-1}a^{\mathrm{T}}.$$

For arbitrary matrices and vectors of the form

$$A_0' = \begin{pmatrix} \alpha' & a'^{\mathrm{T}} \\ b' & B' \end{pmatrix}, \quad u_0 = \begin{pmatrix} \mu \\ u \end{pmatrix} \geq 0, \quad v_0 = \begin{pmatrix} \omega \\ v \end{pmatrix} > 0,$$

$$q(A_0, A_0') = \begin{pmatrix} \varepsilon & e^{\mathrm{T}} \\ f & E \end{pmatrix},$$

satisfying $A_0' \supseteq A_0$ and

$$q(A_0, A_0')u_0 \leq \langle L_0\rangle\langle R_0\rangle u_0 - v_0,$$

we have

$$\varepsilon\mu + e^{\mathrm{T}}u \leq \langle\alpha\rangle\mu - |a|^{\mathrm{T}}u - \omega, \tag{25}$$

$$f\mu + Eu \leq \langle L\rangle\langle R\rangle u - |b|\langle\alpha\rangle^{-1}(\langle\alpha\rangle\mu - |a|^{\mathrm{T}}u) - v. \tag{26}$$

In particular, $\mu(\langle\alpha\rangle - \varepsilon) > 0$ and hence

$$\varepsilon < \langle\alpha\rangle.$$

We compute a bound for distance between A and the Schur complement

$$A' := B' - b'\alpha'^{-1}a'^{\mathrm{T}}$$

of A_0'. Since $A_0' \supseteq A_0$, Lemma 1.7.5(i) implies

$$\begin{aligned} q(A, A') &= q(B, B') + q(b\alpha^{-1}a^{\mathrm{T}}, b'\alpha'^{-1}a'^{\mathrm{T}}) \\ &= q(B, B') - |b|\langle\alpha\rangle^{-1}|a|^{\mathrm{T}} + \beta(b\alpha^{-1}a^{\mathrm{T}}, b'\alpha'^{-1}a'^{\mathrm{T}}) \\ &\leq q(B, B') - |b|\langle\alpha\rangle^{-1}|a|^{\mathrm{T}} + \beta(\alpha^{-1}, \alpha'^{-1})\beta(a, a'^{\mathrm{T}}); \end{aligned}$$

here β is understood componentwise (note that each component of $b\alpha^{-1}a^{\mathrm{T}}$ is a product). Hence, by Lemma 1.7.5(ii),

$$q(A, A') \leq E - |b|\langle\alpha\rangle^{-1}|a|^{\mathrm{T}} + (|b| + f)(\langle\alpha\rangle - \varepsilon)^{-1}(|a| + e)^{\mathrm{T}},$$

and (26) and (25) give

$$\begin{aligned} q(A,A')u &\le Eu - |b|\langle\alpha\rangle^{-1}|a|^{\mathrm{T}}u + (|b|+f)(\langle\alpha\rangle-\varepsilon)^{-1}(|a|+e)^{\mathrm{T}}u \\ &\le \langle L\rangle\langle R\rangle u - v - (|b|+f)\mu + (|b|+f)(\langle\alpha\rangle-\varepsilon)^{-1}(|a|^{\mathrm{T}}u + e^{\mathrm{T}}u) \\ &\le \langle L\rangle\langle R\rangle u - v - (|b|+f)(\langle\alpha\rangle-\varepsilon)^{-1}\omega. \end{aligned}$$

By our induction hypothesis, A' has a triangular decomposition (L', R') with

$$\langle L'\rangle\langle R'\rangle u \ge v + (|b|+f)(\langle\alpha\rangle-\varepsilon)^{-1}\omega.$$

Therefore, with

$$L_0' := \begin{pmatrix} 1 & 0 \\ b'\alpha'^{-1} & L \end{pmatrix}, \quad R_0' := \begin{pmatrix} \alpha & a'^{\mathrm{T}} \\ 0 & R' \end{pmatrix},$$

(L_0', R_0') is a triangular decomposition of A_0', and

$$\begin{aligned} \langle L_0'\rangle\langle R_0'\rangle u_0 &\ge \begin{pmatrix} (\langle\alpha\rangle-\varepsilon)\mu - (|a|+e)^{\mathrm{T}}u \\ \langle L'\rangle\langle R'\rangle u - (|b|+f)(\langle\alpha\rangle-\varepsilon)^{-1}\omega \end{pmatrix} \\ &\ge \begin{pmatrix} \omega \\ v \end{pmatrix} = v_0, \end{aligned}$$

which completes the induction. □

4.5.15 Theorem Let $A, B \in \mathbb{IR}^{n\times n}$ and suppose that A^{G} exists. If

$$\rho(|A^{\mathrm{G}}|q(B,A)) < 1 \tag{27}$$

then B^{G} exists and

$$|B^{\mathrm{G}}| \le (I - |A^{\mathrm{G}}|q(B,A))^{-1}|A^{\mathrm{G}}| = (\langle L\rangle\langle R\rangle - q(B,A))^{-1}, \tag{28}$$

where (L, R) is the triangular decomposition of A.

Proof Suppose that (27) holds and put

$$\Delta := q(B,A), \quad A' := A + [-\Delta, \Delta].$$

Then

$$A, B \subseteq A', \quad q(A, A') = \Delta.$$

By (27), $I - |A^{\mathrm{G}}|\Delta$ has a nonnegative inverse, so that for arbitrary $v > 0$ we have

$$\begin{aligned} u &:= (I - |A^{\mathrm{G}}|\Delta)^{-1}|A^{\mathrm{G}}|v \ge 0, \\ |A^{\mathrm{G}}|\Delta u &= u - |A^{\mathrm{G}}|v, \\ \Delta u &= \langle L\rangle\langle R\rangle|A^{\mathrm{G}}|\Delta u = \langle L\rangle\langle R\rangle u - v. \end{aligned}$$

Hence, by Lemma 4.5.14, A' has a triangular decomposition (L', R'), and

$\langle L'\rangle\langle R'\rangle u \geq v$; in particular, A'^{G} exists. But since $B \subseteq A'$, Theorem 4.5.1 implies that B^{G} exists and $|B^{\mathrm{G}}| \leq |A'^{\mathrm{G}}|$. Hence

$$|B^{\mathrm{G}}|v \leq |A'^{\mathrm{G}}|\langle L'\rangle\langle R'\rangle u = u = (I - |A^{\mathrm{G}}|\Delta)^{-1}|A^{\mathrm{G}}|v.$$

Since $v > 0$ was arbitrary, this implies

$$|B^{\mathrm{G}}| \leq (I - |A^{\mathrm{G}}|\Delta)^{-1}|A^{\mathrm{G}}| = (\langle L\rangle\langle R\rangle - \Delta)^{-1}.$$

Now (28) follows, since $\Delta = q(B, A)$. □

Remark Since $q(B, \check{B}) = \mathrm{rad}(B)$, the theorem implies that B^{G} exists whenever $\check{B}$ is nonsingular and $|\check{B}^{\mathrm{G}}|\mathrm{rad}(B)$ has spectral radius less than one. This sufficient criterion should be compared with the weaker criterion $\rho(|\check{B}^{-1}|\mathrm{rad}(B)) < 1$ for strong regularity of $B \in \mathbb{IR}^{n\times n}$. But note that B^{G} may exist even when $\rho(|\check{B}^{\mathrm{G}}|\mathrm{rad}(B)) \geq 1$; an example is provided by

$$B = \begin{pmatrix} [1, 3] & 4 \\ 4 & 4 \end{pmatrix}.$$

Remarks to Chapter 4

To 4.1 Preconditioning by an approximate midpoint inverse was first suggested by Hansen & Smith (1967). Its superiority to other preconditioning matrices (Theorem 4.1.2) was observed empirically by O. Mayer (1970b) and proved in Neumaier (1984a). An earlier, weaker result is in Krawczyk & Selsmark (1980). The equivalence of strong regularity with (iv) of Proposition 4.1.1 is due to Ris (1972). Theorem 4.1.5 is a slight generalization of a result in Neumaier (1987a). Theorem 4.1.12 is essentially due to Krawczyk (1987a). For a discussion of rounding errors in the computation of an approximate midpoint inverse C and the resulting magnitude of $\|I - C\check{A}\|$ see Wilkinson (1965); the observation that preconditioning yields better results when C^{T} is computed by inverting the transposed midpoint matrix is due to Ris (1972).

Apart from Theorem 4.1.11 (which seems new) there are many other methods for the enclosure of the inverse of an interval matrix. See the references in Alefeld (1977c, 1981b) and also Alefeld & Herzberger (1983), Hansen (1965), Herzberger (1985), Rump (1984), Thieler (1975, 1978).

To 4.2 Krawczyk's method is the specialization of an iterative method for the solution of nonlinear systems introduced by Krawczyk (1969a) (cf. Section 5.2). Theorem 4.2.3 and Theorem 4.2.4 are improvements of bounds in Neumaier (1987b), but the quadratic approximation property of the limit and a weaker version of equation (13) were first proved by Gay (1982).

When A and b are thin then the possibility of obtaining optimal enclosures of $A^{-1}b$ by computing the residual to higher precision was first recognized by Ris (1972). Later it has been popularized by Rump (1980, 1981b, 1982a, 1983, 1984) and Rump & Kaucher (1980) (in conjunction with different techniques based on Brouwer's fixed point theorem). The software packages ACRITH (IBM, 1983a,b) and ARITHMOS (Siemens AG, 1987) with FORTRAN routines for accurate interval computations realize this using the optimal scalar product of Bohlender (1977) and Kulisch & Bohlender (1976).

Enclosures with specified (multiprecision) accuracy are obtained for thin A and b by Demmel & Krückeberg (1985).

To 4.3 Interval Gauss–Seidel iteration was introduced by Ris (1972) (under the assumption that $0 \notin A_{ii}$ for all i). He observed empirically that (with preconditioning) it was superior to Krawczyk's method. The observation that the restriction $0 \notin A_{ii}$ is not necessary is due to Hansen & Sengupta (1981). Theorem 4.3.5 is new and improves a weaker result in Neumaier (1984a); it implies in particular that interval Gauss–Seidel iteration cannot be improved by overrelaxation methods (cf. O. Mayer (1971), Cornelius (1981)). Neumaier (1984a) also contains Theorem 4.3.9.

The speed of interval Gauss–Seidel method improves when it is applied in forward and backward sweeps. The resulting *symmetric Gauss–Seidel method* is discussed in Alefeld (1977a) and (for nonlinear systems) in Schwandt (1981) and Shearer & Wolfe (1985). Another extension, which solves with $O(n^2)$ operations each equation for every variable, is in Neumaier (1988b).

To 4.4 The analysis of linear fixed point equations of the form $x = Ax + b$, where $A \in \mathbb{IR}^{n\times n}$, $b \in \mathbb{IR}^n$, goes back to Apostolatos & Kulisch (1968). The generalization of their results to the more general fixed point equation (1) in Theorem 4.4.1 is due to Neumaier (1987a), where also Theorem 4.4.2 can be found.

Theorem 4.4.4 and a version of Theorem 4.4.5 are in Neumaier (1984a). The optimality Theorem 4.4.8 is due to Barth & Nuding (1974). Theorem 4.4.9 is in Neumaier (1987a). Theorem 4.4.10 appears to be new. Theorem 4.4.11 and Corollary 4.4.13 are based on results in Neumaier (1985a), and Theorem 4.4.12 is an extension of a result of Thiel (1986).

Further results on the speed of interval fixed point iteration for linear interval equations are given by G. Mayer (1986a,b, 1987a,b).

To 4.5 In the early times interval Gauss elimination was notorious for delivering bounds with exponential overestimation (see e.g. Ris, 1972; Schätzle, 1984; Wongwises, 1975). Then Hansen & Smith (1967) showed empirically that it behaves well for preconditioned linear systems. Barth & Nuding (1974) and Beeck (1974) showed that for M-matrices it can yield the hull (Theorem 4.5.8). Kopp (1976) showed empirically that the bounds are acceptable for diagonally dominant systems (cf. Theorem 4.5.11), and Alefeld (1977b) showed that the triangular decomposition exists for all H-matrices (Theorem 4.5.7). Alefeld (1979) and Reichmann (1979b) showed existence for $n = 2$ (Proposition 4.5.4), and Reichmann (1979b) for regular tridiagonal matrices. Reichmann (1979a) gives further existence results for Hessenberg matrices with a special sign structure. See also Hahn, Mohr & Schauer (1985).

Proposition 4.5.2 is in Neumaier (1987a). Theorem 4.5.11 is a generalization of Proposition 2.5 of Krawczyk & Neumaier (1987). The overestimation result in Theorem 4.5.12 is new; the resulting quadratic approximation property was proved by Miller (1972, 1973). Theorem 4.5.15 is in Neumaier (1987a).

Interval Gauss elimination can be coupled with iterative methods by splitting $A = B + E$ and considering the iteration

$$x^{l+1} := B^{G}(b - Ex^{l}). \qquad (*)$$

See Alefeld (1979) (for the case $B = \check{A}$), Neumaier (1987a), Schwandt (1987b), and for more general splittings G. Mayer (1985, 1987c, 1988b). Theorems 4.4.1 and 4.4.2 apply for this iteration, and together with Theorem 4.5.15 one finds that (*) can converge for all $b \in \mathbb{IR}^n$ and all starting boxes x^0 only when A has a triangular decomposition.

Interval versions of *cyclic reduction methods*, and corresponding iterations are discussed in Schwandt (1984, 1987b). *Block interval methods* are treated in Garloff (1990).

Further remarks Linear interval equations are also discussed in some detail in books by Deif (1986) and Kuperman (1971). A survey of solution methods for linear interval equations is in Neumaier (1986b); apart from much of the material covered here it also contains references to methods for *sparse matrices*. Efficient handling of sparsity is possible for M-matrices and diagonally dominant matrices (e.g. through sparse Gauss elimination). The apparently more generally applicable methods in the literature, e.g. Cordes & Kaucher (1987), suffer from exponential overestimation for general matrices. A recent promising approach by Alvarado (1990a,b) to reduce this excessive overestimation is related to techniques by Gambill & Skeel (1988) for reducing the so-called wrapping effect (for enclosures of initial value problems).

Methods for linear interval equations with *circulant matrices* and *Toeplitz matrices* were considered by Garloff (1980, 1986a).

Linear equations whose coefficient matrix depends on parameters varying in specified intervals are treated in Kelch (1988), Neumaier (1987d) and Steckelberg (1980).

Beeck (1977) considers the speed-up obtainable by using *parallel processors* for the solution of linear interval equations.

For an error analysis of preconditioning methods in finite precision interval calculations see Spellucci & Krier (1985). See also Stummel (1984).

Homogeneous linear interval equations are treated in Neumaier (1988b) and Šik (1982).

For techniques for enclosing least squares solutions of *overdetermined linear systems* see Deif (1986), Gay (1988), Jahn (1974), Manteuffel (1981), Neumaier (1986b, 1987c), Rump (1983).

Enclosures for the *determinants* of all $\tilde{A} \in A$ for an interval matrix $A \in \mathbb{IR}^{n \times n}$ are described in Smith (1969).

Linear programming with interval data is considered in Beeck (1978, 1979), Gerlach (1981), Heindl & Reinhart (1976a,b, 1978), Jansson (1985, 1988), Krawczyk (1975), Machost (1970) and Stewart (1973). See also Lodwick (1988).

5

Nonlinear systems of equations

5.1 Existence and uniqueness

In this section we consider the problem of deciding whether a continuous function $F: D_0 \subseteq \mathbb{R}^n \rightarrow \mathbb{R}^n$ has a *zero* x^* in a given subset D of D_0, i.e. whether a vector $x^* \in D \subseteq D_0$ exists such that $F(x^*) = 0$. We also investigate when such a zero is unique, and when it depends continuously on parameter perturbations of F. Then we introduce the Newton operator, the Krawczyk operator and the Hansen–Sengupta operator, which provide simple computational tests for (unique) existence and nonexistence of a zero in a given box.

While the existence problem can be treated assuming only that F is continuous (see Section 5.3), we assume in the present section a form of Lipschitz continuity to also obtain uniqueness results. Since differentiability is not assumed, the results also hold for so-called nonsmooth functions.

Let us begin by fixing the notation. $\|\tilde{x}\|_2 := (\tilde{x}^{\mathrm{T}}\tilde{x})^{1/2}$ denotes the Euclidean norm of a vector $\tilde{x} \in \mathbb{R}^n$ and $\|\tilde{A}\|_2 := \sup\{\|\tilde{A}\tilde{x}\|_2/\|\tilde{x}\|_2 \mid \tilde{x} \neq 0\}$ the corresponding matrix norm. $\bar{D}$, ∂D and $\operatorname{int}(D) = \bar{D}\backslash\partial D$ denote the closure, the boundary, and the interior of a set $D \subseteq \mathbb{R}^n$, respectively. A subset $D \subseteq \mathbb{R}^n$ is called *convex* if

$$\tilde{x}, \tilde{y} \in D \Rightarrow t\tilde{x} + (1-t)\tilde{y} \in D \quad \text{for all } t \in [0, 1]. \tag{1}$$

For example, $\mathbb{R}^n$ itself and all interval boxes in $\mathbb{R}^n$ are convex. The *convex hull* of a set $D \subseteq \mathbb{R}^n$ is the intersection of all convex sets $D' \subseteq \mathbb{R}^n$ with $D \subseteq D'$, and its closure $\operatorname{cch}(D)$ is called the *closed convex hull* of D. Clearly, $\operatorname{cch}(D)$ is closed and convex. We shall need the so-called *separation theorem* for convex sets, which asserts that a point outside a closed convex set is separated from D by a hyperplane:

5.1.1 Proposition Let D be a closed and convex set and $\tilde{z} \in \mathbb{R}^n$. Then $\tilde{z} \notin D$ iff there is a nonzero vector $c \in \mathbb{R}^n$ such that

$$c^{\mathrm{T}}(\tilde{x} - \tilde{z}) \geq 1 \quad \text{for all } \tilde{x} \in D. \tag{2}$$

Proof Suppose that $\tilde{z} \notin D$. Since D is closed, there is a point $x^* \in D$ which minimizes the Euclidean distance $\|\tilde{x} - \tilde{z}\|_2$ between $\tilde{z}$ and points $\tilde{x} \in D$. Now $(1-\varepsilon)x^* + \varepsilon\tilde{x} \in D$ for $0 < \varepsilon \leq 1$, and the optimality of x^* implies $\|x^* - \tilde{z}\|_2^2 \leq \|(1-\varepsilon)x^* + \varepsilon\tilde{x} - \tilde{z}\|_2^2 = \|(x^* - \tilde{z}) + \varepsilon(\tilde{x} - x^*)\|_2^2 = \|x^* - \tilde{z}\|_2^2 + 2\varepsilon(x^* - \tilde{z})^{\mathrm{T}}(\tilde{x} - x^*) + \varepsilon^2\|\tilde{x} - x^*\|_2^2$. Hence $0 \leq (x^* - \tilde{z})^{\mathrm{T}}(\tilde{x} - x^*) + (\varepsilon/2)\|\tilde{x} - x^*\|_2^2$. For $\varepsilon \to 0$ we obtain $(x^* - \tilde{z})^{\mathrm{T}}(\tilde{x} - \tilde{z}) \geq \|x^* - \tilde{z}\|_2^2 > 0$, so that (2) holds with $c = \|x^* - \tilde{z}\|_2^{-2}(x^* - \tilde{z})$. Conversely, if (2) holds then putting $\tilde{x} = \tilde{z}$ we find $\tilde{z} \notin D$. □

In generalization of the definition of a regular interval matrix, let us call a set A of matrices $\tilde{A} \in \mathbb{R}^{n\times n}$ *regular* if A is closed, convex and bounded, and all $\tilde{A} \in A$ are nonsingular.

5.1.2 Lemma If $A \subseteq \mathbb{R}^{n\times n}$ is regular then the set of inverses of matrices in A is bounded, and the mapping A^{H}: $\mathbb{IR}^n \to \mathbb{IR}^n$ defined for $b \in \mathbb{IR}^n$ by

$$A^{\mathrm{H}}b := \square\{\tilde{A}^{-1}\tilde{b} \mid \tilde{A} \in A, \tilde{b} \in b\}$$

is sublinear, with absolute value $|A^{\mathrm{H}}| = A^{-1} := \square\{\tilde{A}^{-1} \mid \tilde{A} \in A\}$.

Proof Since A is closed, $\det(\tilde{A})$ assumes its minimum for some matrix $\tilde{A} \in A$, and since all $\tilde{A} \in A$ are nonsingular, there is a number $\delta > 0$ with $|\det(\tilde{A})| \geq \delta$ for all $\tilde{A} \in A$. By Cramer's rule, $\det(\tilde{A}) \cdot \tilde{A}^{-1}$ is a multilinear function of the coefficients of $\tilde{A}$, and since A is bounded there is a number γ such that $\|\det(\tilde{A}) \cdot \tilde{A}^{-1}\|_2 \leq \gamma$ for all $\tilde{A} \in A$. Hence $\|\tilde{A}^{-1}\|_2 \leq \gamma/\delta$ for all $\tilde{A} \in A$, i.e. the set of inverses of matrices in A is bounded. Therefore $A^{\mathrm{H}}b$ and A^{-1} are well defined. The properties of A^{H} follow as in the case when A is a regular interval matrix (Example 3.5.1(iv)). □

The basic result of this section is a semilocal version of the well-known *implicit function theorem*, which gives a condition under which an equation $G(\tilde{x}, t) = 0$ can be solved for all t from a closed and connected set E by a vector $\tilde{x}$ from a closed and bounded set D when it is known that $F(x^0, t^0) = 0$ for some pair $(x^0, t^0) \in D \times E$. In contrast to the usual local form of the implicit function theorem, D and E can be chosen *before* applying the theorem.

5.1.3 Theorem Let G: $B \subseteq \mathbb{R}^n \times \mathbb{R}^p \to \mathbb{R}^n$ be a continuous function and let $D \subseteq \mathbb{R}^n$, $E \subseteq \mathbb{R}^p$ be closed sets with $D \times E \subseteq B$, D bounded and E connected. Suppose that

$$G(x^0, t^0) = 0 \quad \text{for some } x^0 \in D, t^0 \in E, \tag{3}$$

$$G(\tilde{x}, t) \neq 0 \quad \text{for all } \tilde{x} \in \partial D, t \in E. \tag{4}$$

If there is a regular matrix set A such that for every $\tilde{x}, \tilde{y} \in D$ and every $t \in E$, we have

$$G(\tilde{x}, t) - G(\tilde{y}, t) = \tilde{A}(\tilde{x} - \tilde{y}) \quad \text{for some } \tilde{A} \in A, \tag{5}$$

then there is a unique function $H: E \to D$ such that for all $t \in E$, the vector $\tilde{x} = H(t)$ satisfies $G(\tilde{x}, t) = 0$. Moreover, this function H is continuous.

Proof Let $E' = \{t \in E \mid G(\tilde{x}, t) = 0 \text{ for some } \tilde{x} \in D\}$. If $\tilde{x}, \tilde{y} \in D$ satisfy $G(\tilde{x}, t) = G(\tilde{y}, t) = 0$ then (5) implies $\tilde{x} - \tilde{y} = \tilde{A}^{-1}(G(\tilde{x}, t) - G(\tilde{y}, t)) = 0$ so that $\tilde{x} = \tilde{y}$. Hence, for all $t \in E'$, the equation $G(\tilde{x}, t) = 0$ has a unique solution $\tilde{x} \in D$ which we denote by $H(t)$. By (4), $H(t) \notin \partial D$, so that this defines a function $H: E' \to \text{int}(D)$.

To show continuity of H, suppose that $t^l \in E'$ for $l = 0, 1, 2, \ldots$, and $t = \lim t^l$. Then $t \in E$, since E is closed. Since D is bounded, the sequence of vectors $x^l := H(t^l)$ $(l = 0, 1, 2, \ldots)$ has an accumulation point $\tilde{x} \in D$. By continuity, every accumulation point satisfies $G(\tilde{x}, t) = 0$, so that $t \in E'$, and $\tilde{x} = H(t)$ since the zero is unique. In particular, E' is closed and $H(t)$ is the only accumulation point of the x^l. Hence $\lim H(t^l) = \lim x^l = x = H(t)$, which shows that H is continuous.

It remains to show that $E' = E$. Now E is connected and E' is nonempty since $t^0 \in E'$ by (3). Hence it is sufficient to show that E' is open in E.

In order to show this, let $t^* \in E'$. Since A is a regular matrix set, there are numbers $\alpha, \gamma > 0$ such that $\|\tilde{A}\|_2 \leq \alpha$, $\|\tilde{A}^{-1}\|_2 \leq \gamma$ for all $\tilde{A} \in A$. Since $x^* := H(t^*) \in \text{int}(D)$, there is a number $\delta > 0$ such that every $\tilde{x} \in \mathbb{R}^n$ with $\|\tilde{x} - x^*\|_2 \leq \delta$ is contained in $\text{int}(D)$. Since G is continuous and $G(x^*, t^*) = 0$, there is a neighborhood T of t^* (relative to E) such that $2\gamma\|G(x^*, t)\|_2 \leq \delta$ for all $t \in T$.

Now fix $t \in T$. Since D is closed, there is a point $y^* \in D$ for which the minimum of $\|G(\tilde{y}, t)\|_2$ over $\tilde{y} \in D$ is attained; in particular, $\|G(y^*, t)\|_2 \leq \|G(x^*, t)\|_2$. By assumption, there is a matrix $A^* \in A$ such that $G(x^*, t) - G(y^*, t) = A^*(x^* - y^*)$; therefore

$$\begin{aligned}
\|y^* - x^*\|_2 &= \|A^{*-1}(G(x^*, t) - G(y^*, t))\|_2 \\
&\leq \|A^{*-1}\|_2(\|G(x^*, t)\|_2 + \|G(y^*, t)\|_2) \\
&\leq 2\gamma\|G(x^*, t)\|_2 \leq \delta
\end{aligned}$$

and hence $y^* \in \text{int}(D)$.

Suppose that $G(y^*, t) \neq 0$. Then, since A is regular, the set $R = \{\tilde{A}^{\mathrm{T}}G(y^*, t) \mid \tilde{A} \in A\}$ is closed and convex, and $0 \notin R$. Hence Proposition 5.1.1 implies that there is a nonzero vector $c \in \mathbb{R}^n$ such that $c^{\mathrm{T}}\tilde{A}^{\mathrm{T}}G(y^*, t) \geq 1$ for all $\tilde{A} \in A$. Since $y^* \in \text{int}(D)$, there is a positive number $\varepsilon < 2(\alpha\|c\|_2)^{-2}$ such that $\tilde{y} = y^* - \varepsilon c \in D$, and by (5) we have $G(\tilde{y}, t) = G(y^*, t) - \tilde{A}(y^* - \tilde{y}) = G(y^*, t) - \varepsilon\tilde{A}c$ for some $\tilde{A} \in A$. Hence

$\|G(\tilde{y}, t)\|_2^2 = \|G(y^*, t) - \varepsilon\tilde{A}c\|_2^2 = \|G(y^*, t)\|_2^2 - 2\varepsilon c^{\mathrm{T}}\tilde{A}^{\mathrm{T}}G(y^*, t) + \varepsilon^2\|\tilde{A}c\|_2^2 \leq \|G(y^*, t)\|_2^2 - 2\varepsilon + \varepsilon^2(\alpha\|c\|_2)^2 < \|G(y^*, t)\|_2^2$, contradicting the choice of y^*. Hence $G(y^*, t) = 0$ and $t \in E'$. But this implies that E' contains a neighborhood T of t^* relative to E, so that E' is indeed open in E. Hence $E' = E$, which completes the proof. □

We say that $F: D_0 \subseteq \mathbb{R}^n \rightarrow \mathbb{R}^m$ is *continuously differentiable* in $D \subseteq D_0$ if there is a continuous function $F': D \rightarrow \mathbb{R}^{m\times n}$, the *derivative* (or *Jacobian*) of F, such that, for all $\tilde{x} \in D$,

$$\lim_{\substack{\tilde{y}\to\tilde{x}\\ \tilde{y}\in D}} \frac{\|F(\tilde{y}) - F(\tilde{x}) - F'(\tilde{x})(\tilde{y} - \tilde{x})\|_2}{\|\tilde{y} - \tilde{x}\|_2} = 0.$$

It is easily verified that, when D is contained in the closure of int(D), the components $F'(\tilde{x})_{ik}$ of $F'(\tilde{x})$ are just the partial derivatives $\partial F_i(\tilde{x})/\partial x_k$ of the components of F, where one-sided derivatives are used at the boundary of D.

We say that $G: B \subseteq \mathbb{R}^n \times \mathbb{R}^p \rightarrow \mathbb{R}^m$ is in $D \times E \subseteq B$ *continuously differentiable with respect to the first* n *variables* if there is a continuous function $\partial_1 G: D \times E \rightarrow \mathbb{R}^{m\times n}$, the *partial derivative* with respect to the first n variables such that, for all $\tilde{x} \in D$ and $t \in E$,

$$\lim_{\substack{\tilde{y}\to\tilde{x}\\ \tilde{y}\in D}} \frac{\|G(\tilde{y}, t) - G(\tilde{x}, t) - \partial_1 G(\tilde{x}, t)(\tilde{y} - \tilde{x})\|_2}{\|\tilde{y} - \tilde{x}\|_2} = 0.$$

Similarly, G is said to be in $D \times E$ *continuously differentiable with respect to the last* p *variables* if there is a continuous function $\partial_2 G: D \times E \rightarrow \mathbb{R}^{m\times p}$, the *partial derivative* with respect to the last p variables, such that, for all $\tilde{x} \in D$ and $t \in E$,

$$\lim_{\substack{s\to t\\ s\in E}} \frac{\|G(\tilde{x}, s) - G(\tilde{x}, t) - \partial_2 G(\tilde{x}, t)(s - t)\|_2}{\|s - t\|_2} = 0.$$

Again, $(\partial_1 G(\tilde{x}, t))_{ik} = \partial G_i(\tilde{x}, t)/\partial x_k$ and $(\partial_2 G(\tilde{x}, t))_{ik} = \partial G_i(\tilde{x}, t)/\partial t_k$. The knowledge of partial derivatives allows the construction of a matrix set satisfying (5).

5.1.4 Proposition Let $G: B \subseteq \mathbb{R}^n \times \mathbb{R}^p \rightarrow \mathbb{R}^m$ be in $D \times E \subseteq B$ continuously differentiable with respect to the first n variables. If D is convex and $D \times E$ is compact then the set

$$A := \mathrm{cch}\{\partial_1 G(\tilde{x}, t) \mid \tilde{x} \in D, t \in E\}$$

is a closed, convex and bounded set with the property that (5) holds for every $\tilde{x}, \tilde{y} \in D$ and every $t \in E$.

Proof By continuity of $\partial_1 G$, the set A is bounded since $D \times E$ is compact. Clearly A is closed and convex. To show (5), let $\tilde{x}, \tilde{y} \in D$ and $t \in E$. Since D is convex, the line segment joining $\tilde{x}$ and $\tilde{y}$ is in D. Hence $\varphi(s) := G(\tilde{y} + s(\tilde{x} - \tilde{y}), t)$ defines a function φ: $[0, 1] \to \mathbb{R}^m$. It is easily verified that φ is continuously differentiable with $\varphi'(s) = \partial_1 G(\tilde{y} + s(\tilde{x} - \tilde{y}), t)(\tilde{x} - \tilde{y})$. Hence

$$G(\tilde{x}, t) - G(\tilde{y}, t) = \varphi(1) - \varphi(0) = \int_0^1 \varphi'(s)\, ds = \tilde{A}(\tilde{x} - \tilde{y}),$$

where

$$\tilde{A} = \int_0^1 \partial_1 G(\tilde{y} + s(\tilde{x} - \tilde{y}), t).$$

Since A is closed and convex and $\partial_1 G(\tilde{y} + s(\tilde{x} - \tilde{y}), t) \in A$ for all $s \in [0, 1]$, we have $\tilde{A} \in A$; the required property of the Riemann integral follows from its definition in terms of the limit of a sum. □

For practical application we note the following computable version of this proposition.

5.1.5 Corollary Let G: $B \subseteq \mathbb{R}^n \times \mathbb{R}^p \to \mathbb{R}^m$ be in $D \times E \subseteq B$ continuously differentiable with respect to the first n variables. Then for any interval extension of $\partial_1 F$ and any $x \in \mathbb{I}D$ and $z \in \mathbb{I}E$ such that the interval matrix $A = \partial_1 F(x, z)$ is defined, (5) holds for every $\tilde{x}, \tilde{y} \in x$ and every $t \in z$. □

In particular, if $\partial_1 F(\check{x}, \check{z})$ is nonsingular and the radius of both x and z is sufficiently small then, for any continuous interval extension of $\partial_1 F$, the matrix $A = \partial_1 F(x, z)$ is a regular interval matrix.

If G is not differentiable but Lipschitz continuous then a result similar to Proposition 5.1.4 can be proved in terms of *generalized derivatives*. Since the definitions are somewhat complicated we give no details. For practical purposes one can completely avoid the use of generalized derivatives by working with inclusion algebras. Indeed, the inclusion algebra of Theorem 2.3.10 allows the construction of interval matrices satisfying (5) for all $\tilde{x}, \tilde{y} \in x$ and all $t \in z$ whenever the components of G are defined by arithmetical expressions which are Lipschitz in (x, z).

We now specialize our considerations to the parameter-free case. The condition on F: $D_0 \subseteq \mathbb{R}^n \to \mathbb{R}^m$ corresponding to (5) is

$$F(\tilde{x}) - F(\tilde{y}) = \tilde{A}(\tilde{x} - \tilde{y}) \quad \text{for some } \tilde{A} \in A. \tag{6}$$

We call a closed, convex and bounded set A of $m \times n$ matrices a *Lipschitz set* for F on $D \subseteq D_0$ if (6) holds for every $\tilde{x}, \tilde{y} \in D$, and a *Lipschitz matrix* if, in addition, $A \in \mathbb{IR}^{m\times n}$. In particular, by Corollary 5.1.5, if F is continuously differentiable in D and $x \in \mathbb{I}D$ then, for every interval extension of the derivative, the matrix $A = F'(x)$ is a Lipschitz matrix for F whenever $F'(x)$ is defined.

Note that if F is Lipschitz continuous on D, i.e. $\|F(\tilde{x}) - F(\tilde{y})\| \le \lambda\|\tilde{x} - \tilde{y}\|_2$ for all $\tilde{x}, \tilde{y} \in D$ with some Lipschitz constant λ, then the matrix A with entries $A_{ik} = [-\lambda, \lambda]$ is a Lipschitz matrix for F on D (however, not a very useful one). Conversely, if A is a Lipschitz set for F on D then F is Lipschitz continuous on D with Lipschitz constant $\lambda = \sup\{\|\tilde{A}\|_2 \mid \tilde{A} \in A\}$.

5.1.6 Theorem Let $F: D_0 \subseteq \mathbb{R}^n \to \mathbb{R}^n$ be Lipschitz continuous on $D \subseteq D_0$, and let A be a regular Lipschitz set on D. Then:

(i) For every $b^* \in \mathbb{R}^n$, the equation $F(x^*) = b^*$ has at most one solution $x^* \in D$.

(ii) The inverse function $F^{-1}: F^*(D) \to \mathbb{R}^n$ defined on the range $F^*(D) := \{F(\tilde{x}) \mid \tilde{x} \in D\}$ by

$$F^{-1}(\tilde{b}) = \tilde{x} :\Leftrightarrow F(\tilde{x}) = \tilde{b} \tag{7}$$

is Lipschitz continuous, and A^{-1} is a Lipschitz matrix for F^{-1} on $F^*(D)$.

(iii) If $b^* \in F^*(D)$ then for every $\tilde{x}^0 \in D$,

$$F^{-1}(b^*) \in x^0 + A^{\mathrm{H}}(b^* - F(\tilde{x}^0)). \tag{8}$$

(iv) If D is compact and there is a point $\tilde{x}^0 \in \mathrm{int}(D)$ such that

$$F(\tilde{x}) \neq sF(\tilde{x}^0) + (1-s)b^* \quad \text{for all } \tilde{x} \in \partial D, s \in (0, 1], \tag{9}$$

then $b^* \in F^*(D)$, i.e. the equation $F(x^*) = b^*$ has a unique solution $x^* \in D$.

Proof If $\tilde{x}, \tilde{y} \in D$ satisfy $F(\tilde{x}) = \tilde{b} = F(\tilde{y})$ then (6) implies $\tilde{x} - \tilde{y} = \tilde{A}^{-1}(F(\tilde{x}) - F(\tilde{y})) = 0$, so that $\tilde{x} = \tilde{y}$. Hence every equation $F(\tilde{x}) = \tilde{b}$ has at most one solution $\tilde{x} \in D$. In particular, F^{-1} is well defined. Now let $\tilde{x} = F^{-1}(\tilde{b}), \tilde{y} = F^{-1}(\tilde{a})$, where $\tilde{a}, \tilde{b} \in F^*(D)$. Then (6) gives

$$\begin{aligned} F^{-1}(\tilde{b}) - F^{-1}(\tilde{a}) &= \tilde{x} - \tilde{y} = \tilde{A}^{-1}(F(\tilde{x}) - F(\tilde{y})) \\ &= \tilde{A}^{-1}(\tilde{b} - F(\tilde{y})) = \tilde{A}^{-1}(\tilde{b} - \tilde{a}), \end{aligned}$$

hence

$$F^{-1}(\tilde{b}) - F^{-1}(\tilde{a}) = \tilde{A}^{-1}(\tilde{b} - \tilde{a}) \quad \text{for some } \tilde{A} \in A. \tag{10}$$

Since $\tilde{A}^{-1} \in A^{-1}$, this shows that A^{-1} is a Lipschitz matrix for F^{-1} on $F^*(D)$. Moreover, for $\tilde{a} = b^*$, $\tilde{y} = x^*$ and $\tilde{x} = x^0$, (10) implies (8).

It remains to prove (iv). Pick $\varepsilon \in (0, 1]$ and apply Theorem 5.1.3 with

$p = 1$, $E = [\varepsilon, 1]$ and $G(\tilde{x}, t) := F(\tilde{x}) - tF(\tilde{x}^0) - (1-t)b^*$. We find that $G(\tilde{x}, \varepsilon) = 0$ has a unique solution $\tilde{x} = H(\varepsilon) \in D$, and, by continuity of G, every accumulation point x^* of $H(\varepsilon)$ for $\varepsilon \to 0$ is a solution of the equation $0 = G(x^*, 0) = F(x^*) - b^*$. Hence $b^* \in F^*(D)$, and the uniqueness of x^* follows from (i). □

The above semilocal form of the *inverse function theorem* has two special features not shared by the classical (local) inverse function theorem, and which are very useful for applications. Statement (iii) provides a narrow enclosure for $x^* = F^{-1}(b^*)$ when $F(\tilde{x}^0) \approx b^*$. This can be used for bounding the error of approximate solutions $\tilde{x}^0$ of the equation $F(x^*) = b^*$. Moreover, statement (iv) allows the verification that such a solution really exists.

We now rewrite (iii) and (iv) in a form more suitable for practical calculations. We restrict our attention to the case $b^* = 0$, since the general case is obtained by a simple shift of F.

5.1.7 Theorem (Moore, Nickel) Under the assumption of Theorem 5.1.6, if $\tilde{x} \in x \in \mathbb{I}D$ then every $x' \in \mathbb{IR}^n$ satisfying

$$N(x, \tilde{x}) := \tilde{x} - A^{\mathrm{H}}F(\tilde{x}) \subseteq x' \tag{11}$$

has the following three properties:

(i) Every zero $x^* \in x$ of F satisfies $x^* \in x'$.
(ii) If $x' \cap x = \emptyset$ then F contains no zero in x.
(iii) If $\tilde{x} \in \mathrm{int}(x)$ and $x' \subseteq x$ then F contains a unique zero in x (and hence in x').

Proof We apply Theorem 5.1.6 with $b^* = 0$. Then (8) implies (i), and (ii) is an immediate consequence. To show that F contains a unique zero in x it suffices to verify that

$$F(\tilde{y}) \neq sF(\tilde{x}) \quad \text{for all } \tilde{y} \in \partial x, s \in (0, 1]. \tag{12}$$

Hence suppose that $\tilde{y} \in x$ and $F(\tilde{y}) = sF(\tilde{x})$ for some $s \in (0, 1]$. By (6) we have

$$\begin{aligned} \tilde{y} &= \tilde{x} - \tilde{A}^{-1}(F(\tilde{x}) - F(\tilde{y})) = s\tilde{x} + (1-s)(\tilde{x} - \tilde{A}^{-1}F(\tilde{x})) \\ &\in s\tilde{x} + (1-s)x'. \end{aligned}$$

Since $\tilde{x} \in \mathrm{int}(x)$, $x' \subseteq x$ and $0 < s \leq 1$, this implies $\tilde{y} \in \mathrm{int}(x)$. Thus (12) holds. □

Remark The condition $\tilde{x} \in \mathrm{int}(x)$ in (iii) can be replaced by the more practical condition $x' \subseteq \mathrm{int}(x)$; then existence follows from Theorem 5.4.2 below, and uniqueness as in the present proof.

Note that $N(x, \tilde{x})$ depends on x only indirectly through the condition that A is a regular Lipschitz set for x. $\mathrm{N}(x, \tilde{x})$ is called the *Newton operator* because for $A = F'(x)$ it resembles the improvement step $x^* \approx \tilde{x} - F'(\tilde{x})^{-1}F(x)$ of the classical Newton iteration. But whereas the classical Newton iteration only provides approximations to a zero, the simple replacement of $F'(\tilde{x})$ by $F'(x)$ yields enclosures for the zero (which under our present assumption is unique). Moreover, classical Newton iteration provides no equivalent to the *nonexistence test* (ii) and the existence and uniqueness test (iii) of Theorem 5.1.7. Methods for finding good enclosures for $\mathrm{N}(x, \tilde{x})$ have been discussed in Chapter 4.

While the Newton operator can only be used when A is regular, there are two other interval operators with similar properties which can be defined whenever A is a Lipschitz matrix; since A need not be regular, these operators are more widely applicable. They are the *Krawczyk operator*

$$\mathrm{K}(x, \tilde{x}) := \tilde{x} - CF(\tilde{x}) - (CA - I)(x - \tilde{x}) \tag{13}$$

and the *Hansen–Sengupta operator*

$$\mathrm{H}(x, \tilde{x}) := \tilde{x} + \Gamma(CA, -CF(\tilde{x}), x - \tilde{x}); \tag{14}$$

the latter is a nonlinear version of interval Gauss–Seidel iteration. Both operators involve a preconditioning matrix C. By Theorem 4.3.5, we have

$$\mathrm{H}(x, \tilde{x}) \subseteq \mathrm{K}(x, \tilde{x}) \quad \text{for all } \tilde{x} \in x \subseteq x^0. \tag{15}$$

5.1.8 Theorem (Krawczyk, Kahan; Hansen & Sengupta) Let $F: D_0 \subseteq \mathbb{R}^n \to \mathbb{R}^n$ be Lipschitz continuous on $D \subseteq D_0$, and let $A \in \mathbb{IR}^{n\times n}$ be a Lipschitz matrix for F on D. If $\tilde{x} \in x \in \mathbb{I}D$ and x' denotes either $\mathrm{H}(x, \tilde{x})$ or $\mathrm{K}(x, \tilde{x})$ then:

(i) Every zero $x^* \in x$ of F satisfies $x^* \in x'$.
(ii) If $x' \cap x = \emptyset$ then F contains no zero in x.
(iii) If $\tilde{x} \in \mathrm{int}(x)$ and $\emptyset \neq x' \subseteq \mathrm{int}(x)$ then A is strongly regular and F contains a unique zero in x (and hence in x').

Proof Suppose that $x^* \in x$ is a zero of F. By applying (6) with $\tilde{y} = x^*$ we see that $-F(\tilde{x}) = F(x^*) - F(\tilde{x}) = \tilde{A}(x^* - \tilde{x})$ for some $\tilde{A} \in A$. Therefore

$$\begin{aligned} x^* &= \tilde{x} - CF(\tilde{x}) - (C\tilde{A} - I)(x^* - \tilde{x}) \in \mathrm{K}(x, \tilde{x}), \\ x^* &\in (\tilde{x} + \Sigma(A, -F(\tilde{x}))) \cap x = \tilde{x} + (\Sigma(A, -F(\tilde{x})) \cap (x - \tilde{x})) \\ &\subseteq \mathrm{H}(x, \tilde{x}). \end{aligned}$$

Hence $x^* \in x'$, and (i) and (ii) follow. To prove (iii) we note that $x' \subseteq \mathrm{int}(x)$ implies $\mathrm{H}(x, \tilde{x}) \subseteq \mathrm{int}(x)$ and therefore $\Gamma(CA, -F(x), x - \tilde{x}) \subseteq \mathrm{int}(x - \tilde{x})$. Now Theorem 4.4.5 shows that CA is an H-matrix so that A is strongly regular. By the same theorem, we have

$$\begin{aligned}\mathrm{N}(x, \tilde{x}) &= \tilde{x} - A^{\mathrm{H}}F(\tilde{x}) \subseteq \tilde{x} + (CA)^{\mathrm{F}}(-CF(\tilde{x})) \\ &\subseteq \tilde{x} + \Gamma(CA, -CF(\tilde{x}), x - \tilde{x}) = \mathrm{H}(x, \tilde{x}),\end{aligned}$$

so that $\mathrm{N}(x, \tilde{x}) \subseteq \mathrm{int}(x)$. Now Theorem 5.1.7 shows that F contains a unique zero in x. □

We note that (15) implies that the Hansen–Sengupta operator gives enclosures for a zero of F which are always at least as good as those for the more traditional and formally simpler Krawczyk operator. Moreover, it also gives sharper nonexistence and existence tests (ii) and (iii). On the other hand, the above proof shows that the existence test with $\mathrm{H}(x, \tilde{x})$ is weaker than that with the Newton operator $\mathrm{N}(x, \tilde{x})$. In particular, when A is not regular, the existence tests based on all three operators necessarily fail.

We now prove a result on the optimal choice of $\tilde{x}$ and C for the Krawczyk operator. Let us emphasize the dependence on C by writing

$$\mathrm{K}(x, \tilde{x}; C) := \tilde{x} - CF(\tilde{x}) - (CA - I)(x - \tilde{x}). \tag{13a}$$

5.1.9 Theorem (Krawczyk) Under the assumptions of Theorem 5.1.8 the following statements hold for all $C \in \mathbb{R}^{n\times n}$:

(i) $\mathrm{rad}(\mathrm{K}(x, \check{x}; C)) \le \mathrm{rad}(\mathrm{K}(x, \tilde{x}; C))$.
(ii) If $\mathrm{K}(x, \check{x}; C) \subseteq x$ and $\check{A}$ is regular then $\mathrm{K}(x, \check{x}; \check{A}^{-1}) \subseteq \mathrm{K}(x, \check{x}; C)$.

Proof We have $\mathrm{rad}(\mathrm{K}(x, \tilde{x}; C)) = \mathrm{rad}((CA - I)(x - \tilde{x})) \ge |CA - I| \times \mathrm{rad}(x - \tilde{x}) = |CA - I|\mathrm{rad}(x)$, and, since this holds with equality when $\tilde{x} = \check{x}$, (i) follows. Now suppose that $y := \mathrm{K}(x, \check{x}; C) \subseteq x$. Writing $R := |CA - I|$ we have

$$\check{y} = \check{x} - CF(\check{x}), \quad \mathrm{rad}(y) = R \cdot \mathrm{rad}(x).$$

Hence

$$|CF(\check{x})| = |\check{x} - \check{y}| \le \mathrm{rad}(x) - \mathrm{rad}(y) = (I - R)\mathrm{rad}(x).$$

With

$$R' := |\check{A}^{-1}A - I| = |\check{A}^{-1}|\mathrm{rad}(A) \le |(C\check{A})^{-1}|\,|C|\mathrm{rad}(A)$$

and

$$\Delta := |(C\check{A})^{-1} - I| = |(C\check{A})^{-1}(I - C\check{A})| \le |(C\check{A})^{-1}|\,|I - C\check{A}|$$

we get

$$\begin{aligned}\Delta + R' &\le |(C\check{A})^{-1}|(|I - C\check{A}| + |C|\mathrm{rad}(A)) \\ &= |(C\check{A})^{-1}|R \le (I + \Delta)R.\end{aligned}$$

Therefore $y' := \mathrm{K}(x, \check{x}; \check{A}^{-1})$ satisfies

$$\begin{aligned}|\check{y} - \check{y}'| &= |(\check{A}^{-1} - C)F(\check{x})| = |((C\check{A})^{-1} - I)CF(\check{x})| \\ &\le \Delta|CF(\check{x})| \le \Delta(I - R)\mathrm{rad}(x) \\ &\le (R - R')\mathrm{rad}(x) = \mathrm{rad}(y) - \mathrm{rad}(y').\end{aligned}$$

This implies $y' \subseteq y$; hence (ii) holds. □

Remarks (i) This result shows that the Krawczyk operator gives best results for the choice $\tilde{x} = \check{x}$ and $C = \check{A}^{-1}$.

(ii) Note that essentially the same argument as for (ii) shows that

$$\mathrm{rad}(\mathrm{K}(x, \tilde{x}; C)) < \mathrm{rad}(x) \Rightarrow \mathrm{rad}(\mathrm{K}(x, \check{x}; \check{A}^{-1})) \le \mathrm{rad}(\mathrm{K}(x, \tilde{x}; C)).$$

(iii) It would be interesting to have an analog of these results for the Hansen–Sengupta operator.

Finally, we mention a relation between the Krawczyk operator and a computable form of the Newton operator.

5.1.10 Proposition (Krawczyk) Suppose that $\mathrm{K}(x, \check{x}) \subseteq \mathrm{int}(x)$. Then the spectral radius of $R := |CA - I|$ is <1; the formula

$$A^{\mathrm{I}}b := (I + (I - R)^{-1}R[-1, 1])(Cb) \tag{16}$$

defines an inverse of A; and we have the inclusion

$$\check{x} - A^{\mathrm{I}}F(\check{x}) \subseteq \mathrm{K}(x, \check{x}). \tag{17}$$

Proof The assumption implies $\mathrm{rad}(x) > \mathrm{rad}(\mathrm{K}(x, \check{x})) = \mathrm{rad}((CA - I)) \times (x - \check{x}) \ge |CA - I|\mathrm{rad}(x - \check{x}) = R \cdot \mathrm{rad}(x) \ge 0$; hence $\rho(R) < 1$. In particular, A is strongly regular.

If $\tilde{A} \in A$ then $\tilde{R} := C\tilde{A} - I$ satisfies $|\tilde{R}| \le R$; hence $|(C\tilde{A})^{-1} - I| = |(I - \tilde{R})^{-1}\tilde{R}| \le (I - R)^{-1}R$ so that $(C\tilde{A})^{-1} \in I + (I - R)^{-1}R[-1, 1]$. This shows that $A^{\mathrm{I}}b$ is an inverse of A. Now let $y := \mathrm{K}(x, \check{x})$ and $z := \check{x} - A^{\mathrm{I}}F(\check{x})$. Then

$$\check{y} = \check{z} = \check{x} - CF(\check{x}), \mathrm{rad}(y) = R \cdot \mathrm{rad}(x).$$

Now $|CF(\check{x})| = |\check{x} - \check{y}| \le \mathrm{rad}(x) - \mathrm{rad}(y) = (I - R)\mathrm{rad}(x)$ implies

$$\mathrm{rad}(z) = (I - R)^{-1}R|CF(\check{x})| \le R \cdot \mathrm{rad}(x) = \mathrm{rad}(y).$$

Since $\check{y} = \check{z}$, this shows that $z \subseteq y$. □

5.2 Interval iteration

In this section we study iteration methods for improving interval enclosures of a zero x^* of a function $F\colon D_0 \subseteq \mathbb{R}^n \to \mathbb{R}^n$ by repeated application of the operators N, K or H. We give sufficient conditions for convergence and some results about the convergence speed.

Since $\tilde{x}, x^* \in x$ implies $x^* \in \mathrm{N}(x, \tilde{x})$, $x^* \in \mathrm{K}(x, \tilde{x})$ and $x^* \in \mathrm{H}(x, \tilde{x})$, it is natural to consider the iterations

$$x^0 := x,\ x^{l+1} := \mathrm{N}(x^l, \tilde{x}^l) \cap x^l \quad (l = 0, 1, 2, \ldots) \tag{1}$$
$$x^0 := x,\ x^{l+1} := \mathrm{K}(x^l, \tilde{x}^l) \cap x^l \quad (l = 0, 1, 2, \ldots) \tag{2}$$

and

$$x^0 := x,\ x^{l+1} := \mathrm{H}(x^l, \tilde{x}^l) \qquad (l = 0, 1, 2, \ldots) \tag{3}$$

for suitable $\tilde{x}^l \in x^l$. Here we put $x^{l+1} := \emptyset$ if $x^l = \emptyset$. Due to taking intersections, and in the case of (3) by definition of $\mathrm{H}(x, \tilde{x})$, we have

$$x = x^0 \supseteq x^1 \supseteq \cdots \supseteq x^l \supseteq x^{l+1} \supseteq \cdots, \tag{4}$$
$$x^* \in x, F(x^*) = 0 \Rightarrow x^* \in x^l \text{ for all } l \supseteq 0. \tag{5}$$

In particular, the limit

$$x^\infty := \lim_{l \to \infty} x^l$$

exists for all three sequences and any choice of $\tilde{x}^l \in x^l$. Moreover, if the starting box x contains a zero x^* of F then $x^* \in x^\infty$.

Warning In finite precision calculations one has to make sure that the expression $F(\tilde{x}^l)$ in the above operators is evaluated with outward rounding (i.e. by treating $\tilde{x}^l$ as a thin interval). Otherwise it may happen that, after quick initial convergence, the solution suddenly disappears ($x^{l+1} = \emptyset$) due to rounding errors.

Under suitable conditions it can be shown that the following alternative (SC) holds:

(SC) *Either* F has a unique zero x^* in x and $\lim_{l \to \infty} x^l = x^*$,
or F has no zero in x and $x^l = \emptyset$ for some $l \geq 0$.

If (SC) holds then the iteration is capable of deciding whether F has a zero in x or not, and of enclosing this zero with arbitrary accuracy (assuming sufficiently accurate arithmetic). We shall call a sequence x^l ($l = 0, 1, 2, \ldots$) satisfying (4) and (5) *strongly convergent* if (SC) holds.

5.2.1 Lemma A sequence x^l ($l = 0, 1, 2, \ldots$) satisfying (4) and (5) is strongly convergent if $x^\infty \neq \emptyset$ implies $\mathrm{rad}(x^\infty) = 0$ and $F(x^\infty) = 0$.

Proof If $x^l = \emptyset$ for some $l \geq 0$ then, by (5), F has no zero in x. If $x^l \neq \emptyset$ for all $l \geq 0$ then (4) implies that $x^\infty \neq \emptyset$, hence $\mathrm{rad}(x^\infty) = 0$ by assumption; hence $x^\infty = x^* \in \mathbb{R}^n$ and, again by assumption, $F(x^*) = 0$. □

The convergence analysis is most straightforward for *Krawczyk's iteration* (2). By Theorem 5.1.9(i), a natural choice for $\tilde{x}^l$ in (2) is $\tilde{x}^l = \check{x}^l$. In this case we have the following result. (Here, as in the remainder of this section, $\|\cdot\|$ is a scaled maximum norm.)

5.2.2 Theorem (Krawczyk) Let A be a strongly regular Lipschitz matrix on $x \in \mathbb{I}D_0$ for $F: D_0 \subseteq \mathbb{R}^n \rightarrow \mathbb{R}^n$. Let $C \in \mathbb{R}^{n\times n}$ be such that $\rho(|CA - I|) = \beta^* < 1$. Then Krawczyk's iteration (2) with $\tilde{x}^l = \check{x}^l$ is strongly convergent. Moreover, as long as $x^{l+1} \neq \emptyset$, we have

$$\|\text{rad}(x^{l+1})\| \leq \beta\|\text{rad}(x^l)\|, \quad \text{with } \beta = \|CA - I\|. \tag{6}$$

In particular, the radii converge linearly to zero with asymptotic convergence factor $\leq\beta^*$.

Proof $x^{l+1} \subseteq \text{K}(x^l, \check{x}^l)$ implies $\text{rad}(x^{l+1}) \leq \text{rad}(\text{K}(x^l, \check{x}^l)) = \text{rad}((CA - I) \times (x^l - \check{x}^l)) = |CA - I|\text{rad}(x^l)$. By taking norms, (6) follows.

Now suppose that $x^\infty \neq \emptyset$. By a suitable choice of the norm we may get β arbitrarily close to β^* (Corollary 3.2.3). Hence the radii converge linearly to zero with asymptotic convergence factor $\leq\beta^*$. In particular, $\text{rad}(x^\infty) = 0$ so that $x^\infty = x^* \in \mathbb{R}^n$. Now $x^* \in \text{K}(x^*, x^*) = x^* - CF(x^*)$ implies $CF(x^*) = 0$, and since CA and hence C is regular, $F(x^*) = 0$. By Lemma 5.2.1, (SC) holds. □

For other choices of $\tilde{x}^l$, strong convergence can be proved if the assumptions are strengthened.

5.2.3 Theorem Let A be a strongly regular Lipschitz matrix on $x \in \mathbb{I}D_0$ for $F: D_0 \subseteq \mathbb{R}^n \rightarrow \mathbb{R}^n$. Let $C \in \mathbb{R}^{n\times n}$ be such that $\|CA - I\| \leq \beta < 1$. Then Krawczyk's iteration (2) is strongly convergent for all choices of $\tilde{x}^l \in x^l$ such that

$$\kappa := \sup \frac{\|x^l - \tilde{x}^l\|}{\|\text{rad}(x^l)\|} < \frac{1}{\beta}. \tag{7}$$

In this case (6) holds with $\kappa\beta$ in place of β.

Proof $x^{l+1} \subseteq \text{K}(x^l, \tilde{x}^l)$ implies $\text{rad}(x^{l+1}) \leq \text{rad}(\text{K}(x^l, \tilde{x}^l)) = \text{rad}((CA - I) \times (x^l - \tilde{x}^l)) \leq |CA - I|\,|x^l - \tilde{x}^l|$. Hence $\|\text{rad}(x^{l+1})\| \leq \beta\|x^l - \tilde{x}^l\| \leq \kappa\beta\|\text{rad}(x^l)\|$. Now the remainder of the previous proof applies with $\kappa\beta$ in place of β. □

5.2.4 Corollary (Krawczyk) If $\|CA - I\| < \frac{1}{2}$ then Krawczyk's iteration (2) is strongly convergent for all choices of $\tilde{x}^l \in x^l$.

Proof Since $\tilde{x}^l \in x^l$ implies $0 \in x^l - \tilde{x}^l$, we have $|x^l - \tilde{x}^l| \leq 2\cdot\text{rad}(x^l - \tilde{x}^l) = 2\cdot\text{rad}(x^l)$. Hence (7) is satisfied with $\kappa \leq 2$. □

Since always $H(x, \check{x}) \subseteq K(x, \check{x})$, the preceding results immediately extend to the Hansen–Sengupta iteration (3). For this iteration even much weaker assumptions are sufficient to guarantee strong convergence, although results about rates of convergence have not been obtained under these weaker conditions.

5.2.5 Theorem (Alefeld) Let A be a strongly regular Lipschitz matrix on $x \in ID_0$ for $F: D_0 \subseteq \mathbb{R}^n \to \mathbb{R}^n$. Let $C \in \mathbb{R}^{n \times n}$ be such that CA is an H-matrix. Then the Hansen–Sengupta iteration (3) with $\tilde{x}^l = \check{x}^l$ is strongly convergent.

Proof By taking limits in (3) we get

$$x^\infty = H(x^\infty, \check{x}^\infty). \tag{8}$$

Suppose that $x^\infty \neq \emptyset$. Then $z := x^\infty - \check{x}^\infty$ satisfies $\check{z} = 0$ and

$$z = H(x^\infty, \check{x}^\infty) - \check{x}^\infty = \Gamma(CA, -CF(\check{x}^\infty), z).$$

Therefore $|z| \leq \langle CA \rangle^{-1} \mathrm{rad}(-CF(\check{x}^\infty)) = 0$ by Corollary 4.3.14. This gives $z = 0$, i.e. $\mathrm{rad}(x^\infty) = 0$. Now $0 = \Gamma(CA, -CF(\check{x}^\infty), 0)$ implies $0 = CF(\check{x}^\infty)$. Since C has to be regular, $0 = F(\check{x}^\infty) = F(x^\infty)$. Hence Lemma 5.2.1 applies. $\qquad\square$

5.2.6 Theorem (Thiel) Let A be a strongly regular Lipschitz matrix on $x \in ID_0$ for $F: D_0 \subseteq \mathbb{R}^n \to \mathbb{R}^n$. Let $C \in \mathbb{R}^{n \times n}$ be such that CA is an H-matrix and $(CA)^F$ is regular. Then the Hansen–Sengupta iteration (3) is strongly convergent for all choices of $\tilde{x}^l \in x^l$.

Proof Suppose that $x^\infty \neq \emptyset$ and let $\tilde{x}$ be an accumulation point of the $\tilde{x}^l$ $(l \geq 0)$. By taking limits in (3) we get

$$\tilde{x} \in x^\infty = H(x^\infty, \tilde{x}). \tag{9}$$

Hence $z := z^\infty - \tilde{x}$ satisfies $0 \in z$ and $z = H(x^\infty, \tilde{x}) - \tilde{x} = \Gamma(CA, -CF(\tilde{x}), z)$. Therefore $0 \in z \subseteq (CA)^F(-CF(\tilde{x}))$ by Theorem 4.4.5. Since $(CA)^F$ is regular and $-CF(\tilde{x})$ is thin, this implies $CF(\tilde{x}) = 0$. Since C must be regular, $F(\tilde{x}) = 0$. This yields $z \subseteq (CA)^F 0 = 0$, hence $x^\infty = \tilde{x}$. Now Lemma 5.2.1 applies. $\qquad\square$

Conditions which imply that $(CA)^F$ is regular are given in Theorems 4.4.11 and 4.4.12. In particular, using Corollary 4.4.13, we get the following:

5.2.7 Corollary (Cornelius) If CA is an M-matrix then the Hansen–Sengupta iteration (3) is strongly convergent for all choices of $\tilde{x}^l \in x^l$. $\qquad\square$

We now consider convergence conditions for the Newton iteration (1). However, (1) has to be modified in practice since, in general, the hull inverse is difficult to obtain. Therefore, $A^H F(\tilde{x})$ is usually enclosed by using a preconditioned inverse. Thus we shall generalize (1) to the general *Newton iteration*

$$x^0 := x, \quad x^{l+1} := \mathrm{N_I}(x^l, \tilde{x}^l) \cap x^l \tag{10}$$

with the general *Newton operator*

$$N_\mathrm{I}(x, \tilde{x}) := \tilde{x} - (CA)^\mathrm{I}(CF(\tilde{x})). \tag{11}$$

Here $C \in \mathbb{R}^{n\times n}$ is a preconditioning matrix and $(CA)^\mathrm{I}$ is a suggestive symbol denoting a sublinear mapping such that

$$(CA)^\mathrm{H} b \subseteq (CA)^\mathrm{I} b \quad \text{for all } b \in \mathbb{IR}^n,$$

i.e. $(CA)^\mathrm{I}$ is an inverse of CA. Clearly (4) and (5) also hold for the iteration (10). In the special case where C is the identity matrix and $A^\mathrm{I} = A^\mathrm{H}$, (10) and (11) just reduce to the *optimal Newton iteration* (1) with the optimal Newton operator $\mathrm{N}(x, \tilde{x})$.

5.2.8 Theorem Let A be a strongly regular Lipschitz matrix on $x \in \mathbb{I}D_0$ for F: $D_0 \subseteq \mathbb{R}^n \to \mathbb{R}^n$. Let $C \in \mathbb{R}^{n\times n}$ be such that CA is regular, and let $(CA)^\mathrm{I}$ be an inverse of CA. If $(CA)^\mathrm{I}$ is regular then the Newton iteration (10) is strongly convergent for every choice of $\tilde{x}^l \in x^l$. Moreover, for all $l \geq 0$, we have

$$\text{either } \tilde{x}^l \notin x^{l+1}, \text{ or } x^{l+1} = \tilde{x}^l \text{ and } F(x^l) = 0. \tag{12}$$

Proof Suppose that $x^\infty \neq \emptyset$ and let $\tilde{x}$ be an accumulation point of the $\tilde{x}^l$ $(l \geq 0)$. Since a sublinear mapping is continuous, we have $\tilde{x} \in x^\infty \subseteq \tilde{x} - (CA)^\mathrm{I}(CF(\tilde{x}))$. Therefore $0 \in (CA)^\mathrm{I}(CF(\tilde{x}))$. Since $(CA)^\mathrm{I}$ is regular, this implies $CF(\tilde{x}) = 0$, and hence $x^\infty = \tilde{x}$. Since C is regular we must also have $F(\tilde{x}) = 0$. Now Lemma 5.2.1 shows that (10) is strongly convergent. Finally, if $\tilde{x}^l \in x^{l+1}$ then $0 \in (CA)^\mathrm{I}(CF(\tilde{x}^l))$. Hence, as before, $F(\tilde{x}^l) = 0$, giving $x^{l+1} = \tilde{x}^l \cap x^l = \tilde{x}^l$. □

Relation (12) implies a volume-reducing property of the Newton iteration. Denote by

$$\mathrm{vol}(x) := (\overline{x}_1 - \underline{x}_1)(\overline{x}_2 - \underline{x}_2) \cdot \ldots \cdot (\overline{x}_n - \underline{x}_n) \tag{13}$$

the *volume* of a box $x \in \mathbb{IR}^n$.

5.2.9 Corollary If $(CA)^\mathrm{I}$ is regular then the Newton iteration (10) with $\tilde{x}^l = \check{x}^l$ satisfies

$$\mathrm{vol}(x^{l+1}) \leq \tfrac{1}{2}\,\mathrm{vol}(x^l) \tag{14}$$

for all $l \geq 0$ with $x^{l+1} \neq \emptyset$.

Proof The side lengths $\overline{x}_i^l - \underline{x}_i^l$ are monotone decreasing in l. If $\check{x}_j^l \notin x_j^{l+1}$ for some index j then some side length reduces by a factor of at least $\frac{1}{2}$, giving (14). Otherwise, $\check{x}^l \in x^{l+1}$ and (12) implies $\mathrm{vol}(x^{l+1}) = 0$, and again (14) holds. □

5.2.10 Corollary If, for some $C \in \mathbb{R}^{n\times n}$, CA is an M-matrix or $\|I - CA\| < \frac{1}{2}$ then the optimal Newton iteration (1) is strongly convergent for every choice of $\tilde{x}^l \in x^l$, and (12) and (14) hold.
Proof Under these assumptions A^{H} is regular by Corollary 4.4.13. □

5.2.11 Example Let

$$F(\xi) = \begin{pmatrix} 2\xi_1 - \xi_2 - 2 \\ 3.5\xi_2 - \xi_1^2 - 4\xi_1 + 5 \end{pmatrix}, \quad x = \begin{pmatrix} [-1, 1] \\ [-1, 1] \end{pmatrix}.$$

The unique zero of F in x is $x^* = \begin{pmatrix} 1 \\ 0 \end{pmatrix}$. A Lipschitz matrix for F in x is

$$A := F'(x) = \begin{pmatrix} 2 & -1 \\ [-6, -2] & 3.5 \end{pmatrix}.$$

Since A is an M-matrix, we may iterate without preconditioning, i.e. $C = I$, and have strong convergence for (1) and (3) with arbitrary choice of $\tilde{x}^l$. Let us first consider the Hansen–Sengupta iteration (3). For $\tilde{x}^l = \check{x}^l$ we get

$$x^1 = \begin{pmatrix} [0.5, 1] \\ [-1, \frac{2}{7}] \end{pmatrix}, \quad x^2 = \begin{pmatrix} [0.5, 1] \\ [-\frac{47}{56}, \frac{1}{56}] \end{pmatrix}, \ldots$$

with slow convergence. If we choose $\tilde{x}^l$ close to the zero x^*, the convergence remains slow. Even when $\tilde{x}^l = x^*$ for all $l \geq 0$ we get

$$x_1^l = [1 - \tfrac{1}{2}(\tfrac{6}{7})^{l-1}, 1], \quad x_2^l = [-(\tfrac{6}{7})^l, 0],$$

with slow linear convergence.

The optimal Newton iteration (1) with $\tilde{x}^l = \check{x}^l$ gives

$$x^1 = \begin{pmatrix} [0.4, 1] \\ [-1, 1] \end{pmatrix}, \quad x^2 = \begin{pmatrix} [0.777, 1] \\ [-0.445, 0.18] \end{pmatrix}, \ldots$$

But here the choice $\tilde{x}^l = x^*$ immediately gives $x^1 = x^*$; hence the knowledge of good approximations to the zero can be expected to speed up convergences considerably.

Finally, if we use the Gauss inverse in place of A^{H} then the modified Newton iteration (10) for $\tilde{x}^l = \check{x}^l$ gets stuck with $x^0 = x^1 = x$ since

$$\mathrm{N_G}(x, \check{x}) = \check{x} - A^{\mathrm{G}}F(\check{x}) = \begin{pmatrix} [-2, 2] \\ [-6, 2] \end{pmatrix} \supseteq x.$$

Indeed, the computations for $A^G F(\check{x})$ are just those of Example 4.5.9(ii) which showed that A^G need not be regular for M-matrices. But, again, for $\tilde{x}^0 \approx x^*$ the performance is much better.

Let us draw two conclusions from the example. The Hansen–Sengupta iteration strongly converges under the mildest assumptions, but convergence may be slow and cannot be accelerated by a good approximation to the zero. On the other hand, a Newton iteration makes full use of good zero approximations but can lead to a premature stop of the iteration, in particular, when the hull inverse is replaced by an inverse which is easier to compute. Fortunately, both methods can be combined in such a way that their advantages are preserved and their disadvantages avoided.

5.2.12 Theorem Let A be a strongly regular Lipschitz matrix on $x \in \mathbb{I}D_0$ for $F: D_0 \subseteq \mathbb{R}^n \to \mathbb{R}^n$. Let $C \in \mathbb{R}^{n\times n}$ be such that CA is an H-matrix and let $(CA)^{\mathrm{I}}$ be an inverse of CA. Consider the iteration defined for $l = 0, 1, 2, \ldots$ by

$$x^{l+1/2} := \mathrm{N}_{\mathrm{I}}(x^l, \tilde{x}^l) \cap x^l, \quad x^{l+1} := \mathrm{H}(x^{l+1/2}, \tilde{x}^l), \tag{15a}$$

where

$$x^0 := x \quad \text{and} \quad \tilde{x}^l \in x^l \text{ for } l \geq 0. \tag{15b}$$

Then (4) and (5) hold. Moreover, if either CA is an M-matrix or $\|CA - I\| < \frac{1}{2}$, or $\tilde{x}^l = \check{x}^l$ for infinitely many indices l, then the sequence x^l $(l = 0, 1, 2, \ldots)$ is strongly convergent. Furthermore, if x contains a zero x^* of F then

$$\|x^{l+1} - x^*\| \leq \gamma\|CF(\tilde{x}^l)\| \leq \gamma'\|\tilde{x}^l - x^*\| \tag{16}$$

for all $l \geq 0$, where

$$\gamma := 2\|\mathrm{rad}(\mathrm{cor}(CA)^{\mathrm{I}})\| \quad \text{and} \quad \gamma' = \gamma\|CA\|. \tag{17}$$

Proof Relations (4) and (5) are obvious. Suppose that $x^\infty \neq \emptyset$ and let $\tilde{x}$ be an accumulation point of the $\tilde{x}^l$ $(l \geq 0)$. Taking limits in (15a) gives (9). Hence the same argument as in the proof of Theorem 5.2.6 implies strong convergence when $(CA)^{\mathrm{F}}$ is regular, and in particular when CA is an M-matrix or $\|CA - I\| < \frac{1}{2}$. On the other hand, if $\tilde{x}^l = \check{x}^l$ for infinitely many indices l then taking limits in (15a) gives (8), and the argument of the proof of Theorem 5.2.5 gives strong convergence.

To get (16) we observe that (15a) implies $x^{l+1} \subseteq x^{l+1/2} \subseteq \mathrm{N}_{\mathrm{I}}(x^l, \tilde{x}^l) = \tilde{x}^l - (CA)^{\mathrm{I}}(CF(\tilde{x}^l)) \subseteq \tilde{x}^l - B\tilde{b}$, where

$$B = \mathrm{cor}(CA)^{\mathrm{I}}, \quad \tilde{b} = CF(\tilde{x}^l).$$

Hence

$$|x^{l+1} - x^*| \le |\tilde{x}^l - x^* - B\tilde{b}|. \tag{18}$$

Since A is a Lipschitz matrix we have $F(\tilde{x}^l) = F(\tilde{x}^l) - F(x^*) = \tilde{A}(\tilde{x}^l - x^*)$ for some $\tilde{A} \in A$. Hence $\tilde{x}^l - x^* = (C\tilde{A})^{-1}CF(\tilde{x}^l) \in (CA)^{\mathrm{I}}(CF(\tilde{x}^l)) \subseteq B\tilde{b}$. Together with (18) this yields

$$|x^{l+1} - x^*| \le 2 \cdot \mathrm{rad}(B\tilde{b}) = 2 \cdot \mathrm{rad}(B)|\tilde{b}|. \tag{19}$$

Since $\tilde{b} = CF(\tilde{x}^l) = C\tilde{A}(\tilde{x}^l - x^*) \in CA(\tilde{x}^l - x^*)$, we now get (16) by taking norms in (19). □

In order to exploit the relation (16) it is advisable to determine $\tilde{x}^l$ as an approximation to the zero x^* with small $\|CF(\tilde{x}^l)\|$. For this purpose, any locally fast converging iteration method can be used to approximate x^*, for example a damped Newton or quasi-Newton method. However, we must take account of the possibility that such an approximation $\tilde{x}$ lies outside the current box x^l. Therefore we move $\tilde{x}$ into x^l by means of the cut-off operator defined in equations (2.3.16). In this way we get the following algorithm, which in many cases decides whether a given box $x \in \mathbb{IR}^n$ contains a zero of F and which finds in the affirmative case a good enclosure of this zero.

Algorithm (for enclosing a zero $x^* \in x$ of F)

Step 1 Compute a Lipschitz matrix $A \in \mathbb{IR}^{n\times n}$ for F on x and a preconditioning matrix $C \in \mathbb{R}^{n\times n}$. Put $l := 0$, zero := unknown, $\tilde{x}^0 := \check{x}$, $x^1 := x$.

Step 2 Determine $\tilde{x}^{l+1/2}$ such that $\|CF(\tilde{x}^{l+1/2})\|$ is small, and put $\tilde{x}^{l+1} := \mathrm{cut}(\tilde{x}^{l+1/2}, x^{l+1})$. Replace l by $l+1$.

Step 3 Put $x^{l+1/2} := (\tilde{x} - (CA)^{\mathrm{I}}(CF(\tilde{x}^l))) \cap x^l$ if $(CA)^{\mathrm{I}}$ is defined, and $x^{l+1/2} := x^l$ otherwise. In both cases, put

$$x^{l+1} := \tilde{x}^l + \Gamma(CA, -CF(\tilde{x}^l), x^{l+1/2}).$$

Step 4 If $x^{l+1} = \emptyset$ put zero := no and stop. If $x^{l+1} \subseteq \mathrm{int}(x^l)$ put zero := yes and goto Step 2. If $x^{l+1} = x^l$ stop. In all other cases goto Step 2.

Finite precision arithmetic and the relation $x^{l+1} \subseteq x^l$ imply finite termination of the iteration. If CA is an M-matrix or $\|CA - I\| < \frac{1}{2}$ then, by Theorem 5.2.12, we end up with zero = no or with a narrow enclosure of a zero (and then usually zero = yes), unless a zero lies so close to the boundary of x^0 that existence remains undecided due to rounding errors. In the latter case, trying again with a new small starting box which contains the final box in the interior usually resolves this difficulty. If CA is not an M-matrix and $\|CA - I\| \ge \frac{1}{2}$, then strong convergence is no longer guaranteed and we must solve the equation by global methods (see Section 5.6).

When n is small, the recommended choice for C is $C \approx \check{A}^{-1}$, and $(CA)^{\mathrm{I}} = (CA)^{\mathrm{G}}$ for the inverse in Step 3 (justified by Theorem 4.5.11), or (with somewhat less work) $(CA)^{\mathrm{I}}b = \langle CA\rangle^{-1}|b|[-1, 1]$, with modifications as mentioned after Theorem 4.4.10.

When n is large and A is sparse then it is impractical to choose $C \approx \check{A}^{-1}$; but in the important case where A is an M-matrix, we can choose $C = I$ without losing strong convergence. In this case, $A^{\mathrm{I}}b = \|b\|_v[-u, u]$ is an easily computable inverse, where $u \geq 0$, $\langle A\rangle u \geq v > 0$; if this is used then Step 2 dominates the work of the iteration. Since this step does not involve interval calculations, sophisticated methods for the approximate solution of sparse nonlinear systems can be used to keep the total work small. (See also Theorem 5.6.2 for related results.)

In general, the simplest choice in Step 2 is $\tilde{x}^{l+1/2} = \tilde{x}^l$; but, as we now show, a Newton step (or a simplified Newton step) from $\tilde{x}^l$ yields a vector $\tilde{y}$ with smaller norm $\|CF(\tilde{y})\|$ when $\|CA - I\| < \frac{1}{3}$ (or <1).

5.2.13 Proposition Let A be a strongly regular Lipschitz matrix on $x \in \mathbb{I}D_0$ for $F: D_0 \subseteq \mathbb{R}^n \to \mathbb{R}^n$. Let $C \in \mathbb{R}^{n\times n}$ be such that $\beta := \|CA - I\| < 1$ in some monotone norm. Then:

(i) If $\tilde{x} \in x$ and $\tilde{y} := \tilde{x} - CF(\tilde{x}) \in x$ then $\|CF(\tilde{y})\| \leq \beta\|CF(\tilde{x})\|$.

(ii) If $\mathrm{K}(x, \tilde{x}) \subseteq x$ then the simplified Newton iteration $\tilde{x}^0 := \tilde{x}$, $\tilde{x}^{l+1} := \tilde{x}^l - CF(\tilde{x}^l)$ converges to the unique zero x^* of F in x.

(iii) If $\tilde{x} \in x$ and $\tilde{y} := \tilde{x} - F'(\tilde{x})^{-1}F(\tilde{x}) \in x$ then

$$\|CF(\tilde{y})\| \leq \frac{2\beta}{1-\beta}\|CF(\tilde{x})\|.$$

(iv) If $\mathrm{K}(x, \tilde{x}) \subseteq x$ and $\beta < \frac{1}{3}$ then the Newton iteration $\tilde{x}^0 := \tilde{x}$, $\tilde{x}^{l+1} := \tilde{x}^l - F'(\tilde{x}^l)^{-1}F(\tilde{x}^l)$ converges to the unique zero x^* of F in x.

Proof (i) We have $F(\tilde{y}) = F(\tilde{x}) + \tilde{A}(\tilde{y} - \tilde{x})$ for some $\tilde{A} \in A$, hence

$$\|CF(\tilde{y})\| = \|(I - C\tilde{A})CF(\tilde{x})\| \leq \|I - C\tilde{A}\|\,\|CF(\tilde{x})\| \leq \beta\|CF(\tilde{x})\|.$$

(ii) Apply Schröder's fixed point theorem 3.3.3 (with $\mathbb{R}^n$ in place of $\mathbb{IR}^n$) to $\Phi(\xi) := \xi - CF(\xi)$.

(iii) Since $\tilde{A} := F'(\tilde{x}) \in A$, we have $\|(C\tilde{A})^{-1}\| \leq 1/(1 - \|I - C\tilde{A}\|) \leq 1/(1-\beta)$. Therefore, by subassociativity (3.1.11), we have $\|I - \tilde{A}^{-1}A\| \leq \|(C\tilde{A})^{-1}(C\tilde{A} - CA)\| \leq \|(C\tilde{A})^{-1}\| \cdot (\|C\tilde{A} - I\| + \|I - CA\|) \leq 2\beta/(1-\beta)$. Hence (i) applies with $C' = F'(\tilde{x})^{-1}$ and $\beta' = 2\beta/(1-\beta)$ in place of C and β, respectively.

(iv) We have $\tilde{x}^{l+1} - x^* = \tilde{x}^l - x^* - F'(\tilde{x}^l)^{-1}F(\tilde{x}^l) = (I - F'(\tilde{x}^l)^{-1}\tilde{A})(\tilde{x}^l - x^*)$ so that, as in (iii), $\|\tilde{x}^{l+1} - x^*\| \leq \kappa\|\tilde{x}^l - x^*\|$ with $\kappa = 2\beta/(1-\beta) < 1$. Therefore $\tilde{x}^l \to x^*$ for $l \to \infty$. □

The use of an improved $\tilde{x}^{l+1/2}$ often leads to superlinear convergence. As an example, use of a Newton step usually gives quadratic convergence:

5.2.14 Theorem If CA is an H-matrix and $\tilde{x}^{l+1/2} = \tilde{x}^l - F'(\tilde{x}^l)^{-1}f(\tilde{x}^l)$ in Step 2 of the above scheme then

$$\lim_{l\to\infty} \frac{\|\tilde{x}^{l+1} - x^*\|}{\|\tilde{x}^l - x^*\|} = \lim_{l\to\infty} \|x^l - x^*\|^{1/l} = 0. \tag{20}$$

Moreover, if F' is Lipschitz continuous in x then there are constants $\alpha, \gamma > 0$ such that

$$\|\tilde{x}^{l+1} - x^*\| \le \gamma \|\tilde{x}^l - x^*\|^2, \tag{21}$$

$$\|x^l - x^*\| \le \gamma \cdot \exp(-2^l\alpha). \tag{22}$$

Proof By the mean value theorem, $F(\tilde{x}^l) = F(x^*) + \tilde{A}_l(\tilde{x}^l - x^*)$, where

$$\tilde{A}_l := \int_0^1 F'(x^* + t(\tilde{x}^l - x^*))\,dt \in A.$$

Hence $\|\tilde{x}^{l+1/2} - x^*\| = \|\tilde{x}^l - x^* - F'(\tilde{x}^l)^{-1}\tilde{A}_l(\tilde{x}^l - x^*)\| \le \beta_l\|\tilde{x}^l - x^*\|$, where $\beta_l = \|I - F'(\tilde{x}^l)^{-1}\tilde{A}_l\| \to 0$ for $l \to \infty$. Since $\tilde{x}^{l+1} = \mathrm{cut}(\tilde{x}^{l+1/2}, x)$ and $x^* \in x$ we have $\|\tilde{x}^{l+1} - x^*\| \le \|\tilde{x}^{l+1/2} - x^*\| \le \beta_l\|\tilde{x}^l - x^*\|$. This implies that the first limit in (20) vanishes; the second limit in (20) vanishes because of (16).

If F' is Lipschitz continuous then $\tilde{A}_l = F'(x^*) + \mathrm{O}(\|\tilde{x}^l - x^*\|)$ so that $\beta_l = \mathrm{O}(\|\tilde{x}^l - x^*\|)$. Therefore (21) holds, and again (22) follows from (16). □

In practice, however, it is preferable to improve an approximate zero $\tilde{x}$ by several steps of a Newton or quasi-Newton method until either $\|CF(\tilde{x})\|$ does no longer decrease sufficiently, or a point outside x is found, and then choose $\tilde{x}^{l+1/2}$ as $\tilde{x}$ or $\mathrm{cut}(\tilde{x}, x)$ in Step 2.

Further acceleration of convergence can be achieved by using variable Lipschitz matrices. For reasons given below we give only one particularly neat result of this type.

5.2.15 Theorem (Krawczyk) Let $F\colon D_0 \subseteq \mathbb{R}^n \to \mathbb{R}^n$ be continuously differentiable in $D \subseteq D_0$, and let F' be an interval extension of the derivative of F in D. For $x \in \mathbb{I}D$, we define the special Krawczyk operator

$$\mathrm{K}(x) := \check{x} - \check{A}^{-1}F(\check{x}) - \mathrm{r}(A)(x - \check{x}), \tag{23}$$

where

$$A = F'(x), \quad \mathrm{r}(A) = |\check{A}^{-1}|\mathrm{rad}(A).$$

Then

(i) If $x \in \mathbb{I}D$ satisfies $\mathrm{K}(x) \subseteq x$ and if $F'(x)$ is strongly regular then $\mathrm{K}(\mathrm{K}(x)) \subseteq \mathrm{K}(x)$.

(ii) If $x \in \mathbb{I}D$ satisfies $\mathrm{K}(x) \subseteq \mathrm{int}(x)$ then the iteration

$$x^0 := x, \quad x^{l+1} := \mathrm{K}(x^l) \quad (l = 0, 1, 2, \ldots)$$

defines a nested sequence of intervals x^l converging to the unique zero x^* of F in x, and we have

$$\lim_{l \to \infty} \frac{\|\mathrm{rad}(x^{l+1})\|}{\|\mathrm{rad}(x^l)\|} = 0. \tag{24}$$

Moreover, if F' is the interval evaluation of arithmetical expressions which are Lipschitz at x then

$$\|\mathrm{rad}(x^{l+1})\| = \mathrm{O}(\|\mathrm{rad}(x^l)\|^2). \tag{25}$$

Proof (i) Let $y := \mathrm{K}(x) \subseteq x$ and put $A := F'(x)$ and $B := F'(y)$. Then

$$\check{y} = \check{x} - \check{A}^{-1}F(\check{x}), \quad \mathrm{rad}(y) = \mathrm{r}(A)\mathrm{rad}(x).$$

Since F' is inclusion isotone we have $B \subseteq A$, and, by Corollary 4.1.3, B is strongly regular. Since A is a Lipschitz matrix for x we have

$$F(\check{y}) - F(\check{x}) = \tilde{A}(\check{y} - \check{x}) \quad \text{for some } \tilde{A} \in A.$$

Hence

$$\begin{aligned}
|\check{A}^{-1}F(\check{y})| &= |\check{A}^{-1}F(\check{x}) + \check{A}^{-1}\tilde{A}(\check{y} - \check{x})| \\
&= |(I - \check{A}^{-1}\tilde{A})(\check{x} - \check{y})| \leq \mathrm{r}(A)|\check{x} - \check{y}| \\
&\leq \mathrm{r}(A)(\mathrm{rad}(x) - \mathrm{rad}(y)) = \mathrm{rad}(y) - \mathrm{r}(A)\mathrm{rad}(y).
\end{aligned}$$

Therefore, Theorem 4.1.12(i) implies

$$\begin{aligned}
|\mathrm{mid}(\mathrm{K}(x)) - \mathrm{mid}(\mathrm{K}(y))| &= |\check{y} - \mathrm{mid}(\mathrm{K}(y))| = |\check{B}^{-1}F(\check{y})| \\
&\leq \mathrm{rad}(y) - \mathrm{r}(B)\mathrm{rad}(y) = \mathrm{rad}(\mathrm{K}(x)) - \mathrm{rad}(\mathrm{K}(y)).
\end{aligned}$$

This implies $\mathrm{K}(y) \subseteq \mathrm{K}(x)$.

(ii) Since $\mathrm{rad}(x) > \mathrm{rad}(\mathrm{K}(x)) = \mathrm{r}(F'(x))\mathrm{rad}(x)$, Corollary 3.2.3 implies that $\rho(\mathrm{r}(F'(x))) < 1$, so that $F'(x)$ is strongly regular. Induction using (i) now implies that all $F'(x^l)$ are strongly regular and $x^{l+1} \subseteq x^l$ for $l = 0, 1, 2, \ldots$. In particular, the limit $x^\infty = \lim_{l \to \infty} x^l$ exists. Now $\mathrm{K}(x^\infty) = x^\infty$. Taking midpoints we find $F(\check{x}^\infty) = 0$, so that $\check{x}^\infty = x^*$ is the (unique) zero of F in x. Taking radii we find $\mathrm{r}(F'(x^\infty))\mathrm{rad}(x^\infty) = \mathrm{rad}(x^\infty)$. Since $F'(x^\infty)$ is strongly regular, this implies $\mathrm{rad}(x^\infty) = 0$, hence $x^\infty = x^*$. Now $\mathrm{rad}(x^\infty) = 0$ implies $\mathrm{rad}(F'(x^\infty)) = 0$, hence $\mathrm{r}(F'(x^\infty)) = 0$. Therefore the numbers $\beta_l := \|\mathrm{r}(F'(x^l))\|$ converge to zero. Since

$$\|\mathrm{rad}(x^{l+1})\| = \|\mathrm{r}(F'(x^l))\mathrm{rad}(x^l)\| \leq \beta_l\|\mathrm{rad}(x^l)\|,$$

this implies (24). Moreover, if F' is the interval evaluation of arithmetical expressions which are Lipschitz at x then $\operatorname{rad}(F'(x^l)) = O(\|\operatorname{rad}(x^l)\|)$ by Corollary 2.1.2. Hence

$$\begin{aligned}\beta_l &= \|(\operatorname{mid}(F'(x^l)))^{-1}\operatorname{rad}(F'(x^l))\|\\ &\le \|F'(x^0)^{-1}\|\cdot\|\operatorname{rad}(F'(x^l))\| = O(\|\operatorname{rad}(x^l)\|),\end{aligned}$$

which yields (25). □

The speed-up resulting from the use of a variable Lipschitz matrix like $F'(x)$ must be weighted against the extra work involved, and when $\|I - CA\| \le \frac{1}{2}$, the resulting methods are slower in practice than those with fixed Lipschitz matrix A. Since this inequality will hold when the box is small and $C \approx \check{A}^{-1}$, the above (and other similar) convergence results are of little practical relevance. On the other hand, for most global problems (Section 5.6), variable Lipschitz matrices (or slopes) are essential for success. Also, for large and sparse problems involving M-matrices, the use of variable Lipschitz matrices (or slopes) may noticeably improve speed and limit accuracy in cases where one has to work with $C = I$.

5.3 Set-valued functions

In this section we derive rather general existence results for zeros of continuous parameter-dependent functions which are not necessarily Lipschitz continuous. These weaker assumptions imply, however, that uniqueness results can no longer be proved. Hence the resulting implicit functions are generally set-valued. For the development of the relevant results about such set-valued functions we need some connectivity properties of sets in $\mathbb{R}^n$ which we derive first. (The remainder of this chapter is almost independent of this section; hence one can skip it on first reading.)

$\mathbb{P}(E)$ denotes the set of all subsets of a set E. We say that a set $C \in \mathbb{P}(\mathbb{R}^n)$ *connects* two sets $A, B \in \mathbb{P}(\mathbb{R}^n)$ if there is a *continuum* (i.e. a compact and connected set) $C_0 \subseteq C$ such that $A \cap C_0$ and $B \cap C_0$ are nonempty. For a number $\varepsilon > 0$, we say that a finite sequence $t^0, t^1, \ldots, t^k$ of points of $\mathbb{R}^n$ is an *ε-chain* (*between t^0 and t^k*) if $\|t^l - t^{l-1}\|_\infty \le \varepsilon$ for $l = 1, \ldots, k$.

5.3.1 Lemma Let A and B be closed subsets of $E \subseteq \mathbb{R}^n$.

(i) If the compact set $C \in \mathbb{P}(E)$ connects A and B, then, for every $\varepsilon > 0$, C contains an ε-chain between some point in A and some point in B.

(ii) If E is compact and C_l $(l = 0, 1, 2, \ldots)$ is an infinite sequence of subsets of E such that each C_l connects A and B then the set C of all accumulation points of all sequences t^l $(l = 0, 1, 2, \ldots)$ with $t^l \in C_l$ for $l \ge 0$ connects A and B.

(iii) If $C \in \mathbb{P}(E)$ connects A and B and $\varphi: E \to \mathbb{R}^m$ is continuous then $\varphi(C)$ connects $\varphi(A)$ and $\varphi(B)$.

Proof (i) By assumption we can find a continuum C_0 and two points $\tilde{a} \in A \cap C_0$ and $\tilde{b} \in B \cap C_0$. For any $\varepsilon > 0$, the set $C_\varepsilon(\tilde{a})$ of all points $t \in C_0$ such that C_0 contains an ε-chain between $\tilde{a}$ and t is closed, and $C_0 \backslash C_\varepsilon(\tilde{a})$ is also closed. Since C_0 is connected, this implies that $C_\varepsilon(\tilde{a}) = C_0$. Since $\tilde{b} \in B \cap C_0$, this proves the assertion.

(ii) Let P_l be the set of all pairs $(\tilde{a}, \tilde{b}) \in A \times B$ such that C_l contains some 2^{-l}-chain between $\tilde{a}$ and $\tilde{b}$. Since E is compact and A, B are closed, there is a pair $(\tilde{a}, \tilde{b}) \in A \times B$, a divergent sequence $0 \le i_0 < i_1 < i_2 < \cdots$ of integers and a sequence $(\tilde{a}^l, \tilde{b}^l)$ $(l = 0, 1, 2, \ldots)$ such that

$$(\tilde{a}^l, \tilde{b}^l) \in P_{i_l}, \quad \|\tilde{a}^l - \tilde{a}\|_\infty \le 2^{-l}, \quad \|\tilde{b}^l - \tilde{b}\|_\infty \le 2^{-l}$$

for all $l \ge 0$. By definition of C we have $\tilde{a}, \tilde{b} \in C$. By construction of $(\tilde{a}, \tilde{b})$, the set $C_l^* := C_{i_l} \cup \{\tilde{a}, \tilde{b}\}$ contains some 2^{-l}-chain $\tilde{a} = t(l, 0)$, $t(l, 1), \ldots, t(l, k_l) = \tilde{b}$. Now let C^* be the set of all accumulation points of all sequences $t(l, i_l)$ $(l = 0, 1, 2, \ldots)$ with $i_l \in \{0, \ldots, k_l\}$. Clearly C^* is a closed and hence a compact subset of C.

To show that C^* is connected we suppose that C' and C'' are two disjoint closed subsets of C^* with $C' \cup C'' = C^*$ and $\tilde{a} \in C'$. Since C' and C'' have no common limit point, $\delta := \inf\{\|t' - t''\|_\infty \mid t' \in C', t'' \in C''\} > 0$. Since E is compact, there is a finite subset Δ of E such that the open balls with radius $\varepsilon := \delta/6$ around points of Δ cover E, and w.l.o.g. $\tilde{a}, \tilde{b} \in \Delta$. In particular, for every $t(l, i)$ $(0 \le i \le k_l)$ there is some $s(l, i) \in \Delta$ with $\|t(l, i) - s(l, i)\|_\infty \le \varepsilon$, and we may choose $s(l, 0) = \tilde{a}$ and $s(l, k_l) = \tilde{b}$. In particular, for $l \ge l_0 := \log_2(\varepsilon^{-1})$, the points $s(l, 0), \ldots, s(l, k_l)$ form a 3ε-chain.

Now let $t \in C^*$. By definition of C^* there is a convergent sequence $t(l_\nu, j_\nu)$ $(\nu = 0, 1, 2, \ldots)$ with $l_0 < l_1 < l_2 < \cdots$ and $0 \le j_\nu \le k_{l_\nu}$ whose limit is t. By deleting and/or repeating suitable points of such a 3ε-chain we can obtain 3ε-chains $s(l_\nu, i_{\nu 0}), \ldots, s(l_\nu, i_{\nu k})$ of constant length k (where k is the number of points in Δ) with $i_{\nu 0} = 0$, $i_{\nu k} = j_\nu$. Let $(s^0, t^0, \ldots, s^k, t^k)$ be an accumulation point of the sequence $(s(l_\nu, i_{\nu 0}), t(l_\nu, i_{\nu 0}), \ldots, s(l_\nu, i_{\nu k}), t(l_\nu, i_{\nu k}))$ $(l = 0, 1, 2, \ldots)$. Then $s^0 = t^0 = \tilde{a}$, $t^k = t$, and $s^l \in \Delta$, $t^l \in C^*$, $\|s^l - t^l\|_\infty \le \varepsilon$ for $0 < l \le k$. This implies that $t^0, \ldots, t^k$ is a 5ε-chain in C^*. Since $5\varepsilon < \delta$ and $\tilde{a} \in C'$, the definition of δ implies that $t^l \in C'$ for $l = 0, \ldots, k$. In particular, we find that $t = t^k \in C'$.

Since $t \in C^*$ was arbitrary, this shows that $C' = C^*$ and $C'' = \emptyset$. Therefore C^* is connected. Since $\tilde{a} \in C^* \cap A$, $b^* \in C^* \cap B$ and $C^* \subseteq C$, this implies that C connects A and B.

(iii) This holds since the image of a continuum under a continuous mapping is again a continuum. □

5.3.2 Lemma Let $a, b \in \mathbb{IR}$ and let Σ be a closed subset of $a \times b$. If for every continuous mapping π: $[0, 1] \to a \times b$ with $\pi_2(0) = \underline{b}$, $\pi_2(1) = \overline{b}$ there is a number $s \in [0, 1]$ such that $\pi(s) \in \Sigma$ then Σ connects $\{\underline{a}\} \times \mathbb{R}$ and $\{\overline{a}\} \times \mathbb{R}$.

Proof W.l.o.g. we may assume that $a = b = [0, 1]$. Put $A := \{0\} \times [0, 1]$, $B := \{1\} \times [0, 1]$, $E := [0, 1] \times [0, 1]$.

We fix an arbitrary integer $l > 0$ and write $\langle i, k\rangle$ for the point with coordinates $(i/l, k/l)$ $(i, k \in \mathbb{Z})$. Let Γ be the set consisting of those points $\langle i, k\rangle \in E$ which either have $i = 0$ or satisfy

$$\max(|tl - i|, |sl - k|) \le 2 \quad \text{for some } (s, t) \in \Sigma.$$

It is intuitively clear that there is a piecewise linear path through certain points of Γ which connects A and B. To justify this rigorously we construct a sequence of points of Γ in the following way. Let $u^0 := \langle 0, 0\rangle$, and put $u^1 := \langle 1, 0\rangle$ if $\langle 1, 0\rangle \in \Gamma$, $u^1 := \langle 0, 1\rangle$ otherwise. If $u^j = \langle i, k\rangle$ is already constructed we let u^{j+1} be the first point of Γ which follows u^{j-1} in the sequence $\langle i+1, k\rangle$, $\langle i, k+1\rangle$, $\langle i-1, k\rangle$, $\langle i, k-1\rangle$, $\langle i+1, k\rangle$, $\langle i, k+1\rangle$, $\langle i-1, k\rangle$, $\langle i, k-1\rangle$. (See Fig. 5.1 for an example where the rectangles marked with $\times$ contain some point of Σ, the vertices marked with $\circ$ belong to Γ and the numbers indicate the positions of the u^l.)

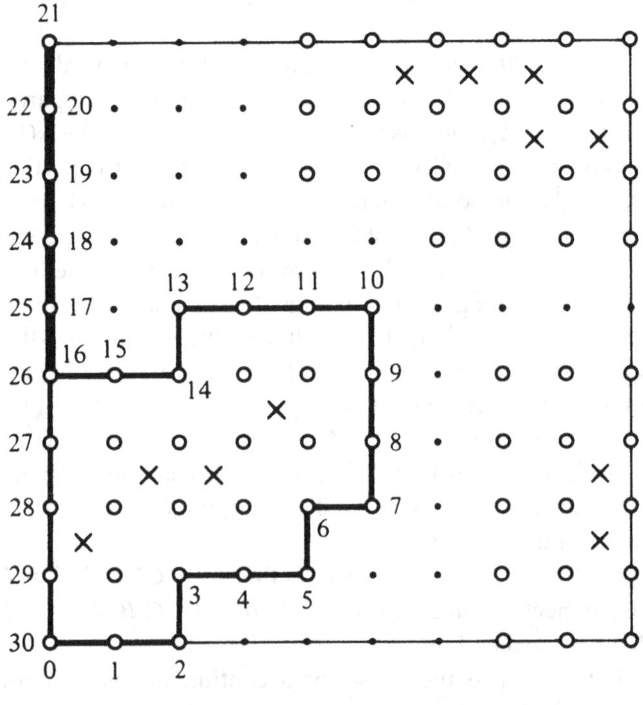

Fig. 5.1. Construction of the path.

Since Γ contains at most $(l+1)^2$ points, there are numbers j and r such that $(u^{j+r}, u^{j+1+r}) = (u^j, u^{j+1})$. The construction now implies that $u^{j-1+r} = u^{j-1}$ and $u^{j+2+r} = u^{j+2}$, so that we have $u^{j+r} = u^j$ for all $j \geq 0$. Now one finds by induction on j that $u^{r-j} = \langle 0, j \rangle$ for $j = 1, \ldots, l$. Hence there is a smallest number q such that $u^q \in [0,1] \times \{1\}$. Let p be the first index such that $u^p \notin [0,1] \times \{0\}$. Then $u^{p-1} \notin \Sigma$ and $u^j \notin [0,1] \times \{0\}$ for $j = p, \ldots, q$. The piecewise linear path formed by the union of the line segments joining u^{j-1} and u^j for $j = p, \ldots, q$ is the image of a continuous mapping π satisfying the hypothesis of our assumption. Hence there is a smallest number j between p and q such that the line segment joining u^{j-1} and u^j intersects Σ. Consideration of the path from u^{j-2} to u^j reveals that u^{j-2}, u^{j-1} and u^j lie on the boundary of E. By construction of p and q this forces $u^j \in \{1\} \times [0,1] = B$. Now let h be the maximal index such that $h < j$ and $u^{h-1} \in \{0\} \times [0,1] = A$. Then the piecewise linear path C_l formed by the union of the line segments joining $u^{\nu-1}$ and u^ν for $\nu = h, \ldots, j$ connects A and B. Moreover, by construction of Γ, for every point $u \in C_l$ there is a point $(t,s) \in \Sigma$ such that $\|u - (t,s)\|_\infty < 2/l$.

Now we apply Lemma 5.3.1(ii) to the sequence $C_1, C_2, C_3, \ldots$ constructed in this way for $l = 1, 2, 3, \ldots$. By construction, every point of C must belong to Σ; hence Σ connects A and B. □

Now let $G: E_0 \subseteq \mathbb{R}^n \to \mathbb{P}(\mathbb{R}^m)$ be a set-valued function. The *graph* of G over $E \subseteq E_0$ is defined as the set

$$\operatorname{graph}(G|E) := \{(t,u) \quad | \quad t \in E, u \in G(t)\}.$$

We call G *connection continuous* or simply c-*continuous* in a set $E \subseteq E_0$ if, for every continuum $C \in \mathbb{P}(E)$ and any $t, t' \in C$, the set $\operatorname{graph}(G|C)$ is compact and connects $\{t\} \times \mathbb{R}^m$ and $\{t'\} \times \mathbb{R}^m$. In particular, we obtain for $C = \{t\}$ that any c-continuous function G has nonempty and compact values $G(t)$. The concept of c-continuity is a natural extension to set-valued functions of the notion of continuity for ordinary functions:

5.3.3 Proposition If $\varphi: E_0 \subseteq \mathbb{R}^n \to \mathbb{R}^m$ is continuous in $E \subseteq E_0$ then the set-valued function $G: E_0 \to \mathbb{P}(\mathbb{R}^m)$ defined by $G(t) := \{\varphi(t)\}$ for $t \in E_0$ is c-continuous in E.

Proof Let $C \subseteq E$ be a continuum and $t, t' \in C$. Since $\operatorname{graph}(G|C)$ is the image of C under the continuous mapping $t \to (t, \varphi(t))$, it is compact and connects $\{(t, \varphi(t))\}$ and $\{(t', \varphi(t'))\}$, hence $\{t\} \times \mathbb{R}^m$ and $\{t'\} \times \mathbb{R}^m$. □

A main tool for our existence theorems is the following generalization of the *intermediate value theorem*.

5.3.4 Theorem Let $a, b \in \mathbb{IR}$ and let $G: a \times b \to \mathbb{P}(\mathbb{R})$ be *c*-continuous in $a \times b$. If

$$\sup(G(\tilde{a}, \underline{b})) \leq 0 \leq \inf(G(\tilde{a}, \overline{b})) \quad \text{for all } \tilde{a} \in a \tag{1}$$

then the set-valued function $H: b \to \mathbb{P}(\mathbb{R})$ defined for $\tilde{a} \in a$ by

$$H(\tilde{a}) := \{\tilde{b} \in b \mid 0 \in G(\tilde{a}, \tilde{b})\}$$

is *c*-continuous in a.

Proof Since the continua in $\mathbb{R}$ are just the closed intervals it is sufficient to show that, for any (closed) subinterval c of a, the set

$$\Sigma := \operatorname{graph}(H|c) = \{(\tilde{c}, \tilde{b}) \in c \times b \mid 0 \in G(\tilde{c}, \tilde{b})\}$$

is compact and connects $\{\underline{c}\} \times \mathbb{R}$ with $\{\overline{c}\} \times \mathbb{R}$. Now Σ is compact since $\operatorname{graph}(G|a \times b)$ is compact. To show that Σ connects $\{\underline{c}\} \times \mathbb{R}$ and $\{\overline{c}\} \times \mathbb{R}$ we verify the hypothesis of Lemma 5.3.2. Let $\pi: [0,1] \to c \times b$ be a continuous mapping satisfying $\pi_2(0) = \underline{b}$, $\pi_2(1) = \overline{b}$. Then $C := \{\pi(s) \mid s \in [0, 1]\}$ is a continuum. Since G is *c*-continuous, $\operatorname{graph}(G|C)$ contains a continuum C' which intersects $\{\pi(0)\} \times \mathbb{R}$ and $\{\pi(1)\} \times \mathbb{R}$. Hence C' contains points of the form $(\pi(0), u_0)$ and $(\pi(1), u_1)$. Since $u_0 \in G(\pi(0))$ and $\pi_2(0) = \underline{b}$, (1) implies $u_0 \leq 0$, and since $u_1 \in G(\pi(1))$ and $\pi_2(1) = \overline{b}$, (1) implies $u_1 \geq 0$. Hence both sets

$$C^+ := \{(t, u) \in C' \mid u \geq 0\}, \quad C^- := \{(t, u) \in C' \mid u \leq 0\}$$

are nonempty. Since C^+ and C^- are closed and cover the connected set C' they must have a nonempty intersection. Hence there is a point $(t, u) \in C^+ \cap C^-$, and this point satisfies $u = 0 \in G(\pi(t))$, i.e. $\pi(t) \in \Sigma$. Now Lemma 5.3.2 applies. □

In order to extend the intermediate value theorem to higher dimensions we need the following result about the composition of *c*-continuous functions.

5.3.5 Proposition Let $G: B \subseteq \mathbb{R}^p \times \mathbb{R}^n \to \mathbb{P}(\mathbb{R}^m)$ be *c*-continuous in $D \times E \subseteq B$. If $H: D_0 \subseteq \mathbb{R}^n \to \mathbb{P}(E)$ is *c*-continuous in $D \subseteq D_0$ then the *composition* $F: D \to \mathbb{P}(\mathbb{R}^m)$ of G and H defined by

$$F(\tilde{x}) = G(\tilde{x}, H(\tilde{x})) := \cup\{G(\tilde{x}, t) \mid t \in H(\tilde{x})\}$$

is *c*-continuous in D.

Proof Let $C \in \mathbb{P}(D)$ be a continuum and $\tilde{y}, \tilde{z} \in C$. Since H is *c*-continuous, $\operatorname{graph}(H|C)$ contains a continuum C' such that C' has some elements of the form $(\tilde{y}, s)$ and $(\tilde{z}, t)$. Since G is *c*-continuous, $\operatorname{graph}(G|C')$ connects $\{(\tilde{y}, s)\} \times \mathbb{R}^m$ and $\{(\tilde{z}, t)\} \times \mathbb{R}^m$. Hence the set

$$\Sigma := \{((\tilde{x}, t), \tilde{a}) \in E \times C \times \mathbb{R}^m \quad | \quad t \in H(\tilde{x}), \tilde{a} \in G(t, \tilde{x})\},$$

which contains graph$(G|C')$, also connects $\{(\tilde{y}, s)\} \times \mathbb{R}^m$ and $\{(\tilde{z}, t)\} \times \mathbb{R}^m$. Since graph$(F|C)$ is the image of Σ under the continuous mapping $((\tilde{x}, t), \tilde{a}) \to (\tilde{x}, \tilde{a})$, Lemma 5.3.1(iii) implies that graph$(F|C)$ connects $\{\tilde{y}\} \times \mathbb{R}^m$ with $\{\tilde{z}\} \times \mathbb{R}^m$.

It remains to show that graph$(F|C)$ is compact. Let $((\tilde{x}^l, t^l), \tilde{a}^l)$ $(l = 0, 1, 2, \ldots)$ be a sequence of elements of Σ. Since graph$(H|C)$ is compact, there is a divergent sequence of integers $l_0 < l_1 < l_2 < \cdots$ such that the sequence (x^{l_ν}, t^{l_ν}) $(\nu = 0, 1, 2, \ldots)$ converges to a point $(\tilde{x}, t)$ with $\tilde{x} \in C, t \in H(C)$. Let $T = \{t\} \cup \{t^{l_\nu} \quad | \quad \nu = 0, 1, 2, \ldots\}$. Then T and hence graph$(G|C \times T)$ are compact. So the sequence $((\tilde{x}^{l_\nu}, t^{l_\nu}), \tilde{a}^{l_\nu})$ $(\nu = 0, 1, 2, \ldots)$ has an accumulation point in graph$(G|C \times T)$ and hence in Σ. Therefore Σ is compact. But this implies that graph$(F|C)$ is also compact. □

5.3.6 Corollary If H: $D_0 \subseteq \mathbb{R}^n \to \mathbb{P}(\mathbb{R}^m)$ is c-continuous in $D \subseteq D_0$ then, for any sequence $I = (i_1, \ldots, i_k)$ of integers $0 < i_1 < i_2 < \cdots < i_k \leq m$, the *projection* H_I: $D_0 \to \mathbb{P}(\mathbb{R}^k)$ defined by

$$H_I(\tilde{x}) := \{\tilde{a} \in \mathbb{R}^k \quad | \quad \tilde{a}_l \in H_{i_l}(\tilde{x}) \quad \text{for } l = 1, \ldots, k\}$$

is c-continuous in D.
Proof Apply Proposition 5.3.5 with $G(x, t) = \{(t_{i_1}, \ldots, t_{i_k})^{\mathrm{T}}\}$. □

We shall use the notation $H_i(\tilde{x})$ for the projection $H_{\{i\}}(\tilde{x})$ to a single coordinate. Then we can formulate the following basic multidimensional existence theorem for set-valued implicit functions.

5.3.7 Theorem Let G: $B \subseteq \mathbb{R}^n \times \mathbb{R}^p \to \mathbb{P}(\mathbb{R}^n)$ be c-continuous in $D \times E \subseteq B$. Let $x \in \mathbb{IR}D$ and suppose that, for $i = 1, \ldots, n$ and all $t \in E$,

$$\sup(G_i(\tilde{x}, t)) \leq 0 \quad \text{if } \tilde{x} \in x \text{ and } \tilde{x}_i = \underline{x}_i,$$
$$\inf(G_i(\tilde{x}, t)) \geq 0 \quad \text{if } \tilde{x} \in x \text{ and } \tilde{x}_i = \overline{x}_i.$$

If E is closed and convex then the set-valued function H: $E \to \mathbb{P}(\mathbb{R}^n)$ defined by

$$H(t) := \{\tilde{x} \in x \quad | \quad 0 \in G(\tilde{x}, t)\}$$

is c-continuous. In particular, $H(t)$ is nonempty for all $t \in E$.

Proof We prove the theorem by induction on n and assume that the assertion holds for some value of n; but for $n = 0$ we assume nothing. To show the validity for $n + 1$ in place of n, let $a \in \mathbb{IR}, x \in \mathbb{IR}^n$ and let $E \in \mathbb{P}(\mathbb{R}^p)$ be closed and convex. Let G: $a \times x \times E \to \mathbb{P}(\mathbb{R} \times \mathbb{R}^n)$ be c-continuous in $a \times x \times E$ and suppose that, for $i = 0, \ldots, n$ and all $t \in E$,

$$\sup(G_i(\tilde{z}, t)) \le 0 \quad \text{if } \tilde{z} \in a \times x \text{ and } \tilde{z}_i = \underline{x}_i, \tag{2}$$
$$\inf(G_i(\tilde{z}, t)) \ge 0 \quad \text{if } \tilde{z} \in a \times x \text{ and } \tilde{z}_i = \overline{x}_i. \tag{3}$$

(Here we refer to the first component of a point of $\mathbb{R} \times \mathbb{R}^n$ by an index 0.) The idea is to eliminate the variable $\tilde{a} = \tilde{z}_0$ from the first 'equation' $0 \in G_0(\tilde{z}, t)$; this reduces the dimension by one, so that we can apply the induction hypothesis.

We show first that the set-valued function $F: x \times E \to \mathbb{P}(\mathbb{R})$ defined by

$$F(\tilde{x}, t) := \{\tilde{a} \in a \mid 0 \in G_0(\tilde{a}, \tilde{x}, t)\}$$

is c-continuous in $x \times E$.

Let C be a continuum in $x \times E$ and let $w = (\tilde{x}, t)$ and $w' = (\tilde{x}', t')$ be two points of C. Let $\varepsilon = 2^{-l}$ and pick an ε-chain $w = w^0, w^1, \ldots, w^n = w'$ in C. We define the piecewise linear function $\pi_l: [0, 1] \to x \times \mathbb{R}^p$ by

$$\pi\left(\frac{i - \sigma}{k}\right) := \sigma w^{i-1} + (1 - \sigma)w^i \quad \text{for } i = 1, \ldots, k \text{ and } 0 \le \sigma \le 1,$$

and put

$$\Pi_l := \{\pi(s) \mid s \in [0, 1]\}.$$

By Proposition 5.3.5, the function $K: a \times [0, 1] \to \mathbb{P}(\mathbb{R})$ defined by

$$K(\tilde{a}, s) := G_0(\tilde{a}, \pi(s))$$

is c-continuous in $a \times [0, 1]$. Now (2) and (3) imply for $i = 0$ that K satisfies the hypothesis of the intermediate value theorem 5.3.4. Hence the mapping $H_l: [0, 1] \to \mathbb{P}(\mathbb{R})$ defined by

$$\begin{aligned} H_l(s) &:= \{\tilde{a} \in a \mid 0 \in K(\tilde{a}, s)\} \\ &= \{\tilde{a} \in a \mid 0 \in G_0(\tilde{a}, \pi(s))\} = F(\pi(s)) \end{aligned}$$

is c-continuous in $[0, 1]$. In particular, $\text{graph}(H_l|[0, 1]) = \{(s, \tilde{a}) \mid \tilde{a} \in F(\pi(s))\}$ connects $\{0\} \times \mathbb{R}$ and $\{1\} \times \mathbb{R}$. Since $\text{graph}(F|\Pi_l)$ is the image under the continuous mapping $(s, \tilde{a}) \to (\tilde{a}, \pi(s))$, Lemma 5.3.1(iii) implies that $\text{graph}(F|\Pi_l)$ connects $\{w\} \times \mathbb{R}$ and $\{w'\} \times \mathbb{R}$.

Now we note that, since C is compact and E is closed and convex, the closed convex hull $\text{cch}(C)$ is a compact subset of $x \times E$, and clearly $\Pi_l \subseteq \text{cch}(C)$. Hence $C_l^* := \text{graph}(F|\Pi_l)$ is a subset of the compact set $\text{graph}(F|\text{cch}(C))$. Since C_l^* connects $\{w\} \times \mathbb{R}$ and $\{w'\} \times \mathbb{R}$, Lemma 5.3.1(ii) implies that the set C^* of all accumulation points of all sequences u^l ($l = 0, 1, 2, \ldots$) with $u^l \in C_l^*$ for $l \ge 0$ connects $\{w\} \times \mathbb{R}$ and $\{w'\} \times \mathbb{R}$. But by construction, each point of C^* belongs to the compact set $\text{graph}(F|C) = \{((\tilde{x}, t), \tilde{a}) \in a \times C \mid 0 \in G_0(\tilde{a}, \tilde{z}, t)\}$. Hence F is c-continuous.

Now Proposition 5.3.5 implies that the function $G^*: x \times E \to \mathbb{P}(\mathbb{R}^n)$ defined by

$$G^*(\tilde{x}, t) := G_N(F(\tilde{x}, t), \tilde{x}, t),$$

where $N = (1, 2, \ldots, n)$, is c-continuous, and (2) and (3) imply that G^* satisfies the induction hypothesis. Hence the function $H^*: E \to \mathbb{P}(\mathbb{R}^n)$ defined by

$$H^*(t) := \{\tilde{x} \in x \quad | \quad 0 \in G^*(\tilde{x}, t)\}$$

is c-continuous.

Now we are in the position to prove that the mapping $H: E \to \mathbb{R} \times \mathbb{R}^n$ defined by

$$H(t) := \{(\tilde{a}, \tilde{x}) \in a \times x \quad | \quad 0 \in G(\tilde{a}, \tilde{x}, t)\}$$

is c-continuous in E. Let $C \in \mathbb{P}(E)$ be a continuum. Since H^* is c-continuous,

$$\begin{aligned} \operatorname{graph}(H^*|C) &= \{(t, \tilde{x}) \quad | \quad 0 \in G^*(\tilde{x}, t)\} \\ &= \{(t, \tilde{x}) \quad | \quad 0 \in G_N(\tilde{a}, \tilde{x}, t) \text{ for some } \tilde{a} \in F(\tilde{x}, t)\} \\ &= \{(t, \tilde{x}) \quad | \quad (\tilde{a}, \tilde{x}, t) \in \operatorname{graph}(H|C) \text{ for some } \tilde{a} \in a\} \end{aligned}$$

connects $\{t\} \times x$ and $\{t'\} \times x$ for all $t, t' \in C$. Hence we can find a continuum $C' \subseteq \operatorname{graph}(H^*|C)$ and two points $\tilde{x}, \tilde{x}' \in x$ such that $(t, \tilde{x})$, $(t', \tilde{x}') \in C'$. Since F is c-continuous, the set

$$C'' := \{((s, \tilde{x}), \tilde{a}) \quad | \quad (s, \tilde{x}) \in C', \tilde{a} \in F(\tilde{x}, s)\} = \operatorname{graph}(F|C')$$

connects $\{(t, \tilde{x})\} \times \mathbb{R}$ and $\{(t', \tilde{x}')\} \times \mathbb{R}$. Hence the set $C''' := \{(s, (\tilde{a}, \tilde{x})) \mid ((s, \tilde{x}), \tilde{a}) \in C''\}$ connects $\{t\} \times \mathbb{R} \times \mathbb{R}^n$ and $\{t'\} \times \mathbb{R} \times \mathbb{R}^n$. Since

$$\begin{aligned} \operatorname{graph}(H|C) &= \{(s, (\tilde{a}, \tilde{x})) \in C \times a \times x \quad | \quad 0 \in G(\tilde{a}, \tilde{x}, s)\} \\ &= \{(s, (\tilde{a}, \tilde{x})) \in C \times a \times x \quad | \quad \tilde{a} \in F(\tilde{x}, t), 0 \in G_N(\tilde{a}, \tilde{x}, \tilde{t})\} \end{aligned}$$

is compact and contains C''', this implies that H is continuous in E. □

5.3.8 Corollary (Miranda) Let $F: D_0 \subseteq \mathbb{R}^n \to \mathbb{R}^n$ be continuous in $x \in \mathbb{I}D_0$. If, for $i = 1, \ldots, n$,

$$\begin{aligned} F_i(\tilde{x}) \leq 0 \quad &\text{for all } \tilde{x} \in x \quad \text{with } \tilde{x}_i = \underline{x}_i, \\ F_i(\tilde{x}) \geq 0 \quad &\text{for all } \tilde{x} \in x \quad \text{with } \tilde{x}_i = \overline{x}_i, \end{aligned}$$

then F has at least one zero $x^* \in x$.

Proof Using Proposition 5.3.3, this follows from Theorem 5.3.7 with $E = \{0\}$ and $G(\tilde{x}, t) = \{F(\tilde{x})\}$. □

From Theorem 5.3.7 we deduce a *global implicit function theorem* for set-valued functions which are 'not too far away' from a nonsingular linear transformation.

5.3.9 Theorem Let E be a closed and convex subset of $\mathbb{R}^p$ and let $G: \mathbb{R}^n \times E \to \mathbb{P}(\mathbb{R}^n)$ be c-continuous in $\mathbb{R}^n \times E$. Suppose that there is a nonsingular matrix $A \in \mathbb{R}^{n \times n}$ and two numbers $\alpha, \beta \in \mathbb{R}$ with $0 \le \alpha$, $0 \le \beta < 1$ such that

$$\tilde{a} \in G(\tilde{x}, t) \Rightarrow \|\tilde{a} - A\tilde{x}\| \le \alpha + \beta\|A\tilde{x}\| \tag{4}$$

in some scaled maximum norm. Then the set-valued function $H: E \to \mathbb{P}(\mathbb{R}^n)$ defined by

$$H(t) := \{\tilde{x} \in \mathbb{R}^n \mid 0 \in G(\tilde{x}, t)\}$$

is c-continuous in E and satisfies

$$\tilde{x} \in H(t) \Rightarrow \|A\tilde{x}\| \le \frac{\alpha}{1-\beta}. \tag{5}$$

Proof Suppose $\|\cdot\| = \|\cdot\|_u$ with $0 < u \in \mathbb{R}^n$. Put $\omega := \alpha/(1-\beta)$ and $z := [-\omega, \omega]u$. Clearly (5) follows from (4) with $\tilde{a} = 0$. Hence $H(t) \subseteq z$ for all $t \in E$. To show c-continuity of H, we define $G^*: z \times E \to \mathbb{P}(\mathbb{R}^n)$ by $G^*(\tilde{z}, t) := G(A \quad \tilde{z}, t)$. If $\tilde{z} \in z$ then $\|\tilde{z}\| \le \omega$ and each $\tilde{a} \in G^*(\tilde{z}, t)$ satisfies $\|\tilde{a} - \tilde{z}\| \le \alpha + \beta\|\tilde{z}\| \le \alpha + \beta\omega = \omega$ by (4). In particular, if $\tilde{z}_i = \underline{z}_i$ then $|\tilde{a}_i + \omega u_i| = |\tilde{a}_i - \tilde{z}_i| \le \|\tilde{a} - \tilde{z}\|u_i \le \omega u_i$ so that $\tilde{a}_i \le 0$. And if $\tilde{z}_i = \overline{z}_i$ then similarly $|\tilde{a}_i - \omega u_i| = |\tilde{a}_i - \tilde{z}_i| \le \omega u_i$ so that $\tilde{a}_i \ge 0$. Hence G^* satisfies the hypothesis of Theorem 5.3.7. Therefore the function $H^*: E \to \mathbb{P}(\mathbb{R}^n)$ defined by $H^*(t) = \{\tilde{z} \in z \mid 0 \in G^*(\tilde{z}, t)\}$ is c-continuous in E. Composition with the linear mapping $\tilde{z} \to A\tilde{z}$ gives H and shows that H is also c-continuous. □

An immediate consequence is the following *global inverse function theorem*:

5.3.10 Theorem Let $F: \mathbb{R}^n \to \mathbb{P}(\mathbb{R}^n)$ be c-continuous. Suppose that there is a nonsingular matrix $A \in \mathbb{R}^{n \times n}$ and two numbers $\alpha, \beta \in \mathbb{R}$ with $0 \le \alpha$, $0 \le \beta < 1$ such that

$$\tilde{a} \in F(\tilde{x}) \Rightarrow \|\tilde{a} - A\tilde{x}\| \le \alpha + \beta\|A\tilde{x}\| \tag{6}$$

in some scaled maximum norm. Then the set-valued function $H: \mathbb{R}^n \to \mathbb{P}(\mathbb{R}^n)$ defined by

$$H(\tilde{a}) := \{\tilde{x} \in \mathbb{R}^n \mid \tilde{a} \in F(\tilde{x})\}$$

is c-continuous in $\mathbb{R}^n$. In particular, $H(\tilde{a})$ is nonempty for every $\tilde{a} \in \mathbb{R}^n$.

Proof Apply Theorem 5.3.9 with $G(\tilde{x}, t) := F(\tilde{x}) - t$, $E := [-\gamma, \gamma]u$ and $\alpha + \gamma$ in place of α. We find that H is c-continuous in E. Since γ is arbitrary this shows that H is c-continuous in $\mathbb{R}^n$. □

5.3.11 Proposition Let $F: D_0 \subseteq \mathbb{R}^n \to \mathbb{P}(\mathbb{R}^n)$ be c-continuous in a continuum $D \subseteq D_0$ and suppose that $0 \in D$. Then the set-valued function $H: \mathbb{R}^n \to \mathbb{P}(\mathbb{R}^n)$ defined by

$$H(\tilde{a}) := \{\tilde{x} \in D \mid \tilde{a} \in F(\tilde{x})\}$$

is c-continuous in

$$D^* := D \backslash \{t\tilde{x} + (1-t)\tilde{a} \mid (\tilde{x}, \tilde{a}) \in \operatorname{graph}(F|\partial D), 0 < t \leq 1\}.$$

Proof We first show that the set-valued function $G: \mathbb{R}^n \to \mathbb{P}(\mathbb{R}^n)$ defined by

$$G(\tilde{x}) := \begin{cases} F(\tilde{x}) & \text{if } \tilde{x} \in \operatorname{int}(D) \\ \{\tilde{x}\} & \text{if } \tilde{x} \notin D \\ \{t\tilde{x} + (1-t)\tilde{a} \mid \tilde{a} \in F(\tilde{x}), t \in [0, 1]\} & \text{if } \tilde{x} \in \partial D \end{cases}$$

is c-continuous in $\mathbb{R}^n$. To see this, let C be a continuum in $\mathbb{R}^n$ and $\tilde{x}, \tilde{x}' \in C$. Fix an integer $l \geq 0$ and define $c_l := \{\tilde{x} + \tilde{c} \mid \tilde{x} \in C, \tilde{e} \in \mathbb{R}^n, \|\tilde{e}\| \leq 2^{-l}\}$. We choose a continuous mapping $\pi_l: [0, 1] \to c_l$ such that $\pi_l(0) = \tilde{x}$, $\pi_l(1) = \tilde{x}'$; such a mapping can be obtained, e.g., by taking a piecewise linear path along a 2^{-l}-chain between $\tilde{x}$ and $\tilde{x}'$. The set $\{t \in [0, 1] \mid \pi_l(t) \in D\}$ is a closed subset of $[0, 1]$, hence a union of mutually disjoint closed intervals $[\underline{t}_{k,l}, \bar{t}_{k,l}]$ $(k \in K_l)$. Since $C_{k,l} := \pi([\underline{t}_{k,l}, \bar{t}_{k,l}])$ is a continuum in D, $\operatorname{graph}(F|C_{k,l})$ contains a continuum $C'_{k,l}$ which intersects $\{\pi_l(\underline{t}_{k,l})\} \times \mathbb{R}^n$ and $\{\pi_l(\bar{t}_{k,l})\} \times \mathbb{R}^n$. Now one easily checks that $C''_l := \operatorname{graph}(G|\pi_l([0, 1])) \cup \bigcup_{k \in K_l} C'_{k,l}$ is a continuum intersecting $\{\pi_l(0)\} \times \mathbb{R}^n$ and $\{\pi_l(1)\} \times \mathbb{R}^n$. Since $C''_l \subseteq \operatorname{graph}(G|C_l)$ this shows that $C'_l := \operatorname{graph}(G|C_l)$ connects $\{\tilde{x}\} \times \mathbb{R}^n$ and $\{\tilde{x}'\} \times \mathbb{R}^n$. Now C'_0 is a closed subset of the bounded set

$$\operatorname{cch}(\operatorname{graph}(G|C_0 \cap D)) \cup \{(\tilde{x}, \tilde{x}) \mid \tilde{x} \in C_0\},$$

hence C'_0 is compact. Since the sets C'_l $(l = 0, 1, 2, \ldots)$ are subsets of C'_0, Lemma 5.3.1(ii) implies that $\operatorname{graph}(G|C)$ connects $\{\tilde{x}\} \times \mathbb{R}^n$ and $\{\tilde{x}'\} \times \mathbb{R}^n$. Hence G is c-continuous in $\mathbb{R}^n$.

Now we apply Theorem 5.3.10 with G in place of F and $A = I$, $\beta = 0$ and $\alpha = \sup\{\|\tilde{x} - \tilde{a}\| \mid (\tilde{x}, \tilde{a}) \in \operatorname{graph}(F|D)\}$. We find that the mapping $H^*: \mathbb{R}^n \to \mathbb{P}(\mathbb{R}^n)$ defined by $H^*(\tilde{a}) := \{\tilde{x} \in \mathbb{R}^n \mid \tilde{a} \in G(\tilde{x})\}$ is c-continuous in $\mathbb{R}^n$ and hence in D^*. But if $\tilde{a} \in D^* \cap G(\tilde{x})$ then $\tilde{a} \in F(\tilde{x})$ by construction of G and D^*; hence H^* agrees on D^* with H, and the assertion follows. □

As a consequence we obtain the following version of an existence theorem of Leray and Schauder, and from it the famous fixed point theorem of Brouwer.

5.3.12 Theorem (Leray & Schauder) Let $F: D_0 \subseteq \mathbb{R}^n \to \mathbb{R}^n$ be continuous in the continuum $D \subseteq D_0$. If $0 \in D$ and if

$$\tilde{x} \in \partial D, \quad t \in \mathbb{R}, \quad F(\tilde{x}) = t\tilde{x} \Rightarrow t \geq 0 \tag{7}$$

then F has at least one zero $x^* \in D$.

Proof Apply Proposition 5.3.11 with $\tilde{a} = 0$, using Proposition 5.3.3. □

5.3.13 Theorem (Brouwer) Let D be a convex and compact subset of $\mathbb{R}^n$ with $\operatorname{int}(D) \neq \emptyset$. Then every continuous mapping $G: D \to D$ has at least one fixed point $x^* \in D$, i.e. a point with $x^* = G(x^*)$.

Proof Let $z \in \operatorname{int}(D)$; then $D' := \{\tilde{z} - z \mid \tilde{z} \in D\}$ is a continuum with $0 \in D'$. We verify (7) with D' in place of D and $F(\tilde{x}) = z + \tilde{x} - G(z + \tilde{x})$. If $\tilde{x} \in \partial D'$, $t \in \mathbb{R}$ and $F(\tilde{x}) = t\tilde{x}$ then $tz + (1 - t)(z + \tilde{x}) = G(z + \tilde{x}) \in D$. But since $z \in \operatorname{int}(D)$ and $z + \tilde{x} \in \partial D$, this forces $t \geq 0$. Now Theorem 5.3.12 applies. □

A slight generalization of the last argument gives the following set-valued analog:

5.3.14 Theorem Let D be a convex and compact subset of $\mathbb{R}^n$ with $\operatorname{int}(D) \neq \emptyset$ and let $G: D \to \mathbb{P}(D)$ be c-continuous. Then there is some point $x^* \in D$ with $x^* \in G(x^*)$.

Proof By Proposition 5.3.5, the set-valued function $F: D \to \mathbb{P}(\mathbb{R}^n)$ defined by $F(\tilde{x}) = \{\tilde{x} - \tilde{a} \mid \tilde{a} \in G(\tilde{x})\}$ is c-continuous in D. Since $0 \in \operatorname{int}(D)$, Proposition 5.3.11 applies. Now $G(\tilde{x}) \subseteq D$ implies that $0 \in D^*$ so that $H(0)$ is a nonempty subset of D. But any $x^* \in H(0)$ satisfies $x^* \in G(x^*)$. □

We end this section by giving a simple sufficient condition for c-continuity of set-valued functions.

5.3.15 Proposition A set-valued function $G: E_0 \subseteq \mathbb{R}^n \to \mathbb{P}(\mathbb{R}^m)$ is c-continuous in every subset E of E_0 with the following two properties:

(C1) for all $t \in E$, the set $G(t)$ is a continuum;
(C2) for every continuum $C \in \mathbb{P}(E)$, $\operatorname{graph}(G|C)$ is compact.

Proof Let $C \in \mathbb{P}(E)$ be a continuum. We show that $\operatorname{graph}(G|C)$ is connected. To see this, let A_1', A_2' be disjoint closed subsets of $C \times \mathbb{R}^m$ with $A_1' \cup A_2' = \operatorname{graph}(G|C)$. If $A_i' \neq \emptyset$ for $i = 1, 2$ then the sets

$$A_i := \{t \in C \mid (t, u) \in A_i' \text{ for some } u \in \mathbb{R}^m\} \quad (i = 1, 2)$$

are nonempty closed subsets of C satisfying $A_1 \cup A_2 = C$. Since C is connected, A_1 and A_2 must have a common point t. But for any $t \in A_1 \cap A_2$ the sets $E_i := \{u \in G(t) \mid (t, u) \in A_i'\}$ $(i = 1, 2)$ are nonempty, closed subsets of $G(t)$ with $G(t) = E_1 \cup E_2$ and $E_1 \cap E_2 = \emptyset$, contradicting the connectedness of $G(t)$. Hence A_1' or A_2' must be empty. Therefore graph$(G|C)$ is connected. Since $G(t) \neq \emptyset$ for all $t \in E$, graph$(G|C)$ connects $\{t\} \times \mathbb{R}^n$ and $\{t'\} \times \mathbb{R}^n$ for any $t, t' \in C$. Since graph$(G|C)$ is compact by hypothesis, G is c-continuous on E. □

5.4 Zeros of continuous functions

Based on a preconditioned version of the Leray–Schauder theorem we show in this section that the use of slopes in place of Lipschitz sets improves the enclosure and iteration methods for zeros discussed in Sections 5.1 and 5.2 in a similar way as the centered forms are sharpened when slopes are used in place of derivatives.

5.4.1 Theorem Let $F: D_0 \subseteq \mathbb{R}^n \to \mathbb{R}^n$ be continuous in $x \in \mathbb{I}D_0$. Let $\check{x} \in x$ and let $C \in \mathbb{R}^{n \times n}$ be a nonsingular matrix such that the implication

$$\tilde{y} \in \partial x, \quad t \in \mathbb{R}, \quad CF(\tilde{y}) = t(\tilde{y} - \check{x}) \Rightarrow t \geq 0 \tag{1}$$

holds. Then F has at least one zero $x^* \in x$.

Proof This follows directly from the Leray–Schauder theorem 5.3.12 with $D = x - \check{x}$ applied to the function $\tilde{z} \to CF(\check{x} + \tilde{z})$ in place of F. We give a second proof which assumes $\check{x} \in \text{int}(x)$ but uses the fixed point theorem of Brouwer (see Theorem 5.3.13, but many other proofs are known).

Put $d := x - \check{x}$. Since $0 \in \text{int}(d)$, there is a number $\varepsilon > 0$ such that d contains all $\tilde{d} \in \mathbb{R}^n$ with $\|\tilde{d}\| \leq \varepsilon$. Hence, for all $\tilde{a} \in \mathbb{R}^n$, the set $T(\tilde{a}) := \{t \geq 0 \mid \tilde{a} \in td\}$ contains $t = \varepsilon^{-1}\|\tilde{a}\|$. Therefore $\tau(\tilde{a}) := \inf T(\tilde{a}) \in T(\tilde{a})$, and we have

$$\tau(\tilde{a}) \leq \varepsilon^{-1}\|\tilde{a}\|. \tag{2}$$

If $s \in T(\tilde{a} - \tilde{b})$ and $t \in T(\tilde{b})$ then $s + t \in T(\tilde{a})$ since $sd + td = (s + t)d$ for $0 \leq s, t \in \mathbb{R}$; hence $\tau(\tilde{a}) \leq \tau(\tilde{a} - \tilde{b}) + \tau(\tilde{b})$. By (2) and symmetry, this implies $|\tau(\tilde{a}) - \tau(\tilde{b})| \leq \varepsilon^{-1}|\tilde{a} - \tilde{b}|$. Therefore τ is continuous.

Now suppose that F has no zero $x^* \in x$. Then $\tau(CF(\tilde{y})) > 0$ for all $\tilde{y} \in x$. Hence the mapping $\Phi: D \to \mathbb{R}^n$ defined for $D = x$ by $\Phi(\tilde{y}) := \check{x} + CF(\tilde{y})/\tau(CF(\tilde{y}))$ is continuous and maps D into itself. So Φ has some fixed point $x^* = \Phi(x^*)$ in D. But then the choice $\tilde{y} = x^*$ and $t = -\tau(CF(\tilde{y}))$ contradicts (1). □

Now let D_0 be a subset of $\mathbb{R}^n$ and let $x, y \in \mathbb{I}D_0$. We say that a closed, convex and bounded set A of $m \times n$ matrices is a *slope set* at $[x, y]$ for the function $F: D_0 \to \mathbb{R}^m$ if for every $\tilde{x} \in x$ and $\tilde{y} \in y$ we have

$$F(\tilde{y}) = F(\tilde{x}) + \tilde{A}(\tilde{y} - \tilde{x}) \quad \text{for some } \tilde{A} \in A; \tag{3}$$

A is called a *slope matrix* if, in addition, $A \in \mathbb{IR}^{m\times n}$.

The concept of a slope set generalizes that of a Lipschitz set since a slope set at $[x, x]$ is the same as a Lipschitz set on $x \in \mathbb{I}D_0$. In practice, slope matrices can easily be computed by using the inclusion algebra of Theorem 2.3.8 for a row-wise construction of A. As shown in Proposition 2.3.12, the resulting slope matrix at $[\tilde{x}, x]$ is always contained in the Lipschitz matrix computed as a slope matrix at $[x, x]$ and has roughly a halved radius. Therefore the following variation of Theorem 5.1.7 has a weaker hypothesis and gives sharper enclosures than Theorem 5.1.7 itself.

5.4.2 Theorem Let $F: D_0 \subseteq \mathbb{R}^n \to \mathbb{R}^n$ be continuous in $x \in \mathbb{I}D_0$. Let $\tilde{x} \in x$ and suppose that A is a regular slope set at $[\tilde{x}, x]$ for F. Then every $x' \in \mathbb{IR}^n$ containing

$$\mathrm{N}(x, \tilde{x}) := \tilde{x} - A^{\mathrm{H}}F(\tilde{x}) \tag{4}$$

has the following three properties:

(i) Every zero $x^* \in x$ of F satisfies $x^* \in x'$.
(ii) If $x' \cap x = \emptyset$ then F contains no zero in x.
(iii) If $x' \subseteq \mathrm{int}(x)$ then F contains at least one zero in x (and hence in x').

Proof Let $x^* \in x$ be a zero of F. Putting $\tilde{y} = x^*$ in (3) gives $x^* = \tilde{x} - \tilde{A}^{-1}F(\tilde{x}) \in \mathrm{N}(x, \tilde{x}) \subseteq x'$. Therefore (i) holds, and (ii) is an immediate consequence. To prove (iii), it suffices to verify the hypothesis of the preconditioned Leray–Schauder theorem 5.4.1 for $D = x$ and $C = \tilde{A}_0^{-1}$, where $\tilde{A}_0 \in A$ is arbitrary. Suppose that $\tilde{y} \in \partial D$ and $CF(\tilde{y}) = t(\tilde{y} - \tilde{x})$ for some $t \in \mathbb{R}$ with $t < 0$. Then $s = 1/(1 - t)$ satisfies

$$0 < s < 1, \quad 1 - s = -st. \tag{5}$$

Combining (3), (5) and the relation $F(\tilde{y}) = t\tilde{A}_0(\tilde{y} - \tilde{x})$ we get

$$(s\tilde{A} + (1 - s)\tilde{A}_0)(\tilde{y} - \tilde{x}) = s(F(\tilde{y}) - F(\tilde{x})) + st\tilde{A}_0(\tilde{y} - \tilde{x}) = -sF(\tilde{x}).$$

Hence

$$\tilde{y} = \tilde{x} - s\tilde{A}_s^{-1}F(\tilde{x}), \tag{6}$$

where $\tilde{A}_s := s\tilde{A} + (1 - s)\tilde{A}_0 \in A$. But $-\tilde{A}_s^{-1}F(\tilde{x}) \in -A^{\mathrm{H}}F(\tilde{x}) \subseteq \mathrm{N}(x, \tilde{x}) - \tilde{x} \subseteq x' - \tilde{x}$, hence (6) and (5) imply $\tilde{y} \in \tilde{x} + s(x - \tilde{x}) \subseteq \mathrm{int}(x)$. But this contradicts the assumption $\tilde{y} \in \partial D$. Hence the hypothesis of Theorem 5.4.1 holds, giving (iii). □

Thus the Newton operator $N(x, \tilde{x})$ defined with a slope set has almost the same properties as when defined with a Lipschitz set; the only difference is that we lose uniqueness of a zero in the case of (iii). Now exactly the same arguments which prove Theorem 5.1.8 show that one can replace the Lipschitz matrices in the Krawczyk operator $K(x, \tilde{x})$ and the Hansen–Sengupta operator $H(x, \tilde{x})$ by slope matrices without affecting the enclosure and existence properties:

5.4.3 Corollary Let $F: D_0 \subseteq \mathbb{R}^n \to \mathbb{R}^n$ be continuous in $x \in \mathbb{I}D_0$. Let $\tilde{x} \in x$ and suppose that $A \in \mathbb{IR}^{n \times n}$ is a slope matrix at $[\tilde{x}, x]$ for F. If x' denotes either $H(x, \tilde{x})$ or $K(x, \tilde{x})$ then:

(i)   Every zero $x^* \in x$ of F satisfies $x^* \in x'$.
(ii) If $x' \cap x = \emptyset$ then F contains no zero in x.
(iii) If $\emptyset \neq x' \subseteq \text{int}(x)$ then A is strongly regular and F contains at least one zero in x (and hence in x'). $\qquad\qquad\qquad\qquad\qquad\qquad$ $\square$

5.5 Local analysis of parameter-dependent nonlinear systems

In this section we consider the sensitivity of solutions of the equation

$$G(\tilde{x}, t) = 0 \qquad\qquad (1)$$

in a neighborhood of an approximate zero $(\tilde{z}, t_0) \in B$, where $G: B \subseteq \mathbb{R}^n \times \mathbb{R}^p \to \mathbb{R}^n$ is a Lipschitz continuous function. Under natural assumptions we show how to construct a narrow box $x \in \mathbb{IR}^n$ with $\tilde{z} \in x$ such that (1) has a unique solution $\tilde{x} = H(t) \in x$ for each t from some given neighborhood E of t_0. We show how any such initial enclosure of the *solution set*

$$\Sigma = \{H(t) \mid t \in E\} \qquad\qquad (2)$$

can be used to construct a better enclosure of Σ which has the quadratic approximation property. Thus interval analysis provides a rigorous and realistic *sensitivity analysis* of the solutions of (1) under perturbations of specified magnitude (i.e. not only asymptotically for 'sufficiently small' perturbations).

Throughout this section we assume that $G: B \subseteq \mathbb{R}^n \times \mathbb{R}^p \to \mathbb{R}^n$ is continuous, $D \subseteq \mathbb{R}^n$ and $E \subseteq \mathbb{R}^p$ are closed and connected sets with $D \times E \subseteq B$, and $A: \mathbb{I}D \to \mathbb{IR}^{n \times n}$ is an interval function such that for all $x \in \mathbb{I}D$, all $\tilde{x}, \tilde{z} \in x$ and all $t \in E$ we have

$$G(\tilde{z}, t) - G(\tilde{x}, t) = \bar{A}(\tilde{z} - \tilde{x}) \quad \text{for some } \bar{A} \in A(x). \qquad (3)$$

We also use a preconditioning matrix $C \in \mathbb{R}^{n \times n}$.

5.5.1 Theorem Let $\tilde{z} \in x \in \mathbb{I}D$ and suppose that

$$(CA(x) - I)(x - \tilde{z}) \subseteq d, \tag{4}$$
$$g^* := \Box\{CG(\tilde{z}, t) \mid t \in E\} \subseteq g, \tag{5}$$

with $d, g \in \mathbb{IR}^n$. Then any of the conditions

$$\tilde{z} - g - d \subseteq \operatorname{int}(x), \tag{6a}$$
$$\emptyset \neq \Gamma(CA(x), g, \tilde{z} - x) \subseteq \operatorname{int}(\tilde{z} - x), \tag{6b}$$
$$(CA(x))^{\mathrm{H}} g \subseteq \tilde{z} - x \quad \text{(where } CA(x) \text{ is assumed regular)} \tag{6c}$$

implies that, for all $t \in E$, (1) has a unique solution $\tilde{x} = H(t) \in x$, and the function H: $E \to \mathbb{R}^n$ defined in this way is continuous. Moreover, the box

$$x' := (\tilde{z} - g - d) \cap x$$

is an enclosure of the solution set (2) satisfying

$$0 \leq \operatorname{rad}(x') - \operatorname{rad}(\Box\Sigma) \leq 2 \cdot \operatorname{rad}(d) + \operatorname{rad}(g) - \operatorname{rad}(g^*). \tag{7}$$

Proof By Theorem 4.4.5 and Corollary 4.4.7, the conditions (6a) and (6b) imply that $CA(x)$ is regular and (6c) holds. By Theorem 5.4.2(iii), (6c) implies that, for each $t \in E$, the equation (1) has a solution $\tilde{x} \in x$. If $\tilde{y} \in x$ also satisfies $G(\tilde{y}, t) = 0$ then (3) shows that $\tilde{y} = \tilde{x}$ so that $\tilde{x}$ is uniquely determined by t. Hence H is well-defined, and continuity of H follows from the implicit function theorem 5.1.3. To show (7) we write $x^* := \Box\Sigma$ and put $\tilde{x} = H(t)$ and $\tilde{y} = \tilde{z} - CG(\tilde{z}, t)$. Then, with $\tilde{A}$ from (3), we have

$$0 = CG(\tilde{x}, t) = C(G(\tilde{z}, t) + \tilde{A}(\tilde{x} - \tilde{z})) = \tilde{z} - \tilde{y} + C\tilde{A}(\tilde{x} - \tilde{z}),$$

so that

$$\tilde{y} - \tilde{x} = (C\tilde{A} - I)(\tilde{x} - \tilde{z}) \in (CA(x) - I)(x - \tilde{z}) \subseteq d.$$

Therefore $\tilde{x} \subseteq \tilde{y} - d \subseteq \tilde{z} - g - d$, giving

$$\Box\Sigma \subseteq (\tilde{z} - g - d) \cap x = x',$$

and $\tilde{y} \subseteq \tilde{x} + d \subseteq \Box\Sigma + d$, giving

$$\tilde{z} - g^* \subseteq \Box\Sigma + d.$$

Taking radii we find $\operatorname{rad}(g^*) \leq \operatorname{rad}(\Box\Sigma) + \operatorname{rad}(d)$, so that

$$\begin{aligned} 0 &\leq \operatorname{rad}(x') - \operatorname{rad}(\Box\Sigma) \\ &\leq (\operatorname{rad}(g) + \operatorname{rad}(d)) - (\operatorname{rad}(g^*) - \operatorname{rad}(d)), \end{aligned}$$

which implies (7). □

Remark More generally, any $x' \in \mathbb{IR}^n$ with $\Box\Sigma \subseteq x' \subseteq x$ satisfies

$$0 \leq \operatorname{rad}(x) - \operatorname{rad}(\Box\Sigma) \leq \operatorname{rad}(x') + \operatorname{rad}(d) - \operatorname{rad}(g^*). \tag{7a}$$

When $\tilde{z}$ is a sufficiently good approximation to a solution of (1) then, usually, the following result provides a narrow box x which contains the solution set.

5.5.2 Proposition Under the assumptions of Theorem 5.5.1, let $0 < u \in \mathbb{R}^n$ and $1 \leq \kappa \in \mathbb{R}$. If the box

$$x := \tilde{z} - g + \kappa\|g\|_u[-u, u] \tag{8}$$

satisfies

$$\|CA(x) - I\| = \beta < \kappa/(\kappa + 1) \tag{9}$$

then both x and

$$x' := \tilde{z} - g + \frac{\beta}{1-\beta}\|g\|_u[-u, u] \tag{8a}$$

contain the solution set (2), and

$$0 \leq \text{rad}(x') - \text{rad}(\Box\Sigma) \leq \frac{2\beta}{1-\beta}\|g\|u + \text{rad}(g) - \text{rad}(g^*). \tag{10}$$

Proof Put $\alpha := \|g\|$. If $\alpha = 0$ then $g = 0$; since C is nonsingular by (9), we have $G(\tilde{z}, t) = 0$ for all $t \in E$, and the assertion trivially holds. Hence suppose that $\alpha > 0$. Then (8) implies $\tilde{z} \in x$ and $\|x - \tilde{z}\|_u \leq (1 + \kappa)\alpha$. Therefore

$$\|(CA(x) - I)(x - \tilde{z})\|_u \leq \beta\|x - \tilde{z}\|_u < \kappa\alpha,$$

so that

$$(CA(x) - I)(x - \tilde{z}) \subseteq \text{int}(\kappa\alpha[-u, u])$$

Hence (4) and (6a) hold with $d = \kappa\alpha[-u, u]$, and Theorem 5.5.1 applies. □

If G is in D continuously differentiable with respect to the first n variables then (3) and (5) can be satisfied with interval extensions of G and $\partial_1 G$ by choosing

$$g := CG(\tilde{z}, e), \quad A(x) := \partial_1 G(x, e),$$

where $e \in \mathbb{IR}^p$ is an interval enclosure of E. In particular, suppose that (x^*, t^*) is a regular zero of (1), i.e.

$$G(x^*, t^*) = 0, \quad \text{rank}\, \partial_1 G(x^*, t^*) = n.$$

If $\partial_1 G$ is given by Lipschitz arithmetical expressions then, for sufficiently narrow $x \ni x^*$ and $e \ni t^*$, the Lipschitz matrix $A(x)$ will have arbitrarily small radius. Since $A(x)$ contains the nonsingular matrix $\partial_1 G(x^*, t^*)$, this implies that the number β in (7) can be made arbitrarily small by choosing C as the midpoint inverse of $A(x)$. Therefore there are neighborhoods U of x^*

and V of t^* such that for every $\tilde{z} \in U$, and every $e \in \mathbb{I}V$, the hypothesis of Proposition 5.5.2 can be satisfied. This means that the proposition is indeed a useful tool for finding initial enclosures of the solution set of nonlinear systems of equations with parameters varying in narrow intervals. The most natural choice of the parameter κ is $\kappa = 1$; this still allows fairly large values of $\beta(<\frac{1}{2})$ in (7).

If we want to make use of the overestimation bounds (7) and (10) it is necessary to have a lower bound for $\text{rad}(g^*)$. Such a bound can be obtained when g is determined by a centered form for $CG(\tilde{x}, t)$, since then Theorem 2.3.3 provides a computable upper bound for $\text{rad}(g) - \text{rad}(g^*)$. Thus (7) and (10) yield a computable overestimation bound for the solution set. In particular, if $E = e \in \mathbb{IR}^p$ is a narrow interval with radius $\text{rad}(e) = O(\varepsilon)$ and g is computed from Lipschitz expressions for G using either the mean value form or the inclusion algebra LI_p (cf. relation (2.3.26)) then we will have $\text{rad}(g) = O(\varepsilon)$ and $\text{rad}(g) - \text{rad}(g^*) = O(\varepsilon^2)$. Since $\beta = O(\varepsilon)$, the estimate (10) shows that the enclosure (8a) of $\square\Sigma$ has the quadratic approximation property.

5.6 Global problems

In this section we consider constructive methods for finding narrow enclosures for *all* solutions $x^* \in D$ of $F(x^*) = 0$, where F: $D_0 \subseteq \mathbb{R}^n \to \mathbb{R}^n$ is continuous (e.g. given by arithmetic expressions). The methods considered so far were constructive only in the case when D was a box and the slope or Lipschitz matrix of F at this box was regular. If we drop these assumptions we have to cope with two difficulties: In order to be able to work with intervals we somehow have to reduce the problem to one with bounded domain; on the other hand we must take care of the possibility that several solutions exist. In the following we shall handle the first problem using *a priori* bounds or suitable problem transformations, and the second using bisection.

Methods to reduce an unbounded problem to one with a bounded domain depend on a previous inspection of the problem 'by hand', since some global information on the unbounded problem must be obtained which allows to show that there are no solutions with large norm. We discuss a few theorems which provide this conclusion under conditions which can often be verified by hand; such conditions generally assume that F is defined for all arguments $\tilde{x} \in \mathbb{R}^n$.

For certain functions with diagonal nonlinearities only, global uniqueness and bounds for the zero can be shown. A function Φ: $\mathbb{R}^n \to \mathbb{R}^n$ is called *diagonal* if each component $\Phi_i(\xi)$ depends on ξ_i only, and *isotone* if $\tilde{x} \geq \tilde{y}$ implies $\Phi(\tilde{x}) \geq \Phi(\tilde{y})$.

5.6.1 Theorem Let $A \in \mathbb{R}^{n\times n}$ be an H-matrix with positive diagonal elements and let $\Phi: \mathbb{R}^n \to \mathbb{R}^n$ be continuous, diagonal and isotone. Then the function $F: \mathbb{R}^n \to \mathbb{R}^n$ defined by $F(\tilde{x}) := A\tilde{x} + \Phi(\tilde{x})$ has a unique zero $x^* \in \mathbb{R}^n$. Moreover, the inequality

$$|x^* - \tilde{x}| \le w \tag{1}$$

holds for every nonnegative vector $w \in \mathbb{R}^n$ satisfying

$$\langle A\rangle w \ge |F(\tilde{x})|. \tag{2}$$

Proof Let $A = D + E$ be the Jacobi splitting of A into diagonal and off-diagonal parts. For all $\tilde{z} \in \mathbb{R}^n$, the equation

$$Dz + \Phi(z) = -E\tilde{z}$$

has a unique solution $z = \Psi(\tilde{z})$, since each component $D_{ii}z_i + \Phi_i(z)$ of the left-hand side is a continuous, strictly isotone function of z_i alone. The function $\Psi: \mathbb{R}^n \to \mathbb{R}^n$ defined in this way satisfies the Lipschitz condition

$$|\Psi(\tilde{y}) - \Psi(\tilde{z})| \le D^{-1}|E|\,|\tilde{y} - \tilde{z}| \tag{3}$$

for all $\tilde{y}, \tilde{z} \in \mathbb{R}^n$. Indeed, let $y = \Psi(\tilde{y})$ and $z = \Psi(\tilde{z})$. Since Φ is diagonal and isotone we have $\Delta\Phi(y) \ge \Delta\Phi(z)$, where Δ is a diagonal matrix with $|\Delta| = I$ and $\Delta(y - z) \ge 0$. Hence

$$\begin{aligned} D|y - z| &= \Delta D(y - z) = \Delta(-E\tilde{y} - \Phi(y) + E\tilde{z} + \Phi(z)) \\ &\le \Delta E(\tilde{z} - \tilde{y}) \le |E|\,|\tilde{z} - \tilde{y}|, \end{aligned}$$

which implies (3).

We now show that for every nonnegative vector $w \in \mathbb{R}^n$ satisfying (2), Ψ maps the box $[\tilde{x} - w, \tilde{x} + w]$ into itself. Indeed, let $\tilde{z} \in [\tilde{x} - w, \tilde{x} + w]$ and $z = \Psi(\tilde{z})$. As before we have $\Delta\Phi(z) \ge \Delta\Phi(\tilde{x})$, where Δ is a diagonal matrix with $|\Delta| = I$ and $\Delta(z - \tilde{x}) \ge 0$. Hence

$$\begin{aligned} D|z - \tilde{x}| = \Delta D(z - \tilde{x}) &= \Delta(-E\tilde{z} - \Phi(z) - D\tilde{x}) \\ &\le \Delta(-E\tilde{z} - \Phi(\tilde{x}) - D\tilde{x}) \\ &= -\Delta(E(\tilde{z} - \tilde{x}) + A\tilde{x} + \Phi(\tilde{x})) \\ &\le |E|\,|\tilde{z} - \tilde{x}| + |A\tilde{x} + \Phi(\tilde{x})| \\ &\le |E|w + \langle A\rangle w = Dw, \end{aligned}$$

so that $\Psi(\tilde{z}) = z \in [\tilde{x} - w, \tilde{x} + w]$ as claimed.

Since A is an H-matrix, $\langle A\rangle u > 0$ for some vector $u > 0$. Hence $P := D^{-1}|E|$ satisfies $Pu = u - \langle A\rangle u < u$, and the spectral radius of P is <1. Hence we may apply Schröder's fixed point theorem 3.3.3 and deduce the existence of a unique fixed point x^* of Ψ in $[\tilde{x} - w, \tilde{x} + w]$. But fixed points of Ψ are just the zeros of F. Since $w = \alpha u$ satisfies (2) for all sufficiently large $\alpha > 0$, the zero must in fact be unique in $\mathbb{R}^n$. □

Remark If $\tilde{x}$ is given, the best bound (1) is obtained when (2) holds with equality, giving

$$|x^* - \tilde{x}| \leq \langle A \rangle^{-1} |F(\tilde{x})|, \tag{4}$$

which can be used as an *a priori* bound for x^* with arbitrary $\tilde{x}$, but also as an *a posteriori* bound for the error of an approximation $\tilde{x}$ to the zero x^*.

We now show that it is possible to adapt Theorem 5.2.12 to the situation of the previous theorem. This provides a globally convergent method for finding arbitrarily narrow enclosures of the zero x^*. We shall use a modification $H_0(x, \tilde{x})$ of the Hansen–Sengupta operator. For arbitrary $x \in \mathbb{IR}^n$ and $\tilde{x} \in \mathbb{R}^n$ (not necessarily $\tilde{x} \in x$), the components y_i of $H_0(x, \tilde{x})$ are recursively defined by

$$y_i := \left(\tilde{x}_i - [0, A_{ii}^{-1}]\left(\sum_{k<i} A_{ik} y_k + A_{ii}\tilde{x}_i + \sum_{k>i} A_{ik} x_k + \Phi_i(\tilde{x})\right)\right) \cap x_i.$$

5.6.2 Theorem Under the assumptions of Theorem 5.6.1, let $u, v \in \mathbb{R}^n$ satisfy

$$u > 0, \quad \langle A \rangle u \geq b > 0, \tag{5}$$

and consider the iteration defined for $l = 0, 1, 2, \ldots$ by

$$x^{l+1/2} := (\tilde{x}^l - \|F(\tilde{x}^l)\|_v [-u, u]) \cap x^l, \tag{6a}$$

$$x^{l+1} := \mathrm{H}_0(x^{l+1/2}, \tilde{x}^l); \tag{6b}$$

here the $\tilde{x}^l \in \mathbb{R}^n$ are arbitrary, and the intersection with x^l in (6a) is performed only for $l > 0$. Then the x^l form a nested sequence of boxes containing the zero x^* of F. Moreover, if there is a number $\beta < 1$ such that

$$|\tilde{x}^l - \check{x}^l| \leq \beta \cdot \mathrm{rad}(x^l) \tag{7}$$

for infinitely many indices l then $\lim_{l\to\infty} x^l = x^*$.

Proof We first show the relation

$$x^* \in x - [0, D^{-1}](Ex^* + D\tilde{x} + \Phi(\tilde{x})), \tag{8}$$

where $A = D + E$ as in the previous proof. Indeed, we have

$$\begin{aligned} Ex^* + D\tilde{x} + \Phi(\tilde{x}) &= Ax^* + D(\tilde{x} - x^*) + \Phi(\tilde{x}) \\ &= D(\tilde{x} - x^*) + \Phi(\tilde{x}) - \Phi(x^*) \\ &= \tilde{D}(\tilde{x} - x^*) \end{aligned}$$

with a diagonal matrix $\tilde{D} \geq D$, and since $\tilde{D}^{-1} \in [0, D^{-1}]$ the inclusion (8) follows. Now, looking at (8) componentwise, we find

$$x^* \in x \Rightarrow x^* \in \mathrm{H}_0(x, \check{x}).$$

Since (2) holds with $w = \|F(\check{x}^l)\|_v u$, Theorem 5.6.1 and an induction argument now shows that $x^* \in x^l$ for all $l \geq 1$. Clearly, the x^l form a nested sequence, hence they have a limit x^∞.

Now suppose that $\mathrm{rad}(x^\infty) > 0$. If (7) holds with $\beta < 1$ for infinitely many indices l then there is an accumulation point $\tilde{x}^\infty$ of the $\tilde{x}^l$ such that $x^\infty = \mathrm{H}(x^\infty, \tilde{x}^\infty)$ and $|\tilde{x}^\infty - \check{x}^\infty| \leq \beta \cdot \mathrm{rad}(x^\infty)$. The first relation implies

$$x^\infty \subseteq \tilde{x}^\infty - [0, D^{-1}](Ex^\infty + D\tilde{x}^\infty + \Phi(\tilde{x})),$$

and the second relation gives $\tilde{x}^\infty \in \mathrm{int}(x^\infty)$. Therefore $d := x^\infty - \tilde{x}^\infty$ satisfies

$$-d \subseteq [0, D^{-1}](F(\tilde{x}^\infty) + Ed), \quad 0 \in \mathrm{int}(d).$$

But this forces $0 \in F(\tilde{x}^\infty) + Ed$, and then

$$\begin{aligned} 0 < \mathrm{rad}(d) &\leq D^{-1}\,\mathrm{rad}(F(\tilde{x}^\infty) + Ed) \\ &= D^{-1}|E|\mathrm{rad}(d) \end{aligned}$$

by Proposition 1.6.8(iv). Since $\rho(D^{-1}|E|) < 1$, this contradicts Corollary 3.2.3. Hence $\mathrm{rad}(x^\infty) = 0$, which implies $x^\infty = x^*$. □

Note that (6a) guarantees

$$\mathrm{rad}(x^{l+1}) \leq \|F(\tilde{x}^l)\|_v u,$$

so that the interval sequence x^{l+1} converges at least as fast as the driving sequence $\tilde{x}^l$. This suggests that we determine $\tilde{x}^l$ in most steps by a locally fast-converging iteration method in ordinary floating point arithmetic, accepting the iterate with smallest norm $\|F(\tilde{x}^l)\|_v$ as $\tilde{x}^l$. However, if this $\tilde{x}^l$ does not satisfy (7) then we use in the *next* step $\tilde{x}^{l+1} = \check{x}^{l+1}$. In this way we can guarantee global convergence of the method (6) even when the global behavior of the method used for the computation of the $\tilde{x}^l$ is erratic. Once we are sufficiently close to x^*, the local method will produce an $\tilde{x}^l$ which approximates x^* with working accuracy, so that x^{l+1} will be a good enclosure of x^*. If $\min\{v_i/u_i \mid i = 1, \ldots, n\}$ is very small, then an extra iteration with a Newton operator may further increase the limit accuracy of the enclosure.

The standard five-point discretization of nonlinear elliptic partial differential equations leads to nonlinear systems of the form $F(\tilde{x}) = A\tilde{x} + \Phi(\tilde{x}) = 0$, where A is a large and sparse M-matrix and Φ is diagonal. In many cases of interest, Φ is also continuous and isotone so that the above results apply. It is remarkable that the work required for each iteration of (6) is roughly equivalent to two function evaluations, so that the bulk of the work appears in the computation of the $\tilde{x}^l$. But for this, efficient methods are available which make full use of the sparsity of the problem. The only nontrivial extra work needed is the determination of u, v satisfying (5). But, as remarked after Proposition 3.7.3, this can be done very efficiently by approximately solving a linear system of equations with the matrix $\langle A \rangle = A$.

Another case where global *a priori* bounds are available is:

5.6.3 Theorem Let $F: \mathbb{R}^n \to \mathbb{R}^n$ be continuous. If there is a nonsingular matrix $A \in \mathbb{R}^{n\times n}$ and two numbers $\alpha, \beta \in \mathbb{R}$ such that $\beta < 1$ and

$$\|F(\tilde{x}) - A\tilde{x}\| \leq \alpha + \beta\|A\tilde{x}\| \quad \text{for all } \tilde{x} \in \mathbb{R}^n \tag{9}$$

in some scaled maximum norm then F has at least one zero, and every zero $x^* \in \mathbb{R}^n$ is bounded by

$$\|x^*\| \leq \frac{\alpha}{1-\beta}\|A^{-1}\|. \tag{10}$$

Proof This follows from Theorem 5.3.9 by taking $G(x, t) = \{F(x)\}$. The bound (10) can be more simply obtained by inserting $\tilde{x} = x^*$ into (9), giving

$$\|x^*\| \leq \|A^{-1}\| \, \|Ax^*\| \leq \|A^{-1}\| \cdot \frac{\alpha}{1-\beta}. \qquad \square$$

Another instance of a class of problems with domain $\mathbb{R}^n$ which can be transformed to a problem with bounded domain is the class of polynomial systems. A mapping $F: \mathbb{R}^n \to \mathbb{R}^n$ is called *polynomial* if every component of F is given by an arithmetical expression $F_i(\xi_1, \ldots, \xi_n)$ which is a polynomial in the variables $\xi_1, \ldots, \xi_n$. If we introduce homogeneous variables $\eta_0, \ldots, \eta_n$ such that $\xi_i = \eta_i/\eta_0$ we can rearrange the expressions $F_i(\eta_1/\eta_0, \ldots, \eta_n/\eta_0)$ into the equivalent form $G_i(\eta_0, \ldots, \eta_n)/\eta_0^{d_i}$ with a polynomial G_i in $\eta_0, \ldots, \eta_n$, where d_i denotes the total degree of F_i. Clearly, the system

$$F(x^*) = 0$$

is equivalent to the system

$$G(y^*) = 0, \quad y_0^* \neq 0,$$

and the solutions transform by

$$x_i^* = y_i^*/y_0^* \quad (i = 1, \ldots, n).$$

The homogeneous form, however, is only determined up to a nonzero multiplier. Thus we may fix the multiplier by some additional restriction. In terms of a positive vector $u \in \mathbb{R}^{n+1}$ (reflecting the scaling) a useful restriction is given by demanding that

$$\max\{|y_i^*|/u_i \mid i = 0, \ldots, n\} = y_l^*/u_l = 1$$

for some $l \in \{0, \ldots, n\}$. Thus the original unbounded polynomial problem has been reduced to a polynomial problem in the $n+1$ bounded domains

$$y^l = \{\tilde{y} \in [-u, u] \mid \tilde{y}_l = \tilde{u}_l\} \quad \{l = 0, \ldots, n\}.$$

The same technique can be used for more general nonlinear systems where some variables are bounded, and all unbounded variables occur polynomially in the system.

In the remainder of this section we assume that only the zeros in a bounded domain are sought. We treat here only the case where this domain is a box $x^0 \in \mathbb{IR}^n$. If this box is narrow enough, the methods of Section 5.2 apply directly. Often, however, x^0 is too wide and bisection methods must complement the earlier techniques. The basic principle of a *bisection method* consists of repeated splitting of boxes until they can be further reduced by means of interval iteration. This requires that a list of currently unprocessed boxes is stored in a so-called *stack* or *queue*. (When parallel processors are available, several boxes can be processed concurrently, and the scheme to be discussed must be modified a little.) The results in Section 5.2 now suggest that one proceeds along the following lines, starting with $x = x^0$.

Algorithm (to solve $F(x^*) = 0$ for $x^* \in x$)

Step 1 If x is empty, goto Step 6. Otherwise find an approximate local minimizer $\tilde{x} \in x$ of $\|F(\tilde{x})\|$ by a quasi-Newton method, say. If an intermediate iterate $\tilde{x}^l$ lies outside x then put $\tilde{x} = \text{cut}(\tilde{x}^l, x)$ and goto Step 4.

Step 2 Perform one iteration with a Newton operator with center $z = \tilde{x}$, followed by one iteration with the Hansen–Sengupta operator. If $\tilde{x} \notin x$ goto Step 1. Otherwise, if $\text{rad}(x)$ has been reduced sufficiently, goto Step 2. Otherwise, if $\text{rad}(x)$ is not sufficiently small, goto Step 4.

Step 3 Try to verify the existence of a solution in a narrow box $x' \supseteq x$ by the methods of Section 5.5. If this is successful, print x' ('contains a solution'); otherwise print x ('possibly contains a solution'). Goto Step 6.

Step 4 Let z be the vertex of x farthest away from $\tilde{x}$. Perform one iteration with the Hansen–Sengupta operator with center z. If $\tilde{x} \notin x$ goto Step 1. Otherwise, if $\text{rad}(x)$ has been reduced sufficiently, goto Step 2.

Step 5 Split the box into two parts chosen such that $\tilde{x}$ is in the part of larger volume. Put the part of smaller volume into the stack and goto Step 2.

Step 6 If the stack is empty, stop. Otherwise get a new box from the stack and goto Step 1.

Note that Step 3 is reached when the current box is already small but cannot be improved much by local methods (due to rounding errors when the limit accuracy is reached). In most cases this implies successful verification, but in case of a numerically singular Jacobian, no verification will result.

Clearly, the algorithm has no difficulties in also finding several (isolated) solutions in the initial box x^0, since sooner or later these solutions will be isolated by bisection. Of course, many modifications of this algorithm are possible. Also, the norm and the terms 'reduced sufficiently', 'sufficiently narrow', 'split the box' must be precisely specified for an implementation. It is not yet clear which strategy yields best results.

Remarks to Chapter 5

To 5.1 The separation theorem Proposition 5.1.1 is a classical result of convex analysis; see e.g. Rockafellar (1970).

Local implicit function theorems for differentiable functions can be found in many books on analysis; for the more general case of nonsmooth Lipschitz continuous functions see e.g. Clarke (1983). A semilocal implicit function theorem for differentiable functions similar to Theorem 5.1.3 is in Neumaier (1985c).

Theorem 5.1.6 is essentially in Neumaier (1985a). Parts (i) and (ii) of Theorem 5.1.7 are due to Moore (1966); part (iii) has been proved by Nickel (1971) (for differentiable functions) using Brouwer's fixed point theorem. Parts (i) and (ii) of Theorem 5.1.8 are due to Krawczyk (1969a) for the Krawczyk operator and to Hansen & Sengupta (1981) for the Hansen–Sengupta operator; part (iii) has been found by Kahan (1968) (and later by Moore (1977)) for the Krawczyk operator and by Moore & Qi (1982) for the Hansen–Sengupta operator. (For some related results see Nickel, 1981, and Qi, 1981a,b, 1982.)

Theorem 5.1.9 is due to Krawczyk (1986a), and Proposition 5.1.10 to Krawczyk (1980).

To 5.2 Moore (1962), Krawczyk (1969a), and Hansen & Sengupta (1981) first considered iteration with a Newton operator, the Krawczyk operator, and the Hansen–Sengupta operator, respectively.

Theorem 5.2.2 has been proved by Krawczyk (1983a), and

Corollary 5.2.4 is essentially already in Krawczyk (1969a). Theorem 5.2.5 is due to Alefeld (1979), and Theorem 5.2.6 is in Thiel (1986, 1989). Corollary 5.2.7 has been obtained earlier by Cornelius (1983) (cf. Cornelius & Alefeld, 1982). Theorems 5.2.8 and 5.2.14 are in Neumaier (1985a); a special case was treated in Alefeld (1984).

Theorem 5.2.12 seems to be new. For treatises of methods to compute approximate zeros of nonlinear systems, in particular for Newton and quasi-Newton methods, see Ortega & Rheinboldt (1970), Schwetlick (1979) and Dennis & Schnabel (1983).

For Proposition 5.2.13(ii) see Moore (1978); convergence results for Newton's method involving only the first derivative go back to Bartle (1955).

Theorem 5.2.15 is due to Krawczyk (1987a). Frommer & Mayer (1990) discuss an example where the derivative at the solution, but not the initial Lipschitz matrix, is an H-matrix; in this example, the Hansen–Sengupta iteration (with $C = I$ and variable Lipschitz matrix) still converges.

More general interval iterations based on splittings are discussed by Krawczyk (1988), Schwandt (1984, 1985a–d, 1986, 1987a) and G. Mayer (1988a, 1990).

To 5.3 Existence theorems for general continuous functions are usually based on the concept of topological degree; see e.g. Ortega & Rheinboldt (1970). I have chosen here a different approach based on the apparently new concept of c-continuous set-valued functions, which leads to the basic existence theorem 5.3.7.

Set-valued functions are discussed, e.g., in the book by Aubin & Ekeland (1984). For a treatment of the topological concepts of connectedness and continua see, e.g., the book by Kuratowski (1966).

Corollary 5.3.8 is due to Miranda (1941). Theorem 5.3.12 is a reformulation of the (finite-dimensional case of the) well-known fixed point theorem of Leray & Schauder (1934), and Theorem 5.3.13 is the fixed point theorem of Brouwer (1912). Using Proposition 5.3.15 it is easily verified that Theorem 5.3.14 is a generalization of the fixed point theorem of Kakutani (1941) for convex-set-valued functions.

To 5.4 Hansen (1978b) and Krawczyk & Neumaier (1985) introduced slopes for the solution of equations. Theorem 5.4.2 is a generalization of a result of Neumaier (1986a).

To 5.5 The results of this section are new; for some related results see Neumaier (1989) and Rump (1990). Other interval methods for the solution of parameter-dependent systems of equations are discussed in Adams (1980a), Adams & Lohner (1983) and Gay (1980, 1983); the latter paper shows the quadratic approximation property for one of Gay's methods.

Neumaier (1988b) and Heizmann (1990) discuss a method which allows the computation of narrow enclosures even for complicated solution curves of systems of equations with one parameter. See also Rheinboldt (1988), Koparkar & Mudur (1985), Mudur and Koparkar (1984) and Toth (1985). If n is large it may be too time consuming to search a big box for all connected components of the solution set. If the user can provide an initial solution, continuation methods are able to determine the connected component of the solution set containing this solution. An interval continuation method is described in Kearfott (1990b).

Krawczyk (1983b, 1984, 1986b,c,d, 1987b), Garloff & Krawczyk (1986) and Krawczyk & Neumaier (1986, 1987) discuss interval methods for zeros of functions $\tilde{f}$: $D \subseteq \mathbb{R}^n \rightarrow \mathbb{R}^n$ only known to lie in a *function strip* $[\underline{f}, \bar{f}]$, i.e. which satisfy

$$\underline{f}(\tilde{x}) \leq \tilde{f}(\tilde{x}) \leq \bar{f}(\tilde{x})$$

for all $\tilde{x} \in D$.

For alternative methods of obtaining an *a posteriori* enclosure of a zero when a (very good) approximation is given see Böhm, Rump & Schuhmacher (1987), Rump (1982b) and Shen & Neumaier (1988).

To 5.6 For the special case when A is an M-matrix, the situation of Theorem 5.6.1 is considered repeatedly in Ortega & Rheinboldt (1970, Chapter 13). The present Theorem 5.6.2 dispenses with their convexity assumptions. (This pleasant property of certain interval methods has first been observed by Alefeld, 1979.)

Theorem 5.6.3 is a slight extension of an existence theorem of Allgower & Georg (1983, Section 6). Their method of proof is constructive; arbitrarily good approximations can be obtained by *simplicial methods* based on piecewise linear approximations of the function and successively finer triangulations of $\mathbb{R}^n$.

Bisection methods for the solution of nonlinear equations go back to Alefeld (1970b), Hansen (1978a) and Krawczyk (1970a) (in the univariate case). Algorithms for systems of equations with various strategies for bisection and choice of interval operator are discussed in Hansen & Greenberg (1983), Hansen & Sengupta (1981), Heizmann (1990), Jones (1980), Kearfott (1987a,b, 1990a–

d), Kearfott & Novoa (1990), Lang (1989), Moore (1980a,b), Moore & Jones (1977), Neumaier (1988b), Rex (1989), Shearer & Wolfe (1985) and Thiel (1989). A promising global strategy along different lines is discussed in Koler & Mladenov (1990).

Further remarks There is a large number of papers on solving equations in a single variable, and in particular for polynomials. For functions of a real variable the best methods are based on the monitoring of *sign changes* (which provide a simple existence test); interval arithmetic then only plays the role of verifying the correctness of the signs. (This simple observation seems not to be in print before Matsumoto, Chua & Ayaki, 1988.) See the survey of Nerinckx & Haegemans (1976) for a comparison of several fast and reliable sign change methods.

For functions of a *complex* variable the winding number must be used which creates some extra difficulties; see e.g. Collins (1977), Henrici (1974), Neumaier (1988a) and Wilf (1978). Many other methods are described thoroughly in Petković (1989); see also Neumaier (1984b) and Wolfe (1988).

Enclosures for zeros of multidimensional *quadratic equations* are given by Alefeld (1986).

Eigenvalue problems can be written in several ways as systems of nonlinear equations, and hence may be treated by the methods discussed here. See Alefeld & Herzberger (1974), Alefeld & Spreuer (1986), Deif (1990), Kalmykov (1978), Krawczyk (1968, 1969b, 1970b), Rohn (1989), Rokne (1971, 1985) and Rump (1981a, 1983). Alternatively, eigenvalue problems can be treated as nonlinear equations in a single variable (Neumaier, 1985b), by transformation methods (Ris, 1972) and by interlacing (Behnke, 1988). See also Hollot & Bartlett (1990).

A survey of results on verifying the *stability* of interval matrices is given in Garloff & Bose (1988). See also Shi & Gao (1988).

For enclosures of *singular values* see Alefeld (1987).

Nonlinear least squares problems are treated by Gay (1988), Kearfott (1987c) and Kolev (1985).

6

Hull computation

6.1 The equation $\check{x} = M|\check{x}| + a$

In this section we prepare the stage for the characterization and computation of the hull inverse. We relate Beeck's characterization of the solution set of a linear system of interval equations to the equation $\tilde{x} = M|\tilde{x}| + a$, and discuss necessary and sufficient conditions for unique solvability of the latter equation. We also obtain methods for the computation of $\tilde{x}$.

Suppose that $A \in \mathbb{IR}^{n\times n}$ and $b \in \mathbb{IR}^n$, and we want to find the set $\Sigma(A, b) = \{\tilde{x} \in \mathbb{R}^n \mid \tilde{A}\tilde{x} = \tilde{b} \text{ for some } \tilde{A} \in A, \tilde{b} \in b\}$. By the characterization of Beeck (Theorem 3.4.3),

$$\tilde{x} \in \Sigma(A, b) \Leftrightarrow |\check{A}\tilde{x} - \check{b}| \leq \text{rad}(A)|\tilde{x}| + \text{rad}(b). \tag{1}$$

We should like to solve this condition for $\tilde{x}$. In order to remove the inequality sign we note that

$$|u| \leq v \Leftrightarrow u = Dv \text{ for some } D \in \mathbb{R}^{n\times n} \text{ with } |D| \leq I; \tag{2}$$

indeed, the diagonal matrix D with diagonal entries $D_{ii} = u_i/v_i$ (and $0/0 = 1$) works. Hence $\tilde{x} \in \Sigma(A, b)$ iff there is a diagonal matrix D with $|D| \leq I$ such that

$$\check{A}\tilde{x} - \check{b} = D(\text{rad}(A)|\tilde{x}| + \text{rad}(b)).$$

In the case that $\check{A}$ is regular, this is equivalent to the fixed point equation

$$\tilde{x} = \check{A}^{-1}D \cdot \text{rad}(A)|\tilde{x}| + \check{A}^{-1}(\check{b} + D \cdot \text{rad}(b)). \tag{3}$$

Therefore we are looking for a method to solve fixed point equations of the form

$$\tilde{x} = M|\tilde{x}| + a \quad (M \in \mathbb{R}^{n\times n}, a \in \mathbb{R}^n). \tag{4}$$

In general, an equation of the form (4) need not have a solution; e.g. if $M = I$

then (4) can be solved only for $a \le 0$ since $\check{x} \le |\check{x}|$ for all $\check{x} \in \mathbb{R}^n$. Therefore we must ask whether the matrix

$$M = \check{A}^{-1}D \cdot \text{rad}(A) \quad (|D| \le I) \tag{5}$$

in (3) has special properties which can be used to analyse (4). Note that every matrix D with $|D| \le I$ is a diagonal matrix.

The situation is very simple if A is strongly regular. For, in this case,

$$|M| = |\check{A}^{-1}D \cdot \text{rad}(A)| \le |\check{A}^{-1}|\,|D|\,|\text{rad}(A)| \le |\check{A}^{-1}|\text{rad}(A),$$

and the spectral radius of $|M|$ is less than one since this holds for the dominating matrix $|\check{A}^{-1}|\text{rad}(A)$. Therefore the following result applies.

6.1.1 Theorem (Rohn) If $M \in \mathbb{R}^{n\times n}$ and $\rho(|M|) < 1$ then, for every $a \in \mathbb{R}^n$, the equation

$$\check{x} = M|\check{x}| + a \tag{4}$$

has a unique solution $\check{x} \in \mathbb{R}^n$, and the iteration

$$\check{x}^{l+1} := M|\check{x}^l| + a \quad (l = 0, 1, 2, \ldots)$$

converges to $\check{x}$ for every choice of the starting vector $\check{x}^0$.

Proof Define $\Phi(x) := M|x| + a$. Then

$$\begin{aligned}|\Phi(x) - \Phi(y)| &= |M|x| - M|y|\,| = |M(|x| - |y|)| \\ &\le |M|\,|\,|x| - |y|\,| \le |M|\,|x - y|,\end{aligned}$$

so that the assertion follows from Schröder's fixed point theorem (Theorem 3.3.3) with $\mathbb{R}^n$ in place of $\mathbb{IR}^n$. □

By a similar argument it can be shown that if $\rho(|M|) < 1$ then the Gauss–Seidel version of the above iteration, namely

$$\check{x}_i^{l+1} := \sum_{j<i} M_{ij}|\check{x}_j^{l+1}| + \sum_{j\ge i} M_{ij}|\check{x}_j^l| + a_i \quad (i = 1, \ldots, n),$$

converges also to the solution of (4), and usually faster. A further speed-up is obtained by iteration according to

$$\check{x}_i^{l+1} := s_i^l/(1 - \text{sgn}(s_i^l)M_{ii}) \quad (i = 1, \ldots, n),$$

where

$$s_i^l := \sum_{j<i} M_{ij}|\check{x}_j^{l+1}| + \sum_{j>i} M_{ij}|\check{x}_j^l| + a_i.$$

For matrices A which are regular, but not strongly regular, the spectral radius of $|M|$ may become ≥ 1, and we need a weaker property of M. We call a matrix $M \in \mathbb{R}^{n\times n}$ *nonexpanding* if, for all $\tilde{x} \in \mathbb{R}^n$,

$$|M\tilde{x}| \geq |\tilde{x}| \Rightarrow \tilde{x} = 0,$$

and *expanding* otherwise. If $\rho(|M|) < 1$ then M is nonexpanding since then $|M\tilde{x}| \geq |\tilde{x}|$ implies $|\tilde{x}| \leq |M|\,|\tilde{x}|$, hence $\tilde{x} = 0$ by Corollary 3.2.3. On the other hand, the matrix

$$M = \frac{2}{3}\begin{pmatrix} 1 & 1 \\ 1 & -1 \end{pmatrix} \tag{6}$$

is nonexpanding (check Theorem 6.1.3(v) below) although $\rho(|M|) = \frac{4}{3} > 1$.

6.1.2 Proposition If $A \in \mathbb{IR}^{n\times n}$ is regular then every matrix of the form (5) is nonexpanding.

Proof Suppose that $|M\tilde{x}| \geq |\tilde{x}|$ for some $\tilde{x} \in \mathbb{R}^n$. Then the vector $\tilde{z} := M\tilde{x}$ satisfies

$$\begin{aligned}|\mathrm{mid}(A\tilde{z})| &= |\check{A}\tilde{z}| = |\check{A}M\tilde{x}| = |D\cdot \mathrm{rad}(A)\tilde{x}| \leq |D|\mathrm{rad}(A)|\tilde{x}| \\ &\leq \mathrm{rad}(A)|\tilde{x}| \leq \mathrm{rad}(A)|M\tilde{x}| = \mathrm{rad}(A)|\tilde{z}| = \mathrm{rad}(A\tilde{z})\end{aligned}$$

so that $0 \in A\tilde{z}$. Hence $\tilde{z} = 0$ by regularity, and $\tilde{x} = 0$ since $|\tilde{x}| \leq |M\tilde{x}| = |\tilde{z}| = 0$. □

Before we prove the solvability of (4) for nonexpanding matrices we first give some characterizations of such matrices.

6.1.3 Theorem (Rohn) For $M \in \mathbb{R}^{n\times n}$, the following conditions are equivalent:

(i) M is nonexpanding;
(ii) M^{T} is nonexpanding;
(iii) $|D|, |D'| \leq I \Rightarrow DMD'$ is nonexpanding;
(iv) $|D| \leq I \Rightarrow I - MD$ is nonsingular;
(v) $|D| = I \Rightarrow \det(I - MD) > 0$;
(vi) $|D| = I \Rightarrow I - MD$ is nonsingular and the diagonal entries of $(I - MD)^{-1}$ are $> \frac{1}{2}$.
(vii) $|D| = I \Rightarrow |\lambda| < 1$ for all real eigenvalues λ of MD.

Proof We first show the equivalence of (i), (iii) and (iv). Clearly (iii) implies (i). (i) $\Rightarrow$ (iii): If $|D| \leq I$ and $|D'| \leq I$ then $|DMD'\tilde{x}| \geq |\tilde{x}|$ implies that $\tilde{z} := D'\tilde{x}$ satisfies $|\tilde{z}| \leq |D'|\,|\tilde{x}| \leq |\tilde{x}| \leq |DMD'\tilde{x}| = |DM\tilde{z}| \leq |D|\,|M\tilde{z}| \leq |M\tilde{z}|$; hence $\tilde{z} = 0$ and $|\tilde{x}| \leq |DMD'\tilde{x}| = |DM\tilde{z}| = 0$ so that $\tilde{x} = 0$.

(i) $\Rightarrow$ (iv): If $(I - MD)\tilde{z} = 0$ then $\tilde{x} := D\tilde{z}$ satisfies $M\tilde{x} = MD\tilde{z} = \tilde{z}$ so that

$|\tilde{x}| = |D\tilde{z}| \leq |D|\,|\tilde{z}| \leq |\tilde{z}| = |M\tilde{x}|$; hence $\tilde{x} = 0$, $\tilde{z} = M\tilde{x} = 0$ and $I - MD$ is regular.

(iv) $\Rightarrow$ (i): If $|M\tilde{x}| \geq |\tilde{x}|$ then $\tilde{z} := M\tilde{x}$ satisfies $|\tilde{z}| \geq |\tilde{x}|$. Now the diagonal matrix D with diagonal entries $D_{ii} = \tilde{x}_i/\tilde{z}_i$ $(i = 1, \ldots, n)$ satisfies $|D| \leq I$ and $\tilde{x} = D\tilde{z}$; therefore $(I - MD)\tilde{z} = 0$ so that $\tilde{z} = 0$ and $\tilde{x} = D\tilde{z} = 0$.

Next we show the equivalence of (iv), (v) and (vi). (iv) $\Rightarrow$ (v): If $|D| = I$ then $I - tMD$ is regular for all $t \in [0, 1]$; hence $\varphi(t) := \det(I - tMD)$ has no zero in $[0, 1]$. Since $\varphi(0) = 1$, we see that $\varphi(t)$ is positive for all $t \in [0, 1]$. In particular, $\det(I - MD) = \varphi(1) > 0$.

(v) $\Rightarrow$ (iv): Let $|D| \leq I$. Since the determinant is a linear function of each row, $\det(I - MD)$ is linear in each D_{ii}. Hence $\det(I - MD) \geq \det(I - MD')$ for some D' with $|D'| = I$, whence $\det(I - MD) > 0$ and $I - MD$ is nonsingular.

(v) $\Leftrightarrow$ (vi): Let $|D| = I$, let D' be obtained from D by changing the sign of D_{ii}. Put $B = I - MD$ and $B' = I - MD'$. Then $D' = D - 2De^{(i)}e^{(i)\mathrm{T}}$ yields

$$\begin{aligned} B' &= I - MD + 2MDe^{(i)}e^{(i)\mathrm{T}} = B + 2(I - B)e^{(i)}e^{(i)\mathrm{T}}, \\ \det(B') &= \det(B) \cdot (1 + 2e^{(i)\mathrm{T}}B^{-1}(I - B)e^{(i)}) \\ &= \det(B) \cdot (2(B^{-1})_{ii} - 1). \end{aligned} \tag{7}$$

If (v) holds then $\det(B)$ and $\det(B')$ are positive so that $(B^{-1})_{ii} > \frac{1}{2}$, and (vi) holds. Conversely, if (vi) holds then $\det(B)$ and $\det(B')$ have the same sign, and, by induction, the sign of $\det(I - MD)$ does not depend on D with $|D| = I$. Since $\det(I - MD)$ is linear in each D_{ii} there is a D with $|D| = I$ and $\det(I - MD) \geq \det(I) = 1$; therefore $\det(I - MD) > 0$ for all D with $|D| = I$.

Finally we prove the remaining equivalences. (i) $\Leftrightarrow$ (vii): If $(\lambda, \tilde{z})$ is a real eigenpair of MD (where $|D| = I$) with $|\lambda| \geq 1$ then $\tilde{x} := D\tilde{z}$ is nonzero and satisfies $|M\tilde{x}| = |MD\tilde{z}| = |\lambda\tilde{z}| \geq |\tilde{z}| = |\tilde{x}|$ so that M is expanding. On the other hand, if all real eigenvalues λ of MD satisfy $|\lambda| < 1$ then $I - MD$ has only positive real or conjugate complex eigenvalues whence $\det(I - MD) > 0$; hence (v) holds and M is nonexpanding.

(i) $\Leftrightarrow$ (ii): Since MD, DM and thus $(DM)^\mathrm{T} = M^\mathrm{T}D$ have the same eigenvalues, (vii) holds for M^T in place of M iff it holds for M itself. Hence M and M^T are simultaneously expanding or nonexpanding. □

We also note some necessary conditions for a matrix to be nonexpanding.

6.1.4 Proposition Let $M \in \mathbb{R}^{n \times n}$ be nonexpanding. Then every principal submatrix of M is nonexpanding. In particular,

$$|M_{ii}| < 1 \quad \text{for } i = 1, \ldots, n.$$

Proof Restrict $\tilde{x}$ to those vectors which have a zero in the deleted rows. The second part follows by taking 1×1 principal submatrices. □

We now proceed to the solution of the equation

$$\tilde{x} = MD|\tilde{x}| + a \quad (|D| = I); \tag{8}$$

this equation contains (4) as the special case $D = I$. The fact that (8) implies $\tilde{x} = MDD'\tilde{x} + a$ for some D' with $|D'| = I$ and $D'\tilde{x} \geq 0$ suggests that we try all diagonal matrices D' with diagonal entries ± 1 in a systematic fashion. This is done by the following algorithm.

Sign accord algorithm (for the solution of (7) for fixed M, D, a)
Step 1 Select D' with $|D'| = I$ (preferably such that $D'a \geq 0$);
Step 2 For $s = 1, \ldots, 2^n$ do:

solve $\tilde{x} = MDD'\tilde{x} + a$;
if $D'\tilde{x} \geq 0$ terminate (success); otherwise compute
$k := \min\{j = 1, \ldots, n \mid D'_{jj}\tilde{x}_j < 0\}$;
change the sign of D'_{kk};

Step 3 Terminate (failure).

In case of successful termination, $\tilde{x}$ is a solution of (8).

The rule for changing the sign is suggested by the aim of finding a nonnegative $D\tilde{x}$ quickly, hopefully in a few steps only. This has the advantage that we are often fast, but, in general, the algorithm may cycle (i.e. return to a diagonal matrix D' already tried). In this case, because of the limit of 2^n trials, the algorithm terminates with failure. Of course, this always happens when (8) has no solution. On the other hand, our next result shows that the algorithm always terminates with success if M is nonexpanding. (Note that the alternative rule of changing the signs of all D'_{jj} for which $D'_{jj}\tilde{x}_j < 0$ is often a little faster in practice; but it has not been shown that this rule cannot lead to cycling when M is nonexpanding.)

6.1.5 Theorem (Rohn) For $M \in \mathbb{R}^{n\times n}$, the following conditions are equivalent:

(i) M is nonexpanding.
(viii) For all $a \in \mathbb{R}^n$, the equation $\tilde{x} = M|\tilde{x}| + a$ has a unique solution $\tilde{x} \in \mathbb{R}^n$.
(ix) For all $a \in \mathbb{R}^n$, the equation $\tilde{x} = M|\tilde{x}| + a$ has at most one solution $\tilde{x} \in \mathbb{R}^n$.
(x) For all $D \in \mathbb{R}^{n\times n}$ with $|D| = I$ and every $a \in \mathbb{R}^n$, the equation $\tilde{x} = MD|\tilde{x}| + a$ has at least one solution $\tilde{x} \in \mathbb{R}^n$.

(xi) The sign accord algorithm terminates with success for every $D \in \mathbb{R}^{n\times n}$ with $|D| = I$, every $a \in \mathbb{R}^n$ and every initial choice of D' with $|D'| = I$.

Proof (i) $\Rightarrow$ (ix): If M is nonexpanding and $\tilde{x}^1$ and $\tilde{x}^2$ are two solutions of (4) then $\tilde{x} := |\tilde{x}^1| - |\tilde{x}^2|$ satisfies $|M\tilde{x}| = |M|\tilde{x}^1| - M|\tilde{x}^2|| = |\tilde{x}^1 - \tilde{x}^2| \geq ||\tilde{x}^1| - |\tilde{x}^2|| = |\tilde{x}|$, so that $\tilde{x} = 0$, $|\tilde{x}^1| = |\tilde{x}^2|$, and hence $\tilde{x}^1 = M|\tilde{x}^1| + a = M|\tilde{x}^2| + a = \tilde{x}^2$.

(ix) $\Rightarrow$ (i): Suppose that $|M\tilde{x}| \geq |\tilde{x}|$, and let D be chosen such that $|M\tilde{x}| = DM\tilde{x}$ and $|D| = I$. Then, for either sign, $\tilde{z} := D\tilde{x} \pm Mx$ satisfies $\pm D\tilde{z} = \pm\tilde{x} + DM\tilde{x} = \pm\tilde{x} + |M\tilde{x}| \geq 0$ so that $|\tilde{z}| = \pm\tilde{x} + |M\tilde{x}|$ and $\tilde{z} - M|\tilde{z}| = D\tilde{x} - M|M\tilde{x}| =: a$. The uniqueness of the solution of $\tilde{z} = M|\tilde{z}| + a$ now implies that $M\tilde{x} = 0$ and therefore $\tilde{x} = 0$. Hence M is nonexpanding.

(i) $\Rightarrow$ (xi): Let M be nonexpanding. By Theorem 6.1.3(iv), the equations in the sign accord algorithm have a unique solution. By reverse induction we show that in the sign accord algorithm the sign of D'_{jj} changes at most 2^{n-j} times. We may suppose that this already holds for all $j > k$, since for $j > n$ the statement is empty.

Suppose that D'_{kk} changes sign in steps s_1 and $s_2 > s_1$, but not in the intermediate steps. Denote by $\tilde{x}^l$ the vector computed in step s_l ($l = 1, 2$). Then, for $l = 1, 2$, we have

$$\tilde{x}^l = MDD^{(l)}\tilde{x}^l + a \text{ with suitable } D^{(l)} \in \mathbb{R}^{n\times n} \text{ with } |D^{(l)}| = I. \tag{9}$$

By construction of k we have $D^{(l)}_{jj}\tilde{x}^l_j \geq 0$ for all $j < k$ and

$$D^{(l)}_{kk}\tilde{x}^l_k < 0. \tag{10}$$

In particular, for all $i \leq k$ we have

$$D^{(1)}_{ii} = D^{(2)}_{ii} \quad \text{or} \quad \tilde{x}^1_i\tilde{x}^2_i \leq 0. \tag{11}$$

Moreover, by choice of the steps we have $D^{(2)}_{kk} = -D^{(1)}_{kk}$, so that (10) implies $\tilde{x}^1_k \neq \tilde{x}^2_k$ and therefore $\tilde{x}^1 \neq \tilde{x}^2$.

We now show by way of contradiction that there is an index $j > k$ such that $D^{(1)}_{jj} \neq D^{(2)}_{jj}$. For, otherwise, the alternative (11) holds for all indices i. But then the vector

$$\tilde{x} := DD^{(1)}\tilde{x}^1 - DD^{(2)}\tilde{x}^2$$

satisfies

$$\begin{aligned} |\tilde{x}_i| &= |D_{ii}D^{(1)}_{ii}(\tilde{x}^1_i - \tilde{x}^2_i)| = |\tilde{x}^1_i - \tilde{x}^2_i| && \text{if } D^{(1)}_{ii} = D^{(2)}_{ii}, \\ |\tilde{x}_i| &\leq |\tilde{x}^1_i| + |\tilde{x}^2_i| = |\tilde{x}^1_i - \tilde{x}^2_i| && \text{if } \tilde{x}^1_i\tilde{x}^2_i \leq 0. \end{aligned}$$

Therefore (11) and (9) imply

$$|\tilde{x}| \leq |\tilde{x}^1 - \tilde{x}^2| = |MDD^{(1)}\tilde{x}^1 - MDD^{(2)}\tilde{x}^2| = |M\tilde{x}|,$$

so that $\tilde{x} = 0$. But then $DD^{(1)}\tilde{x}^1 = DD^{(2)}\tilde{x}^2$ and (9) gives the contradiction $\tilde{x}^1 = \tilde{x}^2$.

Therefore there is an index $j > k$ such that $D_{jj}^{(1)} \neq D_{jj}^{(2)}$. This implies that between any two sign changes of D'_{kk} there is at least one sign change of some D'_{jj} $(j > k)$. Since the number of the latter is at most

$$\sum_{j=k+1}^{n} 2^{n-j} = 2^{n-k} - 1,$$

this shows that D'_{kk} changes sign at most 2^{n-k} times. This completes the induction and shows that the sign accord algorithm terminates in the step following at most

$$\sum_{j=1}^{n} 2^{n-j} = 2^n - 1$$

other steps, and hence always with success.

(xi) $\Rightarrow$ (x) is obvious.

(x) $\Rightarrow$ (i): Let $|M^{\mathrm{T}}\tilde{z}| \geq |\tilde{z}|$ and let D be chosen such that $|M^{\mathrm{T}}\tilde{z}| = DM^{\mathrm{T}}\tilde{z}$. Let $\tilde{x}$ be a solution of $\tilde{x} = MD|\tilde{x}| + \tilde{z}$. Then $\tilde{x}^{\mathrm{T}}\tilde{z} \leq |\tilde{x}|^{\mathrm{T}}|\tilde{z}| \leq |\tilde{x}|^{\mathrm{T}}|M^{\mathrm{T}}\tilde{z}| = |\tilde{x}|^{\mathrm{T}}DM^{\mathrm{T}}\tilde{z} = (MD|\tilde{x}|)^{\mathrm{T}}\tilde{z} = (\tilde{x} - \tilde{z})^{\mathrm{T}}\tilde{z} = \tilde{x}^{\mathrm{T}}\tilde{z} - \tilde{z}^{\mathrm{T}}\tilde{z}$ whence $\tilde{z}^{\mathrm{T}}\tilde{z} \leq 0$. But this implies $\tilde{z} = 0$. Hence M^{T}, and therefore M, is nonexpanding.

Now we have shown that (i), (ix), (x) and (xi) are equivalent. In particular, (ix) implies (x), and by specializing to $D = I$ we see that (ix) is equivalent to (viii). This completes the proof of the theorem. □

Remarks (i) If $\rho(|M|) < 1$ then the recommendation of choosing D' such that $D'a \geq 0$ frequently leads to success in the first step. Indeed, this can be shown under the sharper assumption $D'a > 2|M|\,|a|$. In this case, let α be minimal such that $|\tilde{x} - a| \leq \alpha|a|$. Since $|\tilde{x} - a| = |MDD'\tilde{x}| \leq |M|\,|\tilde{x}| \leq |M|(\alpha + 1)|a| < \frac{1}{2}(\alpha + 1)D'a = \frac{1}{2}(\alpha + 1)|a|$, this forces $\alpha < \frac{1}{2}(\alpha + 1)$ so that $\alpha < 1$, $|D'\tilde{x} - D'a| = |\tilde{x} - a| < |a| = D'a$, and hence $D'\tilde{x} > 0$.

(ii) On the other hand, a diagonal matrix D' such that $D'a \leq 0$ may be a very bad choice. For the solution of (4), where $M_{ik} = 2$ $(i > k)$, $M_{ik} = 0$ $(i \leq k)$ and $r_i = -1$ for all i, the sign accord algorithm started with $D' = I$ takes the full number of 2^n steps. Note that M is nonexpanding by Theorem 6.1.3(vii) since M is strictly lower triangular so that all eigenvalues of M are zero.

(iii) Numerical experiments suggest that, as soon as M is nonexpanding, the iteration of Theorem 6.1.1 (and also the variants mentioned there) converge to the solution of (4). However, for practical purposes the convergence is much too slow, unless $\rho(|M|) \ll 1$.

6.2 Characterization and computation of $A^{\mathrm{H}}b$

In this section we apply the results of the previous section to square linear interval equations. We give necessary and sufficient conditions for the regularity of $A \in \mathbb{IR}^n$ and show how to compute $A^{\mathrm{H}}b$ as the hull of finitely many vectors which can be found by iterative methods or by Rohn's sign accord algorithm.

We begin with a characterization of regular interval matrices.

6.2.1 Theorem Let $A \in \mathbb{IR}^{n\times n}$ and suppose that $\check{A}$ is nonsingular. Then the following statements are equivalent:

(i) A is regular.
(ii) $|D|, |D'| \leq I \Rightarrow \check{A}^{-1}D \cdot \mathrm{rad}(A)D'$ is nonexpanding.
(iii) $|D| = I \Rightarrow \check{A}^{-1}D \cdot \mathrm{rad}(A)$ is nonexpanding.
(iv) $|D| = |D'| = I \Rightarrow$ all real eigenvalues λ of $\check{A}^{-1}D \cdot \mathrm{rad}(A)D'$ satisfy $|\lambda| < 1$.
(v) $|D| = |D'| = I \Rightarrow \det(\check{A} - D \cdot \mathrm{rad}(A)D')$ has the same sign as $\det(\check{A})$.

Proof Let A be regular. Then, for $|D| \leq I$, the matrix $M = \check{A}^{-1}D \cdot \mathrm{rad}(A)$ is nonexpanding by Proposition 6.1.2, hence Theorem 6.1.3 applies and shows that (ii), (iii), (iv) and (v) hold. By the same theorem, (iii), (iv) and (v) are equivalent. Hence, since (ii) implies (iii) trivially, it is sufficient to show that (iii) implies (i). To see this, suppose that (iii) holds and $0 \in A^{\mathrm{T}}\tilde{x}$ for some $\tilde{x} \in \mathbb{R}^n$. If we choose D such that $|D| = I$, $D\tilde{x} \geq 0$ then the matrix $M := \check{A}^{-1}D \cdot \mathrm{rad}(A)$ is nonexpanding, and hence M^{T} is too. Since $0 \in A^{\mathrm{T}}\tilde{x}$, the vector $\tilde{z} := \check{A}^{\mathrm{T}}\tilde{x}$ satisfies $|\tilde{z}| = |\check{A}^{\mathrm{T}}\tilde{x}| \leq \mathrm{rad}(A^{\mathrm{T}})|\tilde{x}| = \mathrm{rad}(A)^{\mathrm{T}}D\tilde{x} = (\check{A}M)^{\mathrm{T}}\tilde{x} = M^{\mathrm{T}}\tilde{z} \leq |M^{\mathrm{T}}\tilde{z}|$. Hence, $\tilde{z} = 0$, and $\tilde{x} = 0$ since $\check{A}$ is regular. By Corollary 3.4.5, this implies that A^{T}, and hence A, is regular. □

Note that Theorem 6.2.1(v) is a criterion that decides in finitely many steps whether A is regular. However, for this, one needs to compute 2^{2n-1} determinants ((D, D') and $(-D, -D')$ give the same determinant), which is far too much to be of practical value, even when the determinants are calculated using rank 1 updates (i.e. by a formula corresponding to equation (6.1.7)) and assuming exact arithmetic. It is not known whether there is a polynomial algorithm for deciding the regularity of an arbitrary interval matrix. Fortunately there are simple $\mathrm{O}(n^3)$-tests which decide the majority of practically relevant cases, namely

(1) if $\rho(|\check{A}^{-1}|\mathrm{rad}(A)) < 1$ then A is (strongly) regular;
(2) if $0 \in A\tilde{x}$ then A is singular. (If $C \approx \check{A}^{-1}$ then the normalized column of C containing the absolutely largest entry of C is a good choice for $\tilde{x}$.)

Our next objective is the determination of extremal points of the solution set

$\Sigma(A, b)$ of a system of linear interval equations. For any set $\Sigma \subseteq \mathbb{R}^n$, a point $\tilde{x} \in \Sigma$ is called *extremal* if no line segment in Σ contains $\tilde{x}$ as an interior point, i.e. if, for $\varepsilon > 0$,

$$\tilde{x} + t\tilde{z} \in \Sigma \text{ for all } t \in [-\varepsilon, \varepsilon] \Rightarrow \tilde{z} = 0.$$

Extremal points are relevant for the determination of the convex hull and the interval hull of a set. Indeed, the convex hull of Σ equals the convex hull of its extremal points, and

$$\square\Sigma = \square\{\tilde{x} \in \Sigma \mid \tilde{x} \text{ extremal in } \Sigma\}.$$

6.2.2 Theorem (Rohn) Let $A \in \mathbb{IR}^{n\times n}$ be regular and $b \in \mathbb{R}^n$. Then:

(i) $\tilde{x} \in \Sigma(A, b) \Leftrightarrow 0 \in A\tilde{x} - b$.

(ii) Every extremal point $\tilde{x}$ of $\Sigma(A, b)$ satisfies the equation

$$\inf(D(A\tilde{x} - b)) = 0 \tag{1}$$

for some $D \in \mathbb{R}^{n\times n}$ with $|D| = I$.

(iii) For every $D \in \mathbb{R}^{n\times n}$ with $|D| = I$, equation (1) has a unique solution $\tilde{x} = x^*_D$, and we have

$$A^{\mathrm{H}}b = \square\{x^*_D \mid |D| = I\}. \tag{2}$$

Proof (i) is essentially Beeck's Theorem 3.4.3; it is proved again here to introduce some notation needed. Put $r := A\tilde{x} - b$. If $\tilde{A}\tilde{x} = \tilde{b}$ for some $\tilde{A} \in A$ and $\tilde{b} \in b$ then $0 = \tilde{A}\tilde{x} - \tilde{b} \in A\tilde{x} - b = r$. Conversely, if $0 \in r$ then $|\check{r}| \leq \mathrm{rad}(r)$ so that $\check{r} = D \cdot \mathrm{rad}(r)$ for some $D \in \mathbb{R}^{n\times n}$ with $|D| \leq I$. If we choose D' such that $|D'| = I$, $D'\tilde{x} \geq 0$ then

$$\tilde{A} := \check{A} - D \cdot \mathrm{rad}(A)D' \in A, \quad \tilde{b} := \check{b} + D \cdot \mathrm{rad}(b) \in b,$$

and $\tilde{A}\tilde{x} = \tilde{b}$ since $\tilde{A}\tilde{x} - \tilde{b} = (\check{A} - D \cdot \mathrm{rad}(A)D')\tilde{x} - (\check{b} + D \cdot \mathrm{rad}(b)) = \check{A}\tilde{x} - \check{b} - D(\mathrm{rad}(A)|\tilde{x}| + \mathrm{rad}(b)) = \check{r} - D \cdot \mathrm{rad}(r) = 0$. In particular, $\tilde{x} \in \Sigma(A, b)$ iff $0 \in r = A\tilde{x} - b$, which yields (i).

Now suppose that $|D_{ii}| < 1$. We show that there is an $\varepsilon > 0$ such that

$$\tilde{x} + t\tilde{A}^{-1}e^{(i)} \in \Sigma(A, b) \quad \text{for all } t \in [-\varepsilon, \varepsilon]. \tag{3}$$

Indeed, let $\tilde{D}$ be obtained from D by changing D_{ii} to $D_{ii} + \delta$. Then

$$\tilde{B} := \check{A} - \tilde{D} \cdot \mathrm{rad}(A)D' = \tilde{A} - (\tilde{D} - D)\mathrm{rad}(A)D' = \tilde{A} - \delta e^{(i)}f^{\mathrm{T}},$$

where $f^{\mathrm{T}} = e^{(i)\mathrm{T}}\,\mathrm{rad}(A)D'$, and $\tilde{B} \in A$ if $|D_{ii} + \delta| \leq 1$. Now $\tilde{z} := \tilde{x} + t\tilde{A}^{-1}e^{(i)}$ satisfies

$$\begin{aligned}\tilde{B}\tilde{z} &= \tilde{A}\tilde{z} - \delta e^{(i)}f^{\mathrm{T}}\tilde{z} = \tilde{A}\tilde{x} + te^{(i)} - \delta(f^{\mathrm{T}}\tilde{z})e^{(i)}\\ &= \tilde{b} + (t - \delta f^{\mathrm{T}}\tilde{z})e^{(i)} = \tilde{b} \in b\end{aligned}$$

provided that $t = \delta f^{\mathrm{T}}\tilde{z} = \delta f^{\mathrm{T}}(\tilde{x} + t\tilde{A}^{-1}e^{(i)})$. By continuity, this can be satisfied by some δ with $|D_{ii} + \delta| \le 1$ for all t sufficiently close to zero, say for $|t| \le \varepsilon$. Hence $\tilde{z} \in \Sigma(A, b)$, i.e. (3) holds.

Now (3) says that $\tilde{x}$ is not extremal in (2). Hence the matrix D corresponding to an extremal $\tilde{x}$ must satisfy $|D_{ii}| \ge 1$ and hence $|D| = I$. Since $\inf(D(A\tilde{x} - b)) = \inf(Dr) = D\check{r} - \mathrm{rad}(r) = 0$, (ii) follows.

To show (iii) we rewrite (1), i.e., $D\check{r} - \mathrm{rad}(r)$ in the equivalent form $\tilde{x} = \check{A}^{-1}(\check{b} + \check{r}) = \check{A}^{-1}D(D\check{b} + \mathrm{rad}(r))$, or

$$\tilde{x} = \check{A}^{-1}D \cdot \mathrm{rad}(A)|\tilde{x}| + \check{A}^{-1}(\check{b} + D \cdot \mathrm{rad}(b)). \tag{4}$$

Proposition 6.1.2 and Theorem 6.1.5 now imply that (4), and hence (1), has a unique solution $\tilde{x} = x_D^*$, and since $A^{\mathrm{H}}b$ is the interval hull of the extremal points of $\Sigma(A, b)$, (2) follows from part (ii). □

We note that (4) implies

$$\check{A}x_D^* = D \cdot \mathrm{rad}(A)|x_D^*| + b_D \quad \text{with } b_D = \check{b} + D \cdot \mathrm{rad}(b). \tag{5a}$$

Moreover, if $|D'| = I$ and $D'\tilde{x} \ge 0$ then $|\tilde{x}| = D'\tilde{x}$ so that

$$x_D^* = (A_{D,D'})^{-1}b_D \quad \text{with } A_{D,D'} = \check{A} - D \cdot \mathrm{rad}(A) \cdot D'. \tag{5b}$$

In many cases, not all 2^n vectors x_D^* ($|D| = I$) are needed in (2). This follows from our next result.

6.2.3 Theorem (Rohn) Let $A \in \mathbb{IR}^{n \times n}$ be regular and $b, c \in \mathbb{IR}^n$. Then there exist $x^*, y^* \in \mathbb{R}^n$ and $D, D' \in \mathbb{R}^{n \times n}$ with $|D| = |D'| = I$ such that

$$\sup\{\tilde{c}^{\mathrm{T}}\tilde{A}^{-1}\tilde{b} \mid \tilde{A} \in A, \tilde{b} \in b, \tilde{c} \in c\} = \sup(c^{\mathrm{T}}x^*) = \sup(y^{*\mathrm{T}}b), \tag{6}$$

$$\inf(D(Ax^* - b)) = 0 \qquad D'x^* \ge 0, \tag{7}$$

$$\inf(D'(A^{\mathrm{T}}y^* - c)) = 0, \qquad Dy^* \ge 0. \tag{8}$$

Proof Denote the left hand side of (6) by γ. Clearly, $\gamma = \sup\{\tilde{c}^{\mathrm{T}}\tilde{x} \mid c \in c, \tilde{x} \in \Sigma(A, b)\}$, and since the convex hull of $\Sigma(A, b)$ is closed, bounded and spanned by the x_D^* ($|D| = I$), the supremum is attained for some $\tilde{c} \in c$ and some $x^* = x_D^*$ ($|D| = I$). Thus (7) holds with a suitable D' with $|D'| = I$, and by (5) we have $x^* = A^{*-1}b^*$, where

$$A^* = \check{A} - D \cdot \mathrm{rad}(A)D', \quad b^* = \check{b} + D \cdot \mathrm{rad}(b).$$

Now put

$$c^* := \check{c} + D'\,\mathrm{rad}(c), \quad y^* := (A^{*\mathrm{T}})^{-1}c^*.$$

Then $\gamma = \tilde{c}^{\mathrm{T}}x^* = \sup(c^{\mathrm{T}}x^*) = \check{c}^{\mathrm{T}}x^* + \mathrm{rad}(c)^{\mathrm{T}}|x^*| = c^{*\mathrm{T}}x^* = c^{*\mathrm{T}}A^{*-1}b^* = y^{*\mathrm{T}}b^*$, whence, for all $\tilde{D}$ with $|\tilde{D}| \le I$ and $\tilde{b} := \check{b} + \tilde{D} \cdot \mathrm{rad}(b) \in b$, we have

$$y^{*\mathrm{T}}\tilde{b} = c^{*\mathrm{T}}A^{*-1}\tilde{b} \le \gamma = y^{*\mathrm{T}}b^*,$$
$$0 \le y^{*\mathrm{T}}(b^* - \tilde{b}) = y^{*\mathrm{T}}(D - \tilde{D})\mathrm{rad}(b).$$

If $\mathrm{rad}(b) > 0$ then, by choosing D' as the matrix obtained from D by reversing the sign of D_{ii}, we find that $D_{ii}y_i^* \ge 0$ for all i. Therefore $Dy^* \ge 0$, and $\inf(D'(A^{\mathrm{T}}y^* - c)) = D'(\check{A}^{\mathrm{T}}y^* - \check{c}) - \mathrm{rad}(A^{\mathrm{T}})|y^*| - \mathrm{rad}(c) = D'(A^{*\mathrm{T}}y^* - c^*) = 0$. Hence the theorem holds if $\mathrm{rad}(b) > 0$. In the general case the assertion now follows by continuity. □

6.2.4 Corollary Let $B \in \mathbb{IR}^{n\times n}$ be an enclosure of A^{-1} and put

$$\mathfrak{D} := \{D \in \mathbb{R}^{n\times n} \mid |D| = I, \sup(DB^{\mathrm{T}}e^{(i)}) \ge 0 \text{ for some } i\}. \qquad (9)$$

Then $A^{\mathrm{H}}b = [\underline{z}, \overline{z}]$ with

$$\underline{z} = \inf\{x^*_{-D} \mid D \in \mathfrak{D}\}, \quad \overline{z} = \sup\{x^*_D \mid D \in \mathfrak{D}\}.$$

Proof Let $z = A^{\mathrm{H}}b$. Applying the previous theorem with $c = e^{(i)}$ we find $\overline{z}_i = x_i^*$. Now (7) gives $x^* = x^*_D$, and (8) shows that $Dy^* \ge 0$. But $y^* \in (A^{\mathrm{T}})^{-1}c \le B^{\mathrm{T}}e^{(i)}$; hence $\sup(DB^{\mathrm{T}}e^{(i)}) \ge Dy^* \ge 0$ so that $D \in \mathfrak{D}$. Similarly, applying the previous theorem with $c = -e^{(i)}$ we find $\underline{z}_i = x_i^*$ with $x^* = x^*_D$, $DB^{\mathrm{T}}e^{(i)} \le 0$, so that $-D \in \mathfrak{D}$. Since all x^*_D are in $A^{\mathrm{H}}b = z$, this implies the assertion. □

If A is strongly regular then a suitable enclosure of A^{-1} is given by Theorem 4.1.11. If necessary, we can improve such an enclosure B_0 by means of the Krawczyk-type iteration

$$B_{l+1} = (C + (I - CA)B_l) \cap B_l \quad (l = 0, 1, 2, \ldots)$$

or, better, by Gauss–Seidel iteration applied separately to each column of B_l. In this way it is often possible to restrict the set of sign matrices considerably.

In particular, if $0 \notin B_{ik}$ for $i, k = 1, \ldots, n$ (so that no $\tilde{A} \in A$ has an inverse with a zero entry – such matrices A are called *inverse stable*) then, for $i = 1, \ldots, n$,

$$(A^{\mathrm{H}}b)_i = [x^*_{-D}, x^*_D]_i \quad \text{with } D_{kk} = \mathrm{sgn}(B_{ik}) \quad (k = 1, \ldots, n).$$

Hence in the inverse stable case the calculation of $A^{\mathrm{H}}b$ only requires the determination of $2n$ vectors x^*_D.

Note that general interval matrices with sufficiently small radii will almost certainly be inverse stable. The inverse even of a sparse but irreducible matrix is 'structurally full' so that zero entries in $\check{A}^{-1}$ appear only with probability zero; and nonzeros of $C = \check{A}^{-1}$ will remain entries of B not containing zero when $\|CA - I\| \ll 1$, i.e. when $\mathrm{rad}(A)$ is sufficiently small.

In the more special case when $B \geq 0$ (so that A is inverse positive) we even have $A^{\mathrm{H}}b = [x^*_{-I}, x^*_I]$ with a further reduction of effort; but we already know methods which compute $A^{\mathrm{H}}b$ in this case (cf. Theorem 3.6.7 and Remark (2) after Theorem 4.4.8). Of course, this applies more generally whenever all columns of B have the same sign distribution.

We end this section with a description of methods for the actual computation of the x^*_D, and hence of $A^{\mathrm{H}}b$. We treat first the strongly regular case where (in particular when A has small radii) iterative methods can be used profitably.

6.2.5 Theorem Let $A \in \mathbb{IR}^{n\times n}$ be strongly regular and let $C \in \mathbb{R}^{n\times n}$ be such that CA is an H-matrix. If $b \in \mathbb{IR}^n$ then, for any enclosure x^0 of $A^{\mathrm{H}}b$ and any $D \in \mathbb{R}^{n\times n}$ with $|D| = I$, the iteration

$$x^{l+1} = \Gamma(C\check{A}, CD \cdot \mathrm{rad}(A)\mathrm{abs}(x^l) + Cb_D, x^l) \tag{10}$$

converges to the solution x^*_D of (1). (Here $\mathrm{abs}(x)$ denotes the interval extension of the absolute value function, not to be confused with $|x|$.)

Proof By (5a), $x^* = x^*_D$ satisfies the equation

$$C\check{A}x^* = CD \cdot \mathrm{rad}(A)|x^*| + Cb_D. \tag{11}$$

Since $x^* \in x^0$, this shows that every iterate x^l contains x^*. Since the x^l form a nested sequence, they have a limit x^∞ containing x^* and satisfying

$$x^\infty = \Gamma(C\check{A}, a, x^\infty), \tag{12}$$

where

$$a = CD \cdot \mathrm{rad}(A)\mathrm{abs}(x^\infty) + Cb_D. \tag{13}$$

Now (12) implies

$$x_i^\infty \subseteq \left(a_i - \sum_{k\neq i} (C\check{A})_{ik} x_k^\infty\right) \Big/ (C\check{A})_{ii}.$$

Since $C\check{A}$ is thin, taking radii gives

$$\mathrm{rad}(x_i^\infty) \leq \left(\mathrm{rad}(a_i) + \sum_{k\neq i} |C\check{A}|_{ik}\, \mathrm{rad}(x_k^\infty)\right) \Big/ \langle C\check{A}\rangle_{ii},$$

so that

$$\langle C\check{A}\rangle \mathrm{rad}(x^\infty) \leq \mathrm{rad}(a). \tag{14}$$

Since Cb_D is thin, $|D| = I$, and $\mathrm{rad}(\mathrm{abs}(x)) \leq \mathrm{rad}(x)$, (13) implies $\mathrm{rad}(a) \leq |C|\mathrm{rad}(A)\mathrm{rad}(x^\infty)$. Combined with (14), this yields $\langle CA\rangle \mathrm{rad}(x^\infty) \leq 0$. Since

CA is an H-matrix, $\text{rad}(x^\infty) \leq 0$ follows. Hence x^∞ is thin, and since it contains x^* we must have $x^\infty = x^*$. □

The following algorithm combines Theorem 6.2.5 with Corollary 6.2.4 to an effective method for hull inverse computation.

Algorithm (for computing $A^{\mathrm{H}}b$ where CA is an H-matrix)

Step 1 Compute enclosures x^0 of $A^{\mathrm{H}}b$ and B of A^{-1}. Put $x = -\infty, y = +\infty$, and choose $D \in \mathfrak{D}$ (given by (9)).

Step 2 Iterate according to (10) until either $x^{l+1} < x$ or $x^{l+1} = x^l$. In the latter case put $\underline{x} := \sup(\underline{x}, \underline{x}^l)$ and $\overline{x} := \sup(\overline{x}, \overline{x}^l)$.

Step 3 Replace D by $-D$. Iterate according to (10) until either $x^{l+1} > y$ or $x^{l+1} = x^l$. In the latter case put $\underline{y} := \inf(\underline{y}, \underline{x}^l)$ and $\overline{y} := \inf(\overline{y}, \overline{x}^l)$.

Step 4 Drop D from $\mathfrak{D}$. If $\mathfrak{D}$ is empty, stop; otherwise choose a new $D \in \mathfrak{D}$ and goto Step 2.

The algorithm stops with two narrow intervals x, y such that $[\overline{y}, \underline{x}] \subseteq A^{\mathrm{H}}b \subseteq [\underline{y}, \overline{x}]$.

The speed of the algorithm is mainly determined by the size of $\mathfrak{D}$ and the degree of diagonal dominance of CA; the latter determines the convergence rate of the iteration (10), which is approximately given by $\rho(|CA - I|)$. If $\|CA - I\|_u \leq 0.1$ for some scaled maximum norm we can expect A to be inverse stable (although this will not always be so), so that $\mathfrak{D}$ consists of n sign matrices only, and each iteration (10) converges after a small number $O(1)$ of steps. Since one step of (10) can be done in $O(n^2)$ operations, the algorithm finds $A^{\mathrm{H}}b$ in $O(n^3)$ operations. When $\|CA - I\|$ becomes larger, the number of sign matrices in $\mathfrak{D}$ and the number of iterations needed in (10) tends to grow rapidly, and the method is no longer economical.

As an alternative to iterative methods we can adapt the sign accord algorithm to the present situation. However, the complexity is at least $O(n^4)$ (for inverse stable matrices) and often exponential in n (when A is not strongly regular), so that this method is only practical for very small n.

Sign accord algorithm (for the computation of $A^{\mathrm{H}}b$)

Step 1 Compute $C \approx \check{A}^{-1}$. If CA is an H-matrix, compute an enclosure B of A^{-1} and define $\mathfrak{D}$ by (9); otherwise let $\mathfrak{D} = \{D \in \mathbb{R}^{n \times n} \mid |D| = I\}$. Put $z := \emptyset$, new := true, and choose $D \in \mathfrak{D}$.

Step 2 Choose $D' \in \mathbb{R}^{n \times n}$ with $|D'| = I$, $D'(Cb) \geq 0$, where $b = \check{b} + D \cdot \text{rad}(b)$. Put $s := 1$.

Step 3 Put $A := \check{A} - D \cdot \text{rad}(A) \cdot D'$ and compute $d := \det(A)$. If (new) then put dold := d, new := false. If $d \cdot \text{dold} \leq 0$ then stop (A singular).

Step 4 Solve $A\tilde{x} = b$. If $D'\tilde{x} \geq 0$, goto Step 5. Otherwise, compute

$k := \min\{j = 1, \ldots, n \mid D'_{jj}\tilde{x}_j < 0\}$ and change the sign of D'_{kk}. If $s = 2^n$, terminate (A singular); otherwise put $s := s + 1$ and goto Step 3.

Step 5 Put $z := \square(z \cup \{\tilde{x}\})$. Drop D from $\mathfrak{D}$. If $\mathfrak{D}$ is empty, stop ($A^{\mathrm{H}}b = [\underline{z}, \overline{z}]$); otherwise choose a new $D \in \mathfrak{D}$ and goto Step 2.

Clearly the algorithm is finite and the properties claimed at stop follow from Theorem 6.1.5, Theorem 6.2.1(v) and equations (4) and (5b). Both the determinant and the solution of $A\tilde{x} = b$ can easily be computed from a triangular decomposition of A. Since the sign change in Step 4 corresponds to a rank 1 change of A, the triangular decomposition can be updated with $O(n^2)$ operations. The same holds in Step 5 when the sign matrices are chosen in an order such that consecutive sign matrices differ in a single component only (this is certainly possible when $\mathfrak{D}$ consists of all 2^n sign matrices).

The above form of the algorithm assumes that exact arithmetic is used throughout. In finite precision arithmetic we obtain only approximations to $\underline{z}, \overline{z}$. Moreover, a wrong sign in Step 4 might be chosen due to rounding errors; this might lead to cycling and hence to a wrong decision for singularity. (In a limited number of experiments this never happened.) It is not clear how the algorithm must be modified to allow a rigorous error control. When using interval arithmetic we must also consider the problems of enclosing the determinant (see the Remarks to Chapter 4) and of updating the required information with $O(n^2)$ operations for each sign change.

Remarks to Chapter 6

To 6.1 Most results in this section are an interval-free version of results of Rohn (1982, 1984, 1985) (published in Rohn, 1988, 1990a); some of his proofs have been simplified.

With the definition $u := \sup(\tilde{x}, 0)$, $v := \sup(-\tilde{x}, 0)$, (4) is equivalent to the linear complementarity problem

$$(I - M)u = (I + M)v + a, \quad \inf(u, v) = 0.$$

See the last chapter in the book by Berman & Plemmons (1979) for a discussion of linear complementarity problems and for references.

To 6.2 Most results are again due to Rohn (1988, 1990a). Theorem 6.2.5 is new. The observation that most inverse matrices are full is due to Duff *et al.* (1988). For numerically stable *rank 1 updates* of triangular decompositions with $O(n^2)$ operations see Fletcher & Matthews (1984).

Earlier partial results on the computation of $A^{H}b$ can be found in Beeck (1975), Braess (1966), Hansen (1969a,b), Hartfiel (1980), Kuperman (1971) and Rossier (1982).

A recent result of Poljak & Rohn (1988) shows that the problem of deciding whether an interval matrix is regular or not is *NP*-complete, and hence very unlikely to be achievable in a polynomial number of operations. (See Garey & Johnson, 1979, for the significance of *NP*-completeness for the complexity of algorithms.)

Further remarks Some results are available for other hull computations. See Garloff (1980) (fast Fourier transform), Garloff & Krawczyk (1986), Krawczyk (1987b) (zero sets of function strips), Rohn (1990b) (eigenvalues of symmetric interval matrices), Deif (1990) (eigenvalues and eigenvectors of general interval matrices), Oelschlägel & Süsse (1983) (convex optimization), Nuding (1981) (matrix exponential of interval *M*-matrices); see also Nuding (1983).

REFERENCES

Adams, E. 1980a. On sets of solutions of collections of nonlinear systems in IR^n. In *Interval Mathematics 1980* (ed. K. Nickel), pp. 247–56. Academic Press, New York.

Adams, E. 1980b. On methods for the construction of the boundaries of sets of solutions for differential equations or finite-dimensional approximations with input sets. *Computing Suppl.* **2**, 1–16.

Adams, E. & Ames, W. F. 1982. Linear or nonlinear hyperbolic wave problems with input sets (Part 1). *J. Engin. Math.* **16**, 23–45.

Adams, E. & Lohner, R. 1983. Error bounds and sensitivity analysis. *Scientific Computing* (eds. R. S. Stepleman *et al.*), pp. 213–22. North-Holland, Amsterdam.

Adams, E. & Spreuer, H. 1975. Konvergente numerische Schrankenkonstruktion mit Spline-Funktionen für nichtlineare gewöhnliche bzw. lineare parabolische Randwertaufgaben. In *Interval Mathematics* (ed. K. Nickel), pp. 118–26. Springer Lecture Notes in Computer Science 29, Berlin.

Alefeld, G. 1970a. Über Eigenschaften und Anwendungsmöglichkeiten einer komplexen Intervallarithmetik. *Z. Angew. Math. Mech.* **50**, 455–65.

Alefeld, G. 1970b. Eine Modifikation des Newtonverfahrens zur Bestimmung der reellen Nullstellen einer reellen Funktion. *Z. Angew. Math. Mech.* **50**, T32–3.

Alefeld, G. 1977a. Das symmetrische Einzelschrittverfahren bei linearen Gleichungen mit Intervallen als Koeffizienten. *Computing* **18**, 329–40.

Alefeld, G. 1977b. Über die Durchführbarkeit des Gaußschen Algorithmus bei Gleichungen mit Intervallen als Koeffizienten. *Computing Suppl.* **1**, 15–19.

Alefeld, G. 1977c. Zur iterativen Einschließung von positiven Inversen. *Computing* **19**, 171–4.

Alefeld, G. 1979. Intervallanalytische Methoden bei nichtlinearen Gleichungen. In *Jahrbuch Überblicke Mathematik 1979* (eds. S. D. Chatterji *et al.*), pp. 63–78. Bibl. Inst., Mannheim.

Alefeld, G. 1981a. Bounding the slope of polynomial operators and some applications. *Computing* **26**, 227–37.

Alefeld, G. 1981b. Konvergenzbeschleunigende Maßnahmen bei einem Verfahren zur Einschließung von positiven Inversen. *Beiträge Numer. Math.* **9**, 7–12.

Alefeld, G. 1984. On the convergence of some interval-arithmetic modifications of Newton's method. *SIAM J. Numer. Anal.* **21**, 363–72.

Alefeld, G. 1986. Componentwise inclusion and exclusion sets for solutions of quadratic equations in finite dimensional spaces. *Numer. Math.* **48**, 391–416.

Alefeld, G. 1987. Rigorous error bounds for singular values of a matrix using the precise scalar product. In *Computerarithmetic* (eds. E. Kaucher, U. Kulisch & Ch. Ullrich), pp. 9–30. Teubner, Stuttgart.

Alefeld, G. & Herzberger, J. 1974. Über die Verbesserung von Schranken für die Eigenwerte symmetrischer Tridiagonalmatrizen. *Angew. Informatik (Elektron. Datenverarbeitung)* **16**, 27–35.

Alefeld, G. & Herzberger, J. 1983. *Introduction to Interval Computations.* Academic Press, New York.

Alefeld, G. & Spreuer, H. 1986. Iterative improvement of componentwise errorbounds for invariant subspaces belonging to a double or nearly double eigenvalue. *Computing* **36**, 321–34.

Allgower, E. & Georg, K. 1983. Predictor-corrector and simplicial methods for approximating fixed points and zero points of nonlinear mappings. In *Mathematical Programming, The State of the Art, Bonn 1982*, pp. 15–56, Springer, Berlin.

Alvarado, F. L. 1990a. Sparsity preservation in matrix interval solutions, to appear.

Alvarado, F. L. 1990b. Sparse W-matrix for interval arithmetic, to appear.

Al-Zanaidi, M. & Grossmann, C. 1989. Monotone iteration discretization algorithm for BVP's. *Computing* **41**, 59–74.

Ams, A. 1987. Anwendung von Intervallarithmetik in Simulations- und Parameterschätzverfahren für dynamische Systeme. Manuskript, Inst. f. Angew. Math., Univ. Karlsruhe.

Apostolatos, N. & Kulisch, U. 1968. Grundzüge einer Intervallrechnung für Matrizen und einige Anwendungen. *Elektron. Rechenanlagen* **10**, 73–83.

Appelt, W. 1973. Fehlereinschließung bei der numerischen Lösung elliptischer Differentialgleichungen unter Verwendung eines intervallarithmetischen Verfahren. Berichte der GMD Bonn 79.

Appelt, W. 1974. Fehlereinschließung für die Lösung einer Klasse elliptischer Randwertaufgaben. *Z. Angew. Math. Mech.* **54**, T207–8.

Aubin, J.-P. & Ekeland, I. 1984. *Applied Nonlinear Analysis.* Wiley, New York.

Barth, W. & Nuding, E. 1974. Optimale Lösung von Intervallgleichungssystemen. *Computing* **12**, 117–25.

Bartle, R. G. 1955. Newton's method in Banach spaces. *Proc. Amer. Math. Soc.* **6**, 827–31.

Bauch, H. 1977. Zur intervallanalytischen Lösungseinschließung bei charakteristischen Anfangswertproblemen mit hyperbolischer Differentialgleichung $z_{st} = f(s,t,z)$. *Z. Angew. Math. Mech.* **57**, 543–7.

Bauch, H. 1980a. On the iterative inclusion of solutions in characteristic initial-value problems with hyperbolic differential equation $z_{st} = f(s,t,z)$. *Computing* **24**, 21–32.

Bauch, H. 1980b. Zur iterativen Lösung von charakteristischen Anfangswertproblemen mit hyperbolischer Differentialgleichung $z_{st} = f(s,t,z)$ mittels Intervallmethoden. *Wiss. Z. Tech. Univ. Dresden* **29**, 117–22.

Bauch, H., Jahn, K.-U., Oelschlägel, D., Süße, H. & Wiebigke, V. 1987. *Intervallmathematik.* Teubner, Leipzig.

Baumann, E. 1986. Globale Optimierung stetig differenzierbarer Funktionen einer Variablen. Freiburger Intervall-Ber. 86/6.

Baumann, E. 1988. Optimal centered forms. *BIT* **28**, 80–7.

Baur, W. & Strassen, V. 1983. The complexity of partial derivatives. *Theor. Comp. Sci.* **22**, 317–30.

Beeck, H. 1971. Über intervallanalytische Methoden bei linearen Gleichungssystemen mit Intervallkoeffizienten und Zusammenhänge mit der Fehleranalysis. Dissertation, Techn. Hochsch. München.

Beeck, H. 1972. Über Struktur und Abschätzungen der Lösungsmengen von linearen Gleichungssystemen mit Intervallkoeffizienten. *Computing* **10**, 231–44.

Beeck, H. 1973. Charakterisierung der Lösungsmenge von Intervallgleichungssystemen. *Z. Angew. Math. Mech.* **53**, T181–2.

Beeck, H. 1974. Zur scharfen Außenabschätzung der Lösungsmenge bei linearen Intervallgleichungssystemen. *Z. Angew. Math. Mech.* **54**, T208–9.

Beeck, H. 1975. Zur Problematik der Hüllenbestimmung von Intervallgleichungssystemen. In *Interval Mathematics* (ed. K. Nickel), pp. 150–9. Springer Lecture Notes in Computer Science 29, Berlin.

Beeck, H. 1977. Parallel algorithms for linear equations with not sharply defined data. In *Parallel Computers – Parallel Mathematics* (ed. M. Feilmeier), pp. 257–61. North-Holland, Amsterdam.

Beeck, H. 1978. Linear programming with inexact data. Institutsbericht Nr. TUM-ISU-7830, Inst. f. Statistik und Unternehmensforschung, Techn. Universität München.

Beeck, H. 1979. Schwankungsbereiche linearer Optimierungsaufgaben. *Oper. Res. Verfahren* **31**, 59–72.

Behnke, H. 1988. Inclusion of eigenvalues of general eigenvalue problems for matrices. *Computing Suppl.* **6**, 69–78.

Berman, A. & Plemmons, R. J. 1979. *Nonnegative Matrices in the Mathematical Sciences.* Academic Press, New York.

Böhm, H., Rump, S. M. & Schuhmacher, G. 1987. E-methods for nonlinear problems. In *Computerarithmetic* (eds. E. Kaucher, U. Kulisch & Ch. Ullrich), pp. 59–80. Teubner, Stuttgart.

Börsken, N. C. 1978. Komplexe Kreis–Standardfunktionen. Freiburger Intervall-Ber. 78/2.

Bohl, E. 1974. *Monotonie: Lösbarkeit und Numerik bei Operatorgleichungen.* Springer, Berlin.

Bohlender, G. 1977. Genaue Summation von Gleitkommazahlen. *Computing Suppl.* **1**, 21–32.

Braess, D. 1966. Die Berechnung der Fehlergrenzen bei linearen Gleichungssystemen mit fehlerhaften Koeffizienten. *Arch. Rat. Mech. Anal.* **19**, 74–80.

Brouwer, L. 1912. Über Abbildungen von Mannigfaltigkeiten. *Math. Ann.* **71**, 97–115.

Caprani, O. & Madsen, K. 1980. Mean value forms in interval analysis. *Computing* **25**, 147–54.

Caprani, O., Madsen, K. & Rall, L. B. 1981. Integration of interval functions. *SIAM J. Math. Anal.* **12**, 321–41.

Clarke, F. H. 1983. *Optimization and Nonsmooth Analysis.* Wiley, New York.

Clemmesen, M. 1984. Interval arithmetic implementations using floating point arithmetic. *ACM Signum Newsletter* **19**, 2–8.

Cody, W. J. 1988. Floating-point standards – theory and practice. In *Reliability in Computing* (ed. R. E. Moore), pp. 99–107. Academic Press, London.

Collatz, L. 1964. *Funktionalanalysis und numerische Mathematik*. Springer, Berlin.

Collins, G. E. 1977. Infallible calculation of polynomial zeros to specified precision. In *Mathematical Software III* (ed. J. R. Rice), pp. 35–68. Academic Press, New York.

Cordes, D. 1987. Verifizierter Stabilitätsnachweis für Lösungen von Systemen periodischer Differentialgleichungen auf dem Rechner mit Anwendungen. Dissertation, Fak. f. Math., Univ. Karlsruhe.

Cordes, D. & Adams, E. 1987. Test for uniform boundedness for dynamic problems admitting parameter and combination resonance. In *Computerarithmetic* (eds. E. Kaucher, U. Kulisch & Ch. Ullrich), pp. 115–32. Teubner, Stuttgart.

Cordes, D. & Kaucher, E. 1987. Self-validating computation for sparse matrix problems. In *Computerarithmetic* (eds. E. Kaucher *et al.*), pp. 133–49. Teubner, Stuttgart.

Corliss, G. 1987a. Computing narrow inclusions for definite integrals. In *Computerarithmetic* (eds. E. Kaucher, U. Kulisch & Ch. Ullrich), pp. 150–69. Teubner, Stuttgart.

Corliss, G. 1987b. Performance of self-validating adaptive quadrature. In *Numerical Integration* (eds. P. Keast & G. Fairweather), pp. 239–59. Reidel, Dordrecht.

Corliss, G. 1988. Applications of differentiation arithmetic. In *Reliability in Computing* (ed. R. E. Moore), pp. 127–48. Academic Press, London.

Corliss, G. & Rall, L. B. 1987. Adaptive, self-validating numerical quadrature. *SIAM J. Sci. Statist. Comput.* **8**, 831–47.

Corliss, G., Krenz, G. S. & Davis, P. 1988. Bibliography on interval methods for the solution of ordinary differential equations. Tech. Rep. 289, Marquette Univ., Milwaukee, Wisc.

Cornelius, H. 1981. Untersuchungen zu einem intervallarithmetischen Iterationsverfahren mit Anwendungen auf eine Klasse nichtlinearer Gleichungssysteme. Dissertation, Techn. Univ. Berlin.

Cornelius, H. 1983. On the acceleration of an interval-arithmetic iteration method. *SIAM J. Numer. Anal.* **20**, 1010–22.

Cornelius, H. & Alefeld, G. 1982. A device for the acceleration of convergence of a monotonously enclosing iteration method. In *Iterative Solution of Nonlinear Systems of Equations* (eds. R. Ansorge, Th. Meis & W. Törnig), pp. 68–79. Springer Lecture Notes in Mathematics 953, Berlin.

Cornelius, H. & Lohner, R. 1984. Computing the range of values of real functions with accuracy higher than second order. *Computing* **33**, 331–47.

Cryer, C. W. 1969. On the computation of rigorous bounds for the solutions of linear integral equations with the aid of interval arithmetic, Computer Science Technical Report #70, University of Wisconsin, Madison.

Deif, A. S. 1986. *Sensitivity Analysis in Linear Systems*. Springer, Berlin.

Deif, A. S. 1990. The interval eigenvalue problem. *Zeitschr. Angew. Math. Mech.*, to appear.

Demmel, J. W. & Krückeberg, F. 1985. An interval algorithm for solving systems of linear equations to prespecified accuracy. *Computing* **34**, 117–29.

Dennis, J. E. & Schnabel, R. B. 1983. *Numerical Methods for Unconstrained Optimization and Nonlinear Equations.* Prentice-Hall, Englewood Cliffs, N.J.

Dobner, H.-J. 1987. Bounds for the solution of hyperbolic problems. *Computing* **38**, 209–18.

Dobner, H.-J. 1989. Computing narrow inclusions for the solution of integral equations. *Numer. Funct. Anal. Optimiz.* **10**, 923–36.

Dobner, H.-J. & Kaucher, E. 1987. Solving characteristic initial value problems with guaranteed errorbounds. In *Computerarithmetic* (eds. E. Kaucher, U. Kulisch & Ch. Ullrich), pp. 170–85. Teubner, Stuttgart.

Dobronets, B. S. 1988. Two-sided solution of ODEs via *a posteriori* error estimates. *J. Comput. Appl. Math.* **23**, 53–62.

Dubois, D. & Prade, H. 1987. Fuzzy numbers: an overview. *Analysis of Fuzzy Information, Math. Logic* **1**, 3–39.

Duff, I. S., Erisman, A. M., Gear, C. W. & Reid, J. K. 1988. Sparsity structure and Gaussian elimination. *ACM SIGNUM Newsletter* **23**, 2–8.

Dwyer, P. S. 1951. *Linear Computations.* Wiley, New York.

Eckmann, J.-P. & Wittwer, P. 1985. *Computer Methods and Borel Summability Applied to Feigenbaum's Equation.* Springer Lecture Notes in Physics 227, Berlin.

Eiermann, M. Ch. 1986. Schranken für lineare Funktionale mit Anwendung auf adaptive Quadratur. Freiburger Intervall-Ber. 86/5.

Eiermann, M. Ch. 1989a. Automatic, guaranteed integration of analytic functions. *BIT* **29**, 270–82.

Eiermann, M. Ch. 1989b. Adaptive Berechnung von Integraltransformationen mit Fehlerschranken. Dissertation, Inst. f. Angew. Math., Freiburg.

Eijgenraam, P. 1981. *The Solution of Initial Value Problems using Interval Arithmetic.* Math. Centre Tracts 144, Amsterdam.

Faass, E. 1975. Beliebig genaue numerische Schranken für die Lösung parabolischer Randwertaufgaben. Dissertation, Univ. Karlsruhe.

Fefferman, C. & de la Llave, R. 1986. Relativistic stability of matter-I. *Revista Mat. Iberoamerica* **2**, 119–213.

Fletcher, R. & Matthews, S. P. J. 1984. Stable modification of explicit LU factors for simplex updates. *Math. Programming* **30**, 267–84.

Frobenius, G. 1912. Über Matrizen aus nicht negativen Elementen. *Sitz. Ber. Preuss. Akad. Wiss. Berlin,* pp. 456–77.

Frommer, A. & Mayer, G. 1990. On the R-order of Newton-like methods for enclosing solutions of nonlinear equations. *SIAM J. Numer. Anal.*, to appear.

Fujii, Y., Ichida, K. & Ozasa, M. 1986. Maximization of multivariable functions using interval analysis. In *Interval Mathematics 1985* (ed. K. Nickel), pp. 17–26. Springer Lecture Notes in Computer Science 212, Berlin.

Gambill, T. N. & Skeel, R. D. 1988. Logarithmic reduction of the wrapping effect with applications to ordinary differential equations. *SIAM J. Numer. Anal.* **25**, 153–62.

Gantmacher, F. R. 1959. *Applications of the Theory of Matrices.* Interscience, New York.

Garey, M. R. & Johnson, D. S. 1979. *Computers and Intractability: A Guide to the Theory of NP-Completeness.* Freeman, San Francisco.

Garloff, J. 1980. Zur intervallmässigen Durchführung der schnellen Fourier-Transformation. *Z. Angew. Math. Mech.* **60**, T291–2.
Garloff, J. 1985. Interval mathematics, a bibliography. Freiburger Intervall-Ber. 85/6.
Garloff, J. 1986a. Solution of linear equations having a Toeplitz interval matrix as coefficient matrix. *Opuscula Math.* **2**, 33–45.
Garloff, J. 1986b. Convergent bounds for the range of multivariate polynomials. In *Interval Mathematics 1985* (ed. K. Nickel), pp. 37–56. Springer Lecture Notes in Computer Science 212, Berlin.
Garloff, J. 1987. Bibliography on interval mathematics, continuation. Freiburger Intervall-Ber. 87/2, pp. 1–50.
Garloff, J. 1990. Block methods for the solution of linear interval equations. *SIAM J. Matrix Anal. Appl.* **11**, 89–106.
Garloff, J. & Bose, N. K. 1988. Boundary implications for stability properties: present status. In *Reliability in Computing* (ed. R. E. Moore), pp. 391–402. Academic Press, London.
Garloff, J. & Krawczyk, R. 1986. Optimal inclusion of a solution set. *SIAM J. Numer. Anal.* **23**, 217–26.
Gay, D. M. 1980. Using scalar and vector majorizing equations to bound solution sets of nonlinear algebraic equation systems. In *Interval Mathematics 1980* (ed. K. Nickel), pp. 329–39. Academic Press, New York.
Gay, D. M. 1982. Solving interval linear equations. *SIAM J. Numer. Anal.* **19**, 858–70.
Gay, D. M. 1983. Computing perturbation bounds for nonlinear algebraic equations. *SIAM J. Numer. Anal.* **20**, 638–51.
Gay, D. M. 1988. Interval least squares – a diagnostic tool. In *Reliability in Computing* (ed. R. E. Moore), pp. 183–206. Academic Press, London.
Gerlach, W. 1981. Zur Lösung linearer Ungleichungssysteme bei Störung der rechten Seite und der Koeffizientenmatrix. *Math. Operations-forsch. Statist. Ser. Optim.* **12**, 41–3.
Grassmann, E. & Rokne, J. 1970. The range of values of a circular complex polynomial over a circular complex interval. *Computing* **23**, 139–69.
Hahn, W., Mohr, K. & Schauer, U. 1985. Some techniques for solving linear equation systems with guarantee. *Computing* **34**, 375–9.
Hansen, E. 1965. Interval arithmetic in matrix computations, Part 1, *SIAM J. Numer. Anal.* **2**, 308–20.
Hansen, E. 1969a. On linear algebraic equations with interval coefficients. In *Topics in Interval Analysis* (ed. E. Hansen), pp. 35–46. Oxford University Press.
Hansen, E. 1969b. On the solution of linear algebraic equations with interval coefficients. *Linear Algebra Appl.* **2**, 153–65.
Hansen, E. 1975. A generalized interval arithmetic. In *Interval Mathematics* (ed. K. Nickel), pp. 7–18. Springer Lecture Notes in Computer Science 29, Berlin.
Hansen, E. 1978a. A globally convergent interval method for computing and bounding real roots. *BIT* **18**, 415–24.
Hansen, E. 1978b. Interval forms of Newton's method. *Computing* **20**, 153–63.
Hansen, E. 1980. Global optimization using interval analysis – the multi-dimensional case. *Numer. Math.* **34**, 247–70.

Hansen, E. 1988. An overview of global optimization using interval analysis. In *Reliability in Computing* (ed. R. E. Moore), pp. 289–308. Academic Press, London.

Hansen, E. & Greenberg, R. I. 1983. An interval Newton method. *Appl. Math. Comput.* **12**, 89–98.

Hansen, E. & Sengupta, S. 1980. Global constrained optimization using interval analysis. In *Interval Mathematics 1980* (ed. K. Nickel), pp. 25–47. Academic Press, New York.

Hansen, E. & Sengupta, S. 1981. Bounding solutions of systems of equations using interval analysis. *BIT* **21**, 203–11.

Hansen, E. & Smith, R. 1967. Interval arithmetic in matrix computations, Part 2. *SIAM J. Numer. Anal.* **4**, 1–9.

Hansen, E. & Walster, G. W. 1982. Global optimization in nonlinear mixed integer problems. In *Proceedings of 10th IMACS World Congress on Syst. Simul. Sci. Comput.*, Vol. 1 (eds. W. F. Ames & R. Vichnevetsky), pp. 379–81. IMACS.

Hansen, E. & Walster, G. W. 1987. Nonlinear equations and optimization. Manuscript.

Hansen, E. & Walster, G. W. 1990a. Equality constrained global optimization. *SIAM J. Contr. Opt.*, to appear.

Hansen, E. & Walster, G. W. 1990b. Bounds for Lagrange multipliers and optimal points. *J. Opt. Theory Appl.*, to appear.

Hartfiel, D. J. 1980. Concerning the solution set of $Ax = b$ where $P \leq A \leq Q$ and $p \leq b \leq q$. *Numer. Math.* **35**, 355–9.

Hauenschild, M. 1974. Arithmetiken für komplexe Kreise. *Computing* **13**, 299–312.

Heindl, G. & Reinhart, E. 1976a. Eine allgemeine Methode zur Berechnung von Minimax-Fehlern, Teil 1: Bei vorliegenden Messungen. *Z. Vermessungswesen* **101**, 126–32.

Heindl, G. & Reinhart, E. 1976b. Eine allgemeine Methode zur Berechnung von Minimax-Fehlern, Teil 2: Vor der Durchführung von Messungen. *Z. Vermessungswesen* **101**, 238–41.

Heindl, G. & Reinhart, E. 1978. Eine allgemeine Methode zur Berechnung von Minimax-Fehlern, Teil 3: Bei teils vorliegenden, teils projektierten Messungen. *Z. Vermessungswesen* **103**, 149–55.

Heizmann, F., 1990. Parametrisierte Einschließung der Lösung von Gleichungssystemen mit Parametern, Diplomarbeit. Inst. f. Angew. Math., Univ. Freiburg.

Henrici, P. 1971. Circular arithmetic and the determination of polynomial zeros. In *Conference on Application of Numerical Analysis,* pp. 86–92. Springer Lecture Notes in Mathematics 228, Berlin.

Henrici, P. 1974. *Applied and Computational Complex Analysis.* Vol. 1, Wiley, New York.

Herzberger, J. 1985. Über ein Iterationsverfahren zur Einschließung der inversen Matrix. *Computing* **35**, 185–8.

Hollot, V. V. & Bartlett, A. C. 1990. On the eigenvalues of interval matrices, to appear.

IBM. 1983a. *High-accuracy Arithmetic, Subroutine Library, General Information Manual.* IBM-FORM GC33-6163 (last revised 1986).

IBM. 1983b. *High-accuracy Arithmetic, Subroutine Library, Program Description and User's Guide.* IBM-FORM SC33-6164 (last revised 1986).

Ichida, K. & Fujii, Y. 1979. An interval arithmetic method for global optimization. *Computing* **23**, 85–97.

Iri, M. 1984. Simultaneous computation of functions, partial derivatives and estimates of rounding errors – complexity and practicality. *Japan J. Appl. Math.* **1**, 223–52.

Jahn, K.-U. 1974. Eine Theorie der Gleichungssysteme mit Intervallkoeffizienten. *Z. Angew. Math. Mech.* **54**, 405–12.

Jansson, Ch. 1985. Zur linearen Optimierung mit unscharfen Daten, Dissertation, Kaiserslautern 1985.

Jansson, Ch. 1988. A self-validating method for solving linear programming problems with interval input data. *Computing Suppl.* **6**, 33–45.

Jones, S. T. 1980. Locating safe starting regions for iterative methods: A heuristic algorithm. In *Interval Mathematics 1980* (ed. K. Nickel), pp. 377–86. Academic Press, New York.

Kahan, W. M. 1968. *A More Complete Interval Arithmetic.* Lecture notes for an engineering summer course in numerical analysis, University of Michigan.

Kakutani, S. 1941. A generalization of Brouwer's fixed point theorem. *Duke Math. J.* **8**, 457–9.

Kalmykov, S. A. 1978. To the problem of determination of the symmetric matrix eigenvalues by means of the interval method. In *Numerical Analysis*, pp. 55–9. Collect. Sci. Works Sov. Acad. Sci., Sib. Branch, Inst. Theor. Appl. Mech., Novosibirsk. (In Russian.)

Kaucher, E. 1984. Self-validating numerical methods. In *Proceedings of the International Amse Conference on Modelling and Simulation*, Vol. 1.3, pp. 9–32. Athens, June 27–9, 1984.

Kaucher, E. 1987. Self-validating computation of ordinary and partial differential equations. In *Computerarithmetic* (eds. E. Kaucher, U. Kulisch & Ch. Ullrich), pp. 221–54. Teubner, Stuttgart.

Kaucher, E. & Miranker, W. L. 1984. *Self-Validating Numerics for Function Space Problems.* Academic Press, Orlando, Florida.

Kaucher, E. & Miranker, W. L. 1988. Validating computation in a function space. In *Reliability in Computing* (ed. R. E. Moore), pp. 403–26. Academic Press, London.

Kearfott, R. B. 1987a. Abstract generalized bisection and a cost bound. *Math. Comp.* **49**, 187–202.

Kearfott, R. B. 1987b. Some tests of generalized bisection. *ACM Trans. Math. Software* **13**, 197–220.

Kearfott, R. B. 1987c. An interval Newton method for nonlinear least squares. Manuscript.

Kearfott, R. B. 1990a. Interval Newton/generalized bisection when there are singularities near roots. *Ann. Oper. Res.*, to appear.

Kearfott, R. B. 1990b. Preconditioners for the interval Gauss–Seidel method. *SIAM J. Numer. Anal.*, **27**, to appear.

Kearfott, R. B. 1990c. An interval step control for continuation methods. Manuscript, Lafayette, LA.

Kearfott, R. B. 1990d. Interval arithmetic techniques in the computational solution of nonlinear systems of equations. In *Computational Solution of*

Nonlinear Systems of Equations (eds. E. L. Allgower & K. Georg), Amer. Math. Soc., Providence, R.I.

Kearfott, R. B. & Novoa, M., III. 1990. INTBIS, A portable interval Newton/bisection package. *ACM Trans. Math. Softw.*, to appear.

Kelch, R. 1989. Kontinuierliche Einschließung der Lösungsfunktion funktionaler und parameterabhängiger Probleme auf dem Computer. *Z. Angew. Math. Mech.* **69**, T42–4.

Kenney, Ch., Linz, P. & Wang, R. L. C. 1989. Effective error estimates for the numerical solution of Fredholm integral equations. *Computing* **42**, 353–62.

Klatte, R. & Wolff von Gudenberg, J. 1986. Forschungsschwerpunkt 'Computerarithmetik und Programmiersysteme für ingenieurwissenschaftliche Anwendungen'. Inst. f. Angew. Math., Univ. Karlsruhe.

Klein, H.-O. 1973. Eine intervallanalytische Methode zur Lösung von Integralgleichungen mit mehrparametrigen Funktionen. Berichte der GMD Bonn 69.

Kolev, L. V. 1985. Global solution of the least squares nonlinear regression problem. *Automat. Izchisl. Tekhn.* **2**, 31–6. (In Bulgarian.)

Kolev, L. V. & Mladenov, V. M. 1990. An interval method for finding all operating points of non-linear resistive circuits. *Int. J. Circuit Th. Appl.* **18**, to appear.

Koparkar, P. A. & Mudur, S. P. 1985. Subdivision techniques for processing geometric objects. In *Fundamental Algorithms for Computer Graphics* (ed. R. A. Earnshaw), pp. 751–801. Springer, Berlin.

Kopp, G. 1976. Die numerische Behandlung von reellen linearen Gleichungssystemen mit Fehlererfassung für M-Matrizen sowie für diagonaldominante und invers-isotone Matrizen, Diplomarbeit, Inst. f. prakt. Math, Univ. Karlsruhe.

Krawczyk, R. 1968. Iterative Verbesserung von Schranken für Eigenwerte und Eigenvektoren reeller Matrizen. *Z. Angew. Math. Mech.* **48**, T80–3.

Krawczyk, R. 1969a. Newton-Algorithmen zur Bestimmung von Nullstellen mit Fehlerschranken. *Computing* **4**, 187–201.

Krawczyk, R. 1969b. Fehlerabschätzung reeller Eigenwerte und Eigenvektoren von Matrizen. *Computing* **4**, 281–93.

Krawczyk, R. 1970a. Einschließung von Nullstellen mit Hilfe einer Intervallarithmetik. *Computing* **5**, 356–70.

Krawczyk, R. 1970b. Verbesserung von Schranken für Eigenwerte und Eigenvektoren von Matrizen. *Computing* **5**, 200–6.

Krawczyk, R. 1975. Fehlerabschätzung bei linearer Optimierung. In *Interval Mathematics* (ed. K. Nickel), pp. 215–22. Springer Lecture Notes in Computer Science 29, Berlin.

Krawczyk, R. 1980. Zur Konvergenz iterierter Mengen. Freiburger Intervall-Ber. 80/3, pp. 1–32.

Krawczyk, R. 1983a. A remark about the convergence of interval sequences. *Computing* **31**, 255–9.

Krawczyk, R. 1983b. Interval operators of a function strip of which the Lipschitz-matrix is an interval M-matrix. *Computing* **31**, 245–53.

Krawczyk, R. 1984. Interval iteration for including a set of solutions. *Computing* **32**, 13–31.

Krawczyk, R. 1986a. A class of interval-Newton-operators. *Computing* **37**, 179–83.
Krawczyk, R. 1986b. A Lipschitz operator for function strips. *Computing* **36**, 169–74.
Krawczyk, R. 1986c. Properties of interval operators. *Computing* **37**, 227–45.
Krawczyk, R. 1986d. Interval operators and fixed intervals. In *Interval Mathematics 1985* (ed. K. Nickel), pp. 81–94. Springer Lecture Notes in Computer Science 212, Berlin.
Krawczyk, R. 1987a. Conditionally isotone interval operators. *Computing* **39**, 261–70.
Krawczyk, R. 1987b. Optimal enclosure of a generalized zero set of a function strip. *SIAM J. Numer. Anal.* **24**, 1202–11.
Krawczyk, R. 1988. On interval operators obtained by splitting the Lipschitz matrix. *J. Math. Anal. Appl.* **136**, 609–31.
Krawczyk, R. & Neumaier, A. 1985. Interval slopes for rational functions and associated centered forms. *SIAM J. Numer. Anal.* **22**, 604–16.
Krawczyk, R. & Neumaier, A. 1986. An improved interval Newton operator. *J. Math. Anal. Appl.* **118**, 194–207.
Krawczyk, R. & Neumaier, A. 1987. Interval Newton operators for function strips. *J. Math. Anal. Appl.* **124**, 52–72.
Krawczyk, R. & Selsmark, F. 1980. Order convergence and iterative interval methods. *J. Math. Anal. Appl.* **73**, 1–23.
Krenz, G. S. 1987. Using weight functions in self-validating quadrature. In *Numerical Integration* (eds. P. Keast & G. Fairweather), pp. 261–8. Reidel, Dordrecht.
Krier, N. 1974. Komplexe Kreisarithmetik. *Z. Angew. Math. Mech.* **54**, T225–6.
Krückeberg, F. & Leisen, R. 1985. Solving initial value problems of ordinary differential equations to arbitrary accuracy with variable precision arithmetic. In *Proc. 11th IMACS Congr. Syst. Simul. Sci. Comput.*, Vol. 1, pp. 111–14. IMACS.
Kulisch, U. 1969. Grundzüge der Intervallrechnung. In *Überblicke Mathematik 2* (ed. L. Laugwitz), pp. 51–98. Bibl. Inst., Mannheim.
Kulisch, U. 1976. *Grundlagen des Numerischen Rechnens*, Reihe Informatik 19. Bibl. Inst., Mannheim.
Kulisch, U. 1983. A new arithmetic for scientific computation. In *A New Approach to Scientific Computation* (eds. U. W. Kulisch & W. L. Miranker), pp. 1–26. Academic Press, New York.
Kulisch, U. (ed.) 1987a. *PASCAL-SC, A PASCAL-Extension for Scientific Computation.* Information manual and floppy disks, IBM-PC, Wiley-Teubner, Stuttgart.
Kulisch, U. (ed.) 1987b. *PASCAL-SC, A PASCAL-Extension for Scientific Computation.* Information manual and floppy disks, ATARI ST, Teubner, Stuttgart.
Kulisch, U. & Bohlender, G. 1976. Formalization and implementations of floating-point matrix operations. *Computing* **16**, 239–61.
Kuperman, I. B. 1971. *Approximate Linear Algebraic Equations.* van Nostrand, London.
Kuratowski, K. 1966. *Topologie*, Vol. 2, 4th edn. Academic Press, New York.
Kuttler, J. 1971. A fourth order finite-difference approximation for the fixed membrane eigenproblem. *Math. Comput.* **25**, 237–56.

Lanford, O. E., III. 1986. *Computer Assisted Proofs in Analysis.* Abstract of the International Congress of Mathematics, Berkeley, Aug. 3–11, 1986, pp. 295–6.

Lang, B. 1989. Lokalisierung und Darstellung von Nullstellenmengen einer Funktion. Diplomarbeit, Univ. Karlsruhe.

Lehmann, N. J. 1985. Die analytische Maschine, Grundlagen einer Computer-Analytik, *Sitzungsber. Sächs. Akad. Wiss. Leipzig, Math.-Naturw. Kl.* **118**, (4).

Leray, J. & Schauder, J. 1934. Topologie et équations fonctionelles. *Ann. Sci. École Norm. Sup.* **51**, 45–78.

Lodwick, W. A. 1988. The use of interval arithmetic in uncovering structure of linear systems. In *Reliability in Computing* (ed. R. E. Moore), pp. 341–53. Academic Press, London.

Lohner, R. 1987. Enclosing the solutions of ordinary initial- and boundary-value problems. In *Computerarithmetic* (eds. E. Kaucher, U. Kulisch & Ch. Ullrich), pp. 255–86. Teubner, Stuttgart.

Lohner, R. 1988. Einschließung der Lösung gewöhnlicher Anfangs- und Randwertaufgaben und Anwendungen. Dissertation, Univ. Karlsruhe.

Lohner, R., Adams, E. & Ames, W. F. 1985. Untersuchung der praktischen Stabilität von Lösungen nichtlinearer hyperbolischer Anfangswertaufgaben. *Z. Angew. Math. Mech.* **65**, T76–8.

Machost, B. 1970. Numerische Behandlung des Simplexverfahrens mit intervallanalytischen Methoden. Berichte der GMD Bonn 30.

Manteuffel, T. A. 1981. An interval analysis approach to rank determination in linear least squares problems. *SIAM J. Sci. Stat. Comput.* **2**, 335–48.

Matsumoto, T., Chua, L. O. & Ayaki, K. 1988. Reality of chaos in the double scroll circuit: a computer-assisted proof. *IEEE Trans. Circ. Syst.* **35**, 909.

Mayer, G. 1985. Enclosing the solution set of linear systems with inaccurate data by iterative methods based on incomplete LU-decompositions. *Computing* **35**, 189–206.

Mayer, G. 1986a. On a theorem of Stein–Rosenberg type in interval analysis. *Numer. Math.* **50**, 17–26.

Mayer, G. 1986b. On the speed of convergence of some iterative processes. In *Numerical Methods*, pp. 207–28. Coll. Math. Soc. János Bolyai 50, Miscolc.

Mayer, G. 1987a. On the asymptotic convergence factor of the total step method in interval computation. *Linear Algebra Appl.* **85**, 153–64.

Mayer, G. 1987b. Über den asymptotischen Konvergenzfaktor des Einzelschrittverfahrens in der Intervallrechnung. *Z. Angew. Math. Mech.* **67**, T490–2.

Mayer, G. 1987c. Comparison theorems for an iterative method based on strong splittings. *SIAM J. Numer. Anal.* **24**, 215–27.

Mayer, G. 1988a. Zur Lösungseinschließung bei nichtlinearen Gleichungssystemen. *Z. Angew. Math. Mech.* **68**, T499–500.

Mayer, G. 1988b. Enclosing the solutions of systems of linear equations for interval iterative processes. *Computing Suppl.* **6**, 47–58.

Mayer, O. 1970a. Algebraische und metrische Strukturen in der Intervallrechnung und einige Anwendungen. *Computing* **5**, 144–62.

Mayer, O. 1970b. Über die Bestimmung von Einschließungsmengen für die Lösungen linearer Gleichungssysteme mit fehlerbehafteten Koeffizienten. *Angew. Informatik* (*Elektron. Datenverarbeitung*) **12**, 164–7.

Mayer, O. 1971. Über intervallmäßige Iterationsverfahren bei linearen Gleichungssystemen und allgemeineren Intervallgleichungssystemen. *Z. Angew. Math. Mech.* **51**, 117–24.

Metzger, M. 1988. FORTRAN-SC, A FORTRAN extension for engineering/-scientific computation with access to ACRITH: demonstration of the compiler and sample programs. In *Reliability in Computing* (ed. R. E. Moore), pp. 63–80. Academic Press, London.

Mfayokurera, A. 1989. Nonconvex phase equilibria computations by global minimization. MSc thesis, Dept. of Chem. Eng., Univ. Wisconsin-Madison.

Miller, W. 1972. Quadratic convergence in interval arithmetic, Part 2. *BIT* **12**, 291–8.

Miller, W. 1973. More on quadratic convergence in interval arithmetic. *BIT* **13**, 76–83.

Miranda, C. 1941. Un 'osservazione su un teorema di Brouwer. *Bol. Un. Mat. Ital. Ser. II* **3**, 5–7.

Mohd, I. bin, 1986. Global optimization using interval arithmetic. Ph.D. Thesis, University of St Andrews.

Moore, R. E. 1959. Automatic Error Analysis in Digital Computation, Technical Report LMSD-48421, Lockheed Missiles and Space Division, Sunnyvale, CA.

Moore, R. E. 1962. Interval arithmetic and automatic error analysis in digital computing. Ph.D. Thesis, Appl. Math. Statist. Lab. Rep. 25, Stanford University.

Moore, R. E. 1966. *Interval Analysis*. Prentice-Hall, Englewood Cliffs, N.J.

Moore, R. E. 1977. A test for existence of solution to nonlinear systems. *SIAM J. Numer. Anal.* **14**, 611–15.

Moore, R. E. 1978. A computational test for convergence of iterative methods for nonlinear systems. *SIAM J. Numer. Anal.* **15**, 1194–6.

Moore, R. E. 1979. *Methods and Applications of Interval Analysis*. SIAM, Philadelphia.

Moore, R. E. 1980a. Interval methods for nonlinear systems. *Computing Suppl.* **2**, 113–20.

Moore, R. E. 1980b. New results on nonlinear systems. In *Interval Mathematics 1980* (ed. K. Nickel), pp. 165–80. Academic Press, New York.

Moore, R. E. 1984. Upper and lower bounds on solutions of differential and integral equations without using monotonicity, convexity, or maximum principles. In *Advances in Computer Methods for Partial Differential Equations* (eds. R. Vichnevetsky & R. S. Stepleman), pp. 458–62. IMACS.

Moore, R. E. & Jones, S. T. 1977. Safe starting regions for iterative methods. *SIAM J. Numer. Anal.* **14**, 1051–65.

Moore, R. E. & Qi, L. 1982. A successive interval test for nonlinear systems. *SIAM J. Numer. Anal.* **19**, 845–50.

Mudur, S. P. & Koparkar, P. A. 1984. Interval methods for processing geometric objects. *Comput. Graphics Appl.* **3**, 7–17.

Nakao, M. T. 1988. A numerical approach to the proof of existence of solutions for elliptic problems. *Jpn. J. Appl. Math.* **5**, 313–32.

Nakao, M. T. 1989. A computational verification method of existence of solutions for nonlinear elliptic equations. In *Recent topics in nonlinear PDE IV, Kyoto 1988, Lecture Notes in Num. Appl. Anal.* **10**, 101–20.

Nazarenko, T. I. & Marchenko, L. V. 1983. On optimal interval solutions of some differential and integral equations. In *Investigation in Continuum Mechanics*, pp. 170–8. Irkutsk. (In Russian.)

Nerinckx, D. & Haegemans, A. 1976. A comparison of non-linear equation solvers. *J. Comput. Appl. Math.* **2**, 145–8.

Neuland, W. 1969. Die numerische Behandlung von Integralgleichungen mit intervall-analytischen Methoden. Berichte der GMD Bonn 21.

Neumaier, A. 1984a. New techniques for the analysis of linear interval equations. *Linear Algebra Appl.* **58**, 273–325.

Neumaier, A. 1984b. An interval version of the secant method. *BIT* **24**, 366–72.

Neumaier, A. 1985a. Interval iteration for zeros of systems of equations. *BIT* **25**, 256–73.

Neumaier, A. 1985b. Einführung in die numerische Mathematik, Vorlesungsskript. Inst. f. Angew. Math., Univ. Freiburg. To appear as *Introduction to Numerical Analysis*, Cambridge University Press.

Neumaier, A. 1985c. Existence regions and error bounds for implicit and inverse function. *Z. Angew. Math. Mech.* **65**, 49–55.

Neumaier, A. 1986a. Existence of solutions of piecewise differentiable systems of equations. *Arch. Math.* **47**, 443–7.

Neumaier, A. 1986b. Linear interval equations. In *Interval Mathematics 1985* (ed. K. Nickel), pp. 109–20. Springer Lecture Notes in Computer Science 212, Berlin.

Neumaier, A. 1986c. Tolerance analysis with interval arithmetic. Freiburger Intervall-Ber. 86/9, pp. 5–19.

Neumaier, A. 1987a. Further results on linear interval equations. *Linear Algebra Appl.* **87**, 155–79.

Neumaier, A. 1987b. Overestimation in linear interval equations. *SIAM J. Numer. Anal.* **24**, 207–14.

Neumaier, A. 1987c. Solving linear least squares problems by interval arithmetic. Freiburger Intervall-Ber. 87/6, pp. 37–42.

Neumaier, A. 1988a. An existence test for root clusters and multiple roots. *Z. Angew. Math. Mech.* **68**, 256.

Neumaier, A. 1988b. The enclosure of solutions of parameter-dependent systems of equations. In *Reliability in Computing* (ed. R. E. Moore), pp. 269–86. Academic Press, London.

Neumaier, A. 1989. Rigorous sensitivity analysis for parameter-dependent systems of equations. *J. Math. Anal. Appl.* **144**, 16–25.

Nickel, K. 1968. Quadraturverfahren mit Fehlerschranken. *Computing* **3**, 47–64.

Nickel, K. 1971. *On the Newton Method in Interval Analysis.* MRC Technical Summary Report #1136, University of Wisconsin, Madison.

Nickel, K. 1979. The construction of *a priori* bounds for the solution of a two point boundary value problem with finite elements, I. *Computing* **23**, 247–65.

Nickel, K. 1981. A globally convergent ball Newton method. *SIAM J. Numer. Anal.* **18**, 988–1003.

Nickel, K. 1982. Die Auflösbarkeit linearer Kreisscheiben- und Intervall-Gleichungssysteme. *Linear Algebra Appl.* **44**, 19–40.

Nickel, K. 1986a. Using interval methods for the numerical solution of ODE's. *Z. Angew. Math. Mech.* **66**, 513–23.

Nickel, K. 1986b. Optimization using interval mathematics. Freiburger Intervall-Ber. 86/7, pp. 55–83.

Nuding, E. 1981. Schrankentreue Berechnung der Exponentialfunktion wesentlich-nichtnegativer Matrizen. *Computing* **26**, 57–66.

Nuding, E. 1983. Schrankentreuer Algorithmen. *Beiträge Numer. Math.* **11**, 115–37.

Oelschlagel, D. & Süsse, H. 1983. Die Berechnung von Intervall-Hüllen in der Optimierung. *Z. Angew. Math. Mech.* **63**, 379–86.

Oettli, W. 1965. On the solution set of a linear system with inaccurate coefficients. *SIAM J. Numer. Anal.* **2**, 115–18.

Oettli, W. & Prager, W. 1964. Compatibility of approximate solution of linear equations with given error bounds for coefficients and right-hand sides. *Numer. Math.* **6**, 405–9.

Oettli, W., Prager, W. & Wilkinson, J. H. 1965. Admissible solutions of linear systems with not sharply defined coefficients. *SIAM J. Numer. Anal.* **2**, 291–9.

Ohsmann, M. 1988. Verified inclusion for eigenvalues of certain difference and differential equations. *Computing Suppl.* **6**, 79–88.

Ortega, J. M. & Rheinboldt, W. C. 1970. *Iterative Solution of Nonlinear Equations in Several Variables.* Academic Press, New York.

Ostrowski, A. M. 1937. Über die Determinanten mit überwiegender Hauptdiagonale. *Comment. Math. Helv.* **10**, 69–96.

Pardalos, P. M. & Rosen, J. B. 1987. Constrained global optimization: algorithms and applications. Springer Lecture Notes in Computer Science 268, Berlin.

Perron, O. 1907. Zur Theorie der Matrizen. *Math. Ann.* **64**, 248–63.

Petković, L. D. 1986. A note on the evaluation in circular arithmetics. *Z. Angew. Math. Mech.* **66**, 371–3.

Petković, L. D. & Petković, M. S. 1984. On the k-th root in circular arithmetic. *Computing* **33**, 27–35.

Petković, L. D. & Petković, M. S. 1985. Error estimates of interpolation and numerical integration using circular complex functions. Freiburger Intervall-Ber. 85/3, pp. 51–60.

Petković, L. D. & Petković, M. S. 1988/9. On some applications of circular complex functions. *Computing* **41**, 141–8.

Petković, M. S. 1981. On a generalization of the root iterations for polynomial complex zeros in circular interval arithmetic. *Computing* **27**, 37–55.

Petković, M. S. 1987. Some interval iterations for finding a zero of a polynomial with error bounds. *Comp. Math. Appl.* **14**, 479–95.

Petković, M. S. 1989. *Iterative Methods for Simultaneous Inclusion of Polynomial Zeros.* Springer Lecture Notes in Mathematics 1387, Springer, Berlin.

Petković, M. S. & Stefanović, L. V. 1984. The numerical stability of the generalized root iterations for polynomial zeros. *Comput. Math. Appl.* **10**, 97–106.

Poljak, S. and Rohn, J. 1988. Radius of nonsingularity. Manuscript, Department of Mathematics and Physics, Charles University, Praha.

Qi, L. 1981a. An interval test using the new Krawczyk operator. Freiburger Intervall-Ber. 81/4, pp. 25–37.

Qi, L. 1981b. Interval boxes of solutions of nonlinear systems. *Computing* **27**, 137–44.

Qi, L. 1982. A note on the Moore test for nonlinear system. *SIAM J. Numer. Anal.* **19**, 851–7.

Rall, L. B. 1981. *Automatic Differentiation: Techniques and Applications.* Springer Lecture Notes on Computer Science 120, Berlin.

Rall, L. B. 1983. Differentiation and generation of Taylor coefficients in PASCAL-SC. In *A New Approach to Scientific Computation* (eds. U. Kulisch *et al.*), pp. 291–309. Academic Press, New York.

Rall, L. B. 1984a. Differentiation in PASCAL-SC: type gradient. *ACM. Trans. Math. Softw.* **10**, 161–84.

Rall, L. B. 1984b. Application of interval integration to the solution of integral equations. *J. Integral Equations* **6**, 127–41.

Rall, L. B. 1985. Computable bounds for solutions of integral equations. In *Proc. 11th IMACS Congr. Syst. Simul. Sci. Comput.*, Vol. 1, pp. 119–21. IMACS.

Rall, L. B. 1986. The arithmetic of differentiation. *Math. Magazine* **59**, 275–82.

Rall, L. B. 1987. Optimal implementation of differentiation arithmetic. In *Computerarithmetic* (eds. E. Kaucher, U. Kulisch & Ch. Ullrich), pp. 287–95. Teubner, Stuttgart.

Ratschek, H. 1971. Die Subdistributivität der Intervallarithmetik. *Z. Angew. Math. Mech.* **51**, 189–92.

Ratschek, H. 1988. Some recent aspects of interval algorithms for global optimization. In *Reliability in Computing* (ed. R. E. Moore), pp. 325–40. Academic Press, London.

Ratschek, H. & Rokne, J. 1984. *Computer Methods for the Range of Functions.* Ellis Horwood Ser. Math. Appl.

Ratschek, H. & Rokne, J. 1987. Efficiency of a global optizimation algorithm. *SIAM J. Numer. Anal.* **24**, 1191–201.

Ratschek, H. & Rokne, J. 1988. *New Computer Methods for Global Optimization.* Ellis Horwood Ltd, Chichester (UK), and Wiley, New York.

Ratschek, K. & Sauer, W. 1982. Linear interval equations. *Computing* **28**, 105–15.

Reichmann, K. 1979a. Ein hinreichendes Kriterium für die Durchführbarkeit des Intervall-Gauß-Algorithmus bie Intervall-Hessenberg-matrizen ohne Pivotsuche. *Z. Angew. Math. Mech.* **59**, 373–9.

Reichmann, K. 1979b. Abbruch beim Intervall-Gauß-Algorithmus. *Computing* **22**, 355–61.

Rex, G. 1989. E-Verfahren und parameterabhängige nichtlineare Gleichungssysteme. Preprint 89/1, Karl-Marx-Universität, Leipzig.

Rheinboldt, W. C. 1988. Error questions in the computation of solution manifolds of parametrized equations. In *Reliability in Computing* (ed. R. E. Moore), pp. 249–68. Academic Press, London.

Ris, F. N. 1972. Interval analysis and applications to linear algebra, D.Phil. Thesis, Oxford.

Rockafellar, R. T. 1970. *Convex Analysis.* Princeton University Press.

Rohn, J. 1982. An algorithm for solving interval linear systems and inverting interval matrices. Freiburger Intervall-Ber. 82/5, pp. 23–36.

Rohn, J. 1984. Solving interval linear systems; Proofs to 'Solving interval linear systems'. Freiburger Intervall-Ber. 84/7, pp. 1–14; 17–30; 33–58.

Rohn, J. 1985. Some results on interval linear systems. Freiburger Intervall-Ber. 85/4, pp. 93–116.

Rohn, J. 1988. Solving systems of linear interval equations. In *Reliability in Computing* (ed. R. E. Moore), pp. 171–82. Academic Press, London.

Rohn, J. 1989. A two-sequence method for linear interval equations. *Computing* **41**, 137–40.

Rohn, J. 1990a. Systems of linear interval equations. *Linear Algebra Appl.*, to appear.

Rohn, J. 1990b. Real eigenvalues of an interval matrix with rank one radius. *Z. Angew. Math. Mech.*, to appear.

Rokne, J. 1971. Fehlererfassung bei Eigenwertproblemen von Matrizen. *Computing* **7**, 145–52.

Rokne, J. 1985. Including iterations for the lambda-matrix eigenproblem. *Computing* **35**, 207–18.

Rokne, J. 1986. Low complexity k-dimensional centered forms. *Computing* **37**, 247–52.

Rossier, E. 1982. L'inverse d'une matrice d'intervalles. *Rairo Rech. Oper.* **16**, 99–124.

Rump, S. M. 1980. Kleine Fehlerschranken bei Matrixproblemen. Dissertation, Inst. f. Angew. Math., Univ. Karlsruhe.

Rump, S. M. 1981a. Exakte Fehlerschranken für Eigenwerte und Eigenvektoren. *Z. Angew. Math. Mech.* **61**, T311–13.

Rump, S. M. 1981b. Kleine, exakte Fehlerschranken für die Lösung linearer Gleichungssysteme. *Z. Angew. Math. Mech.* **61**, T313–15.

Rump, S. M. 1982a. Lösung linearer und nichtlinearer Gleichungssysteme mit maximaler Genauigkeit. In *Wissenschaftliches Rechnen und Programmiersprachen* (eds. U. Kulisch & Ch. Ullrich), pp. 147–74. Teubner, Stuttgart.

Rump, S. M. 1982b. Solving nonlinear systems with least significant bit accuracy. *Computing* **29**, 183–200.

Rump, S. M. 1983. Solving algebraic problems with high accuracy. In *A New Approach to Scientific Computation* (eds. U. W. Kulisch & W. L. Miranker), pp. 51–120. Academic Press, New York.

Rump, S. M. 1984. Solution of linear and nonlinear algebraic problems with sharp, guaranteed bounds. *Computing Suppl.* **5**, 147–68.

Rump, S. M. 1990. Rigorous sensitivity analysis for systems of linear and nonlinear equations. *Math. Comput.*, to appear.

Rump, S. M. & Kaucher, E. 1980. Small bounds for the solution of systems of linear equations. *Computing Suppl.* **2**, 157–64.

Scharf, V. 1968. Ein Verfahren zur Lösung des Cauchy-Problems für lineare Systeme von partiellen Differentialgleichungen, Forschungsbericht des Landes Nordrhein-Westfalen 1904, West-deutscher Verlag, Köln, Opladen; Schriften d. Rheinisch-Westfälischen Inst. f. Instrumentelle Math., Univ. Bonn, Ser. A, 21.

Schätzle, F. 1984. Überschätzung beim Gauß-Algorithmus für lineare Intervallgleichungssysteme. Freiburger Intervall-Ber. 84/3.

Scheu, G. 1975. Schrankenkonstruktion für die Lösung der elliptischen Randwertaufgabe mit konstanten Koeffizienten. *Z. Angew. Math. Mech.* **55**, T221–3.

Scheu, G. & Adams, E. 1975. Schrankenkonstruktion für die Randwertaufgabe der linearen, elliptischen Differentialgleichung, Deutsche Luft- und Raumfahrt, Forschungsbericht 75-47.

Schröder, J. 1956. Das Iterationsverfahren bei allgemeinem Abstandsbegriff. *Math. Z.* **66**, 111–16.

Schröder, J. 1990. Operator inequalities and applications. In *Proc. Conf. Inequalities: Fifty years on from Hardy, Littlewood and Polya, Birmingham 1987*, to appear.

Schroder, J. 1988. A method for producing verified results for two-point boundary value problems. *Computing Suppl.* **6**, 9–22.

Schwandt, H. 1981. Schnelle fast global konvergente Verfahren für die Fünf-Punkt-Diskretisierung der Poissongleichung mit Dirichletschen Randbedingungen bei Rechteckgebieten. Dissertation, Techn. Univ. Berlin.

Schwandt, H. 1984. An interval arithmetic approach for the construction of an almost globally convergent method for the solution of the nonlinear Poisson equation on the unit square. *SIAM J. Sci. Statist. Comput.* **5**, 427–52.

Schwandt, H. 1985a. Improvement of an interval arithmetic method for the solution of systems of nonlinear equations. *Z. Angew. Math. Mech.* **65**, T403–4.

Schwandt, H. 1985b. Newton-like interval methods for large nonlinear systems of equations on vector computers. *Comput. Phys. Comm.* **37**, 223–32.

Schwandt, H. 1985c. The solution of nonlinear elliptic Dirichlet problems on rectangles by almost globally convergent interval methods. *SIAM J. Sci. Statist. Comput.* **6**, 617–38.

Schwandt, H. 1985d. Accelerating Krawczyk-like interval algorithms for the solution of nonlinear systems of equations by using second derivatives. *Computing* **35**, 355–67.

Schwandt, H. 1986. Almost globally convergent interval methods for discretizations of nonlinear elliptic partial differential equations. *SIAM J. Numer. Anal.* **23**, 304–24.

Schwandt, H. 1987a. Interval arithmetic multistep methods for nonlinear systems of equations. *Japan J. Appl. Math.* **4**, 139–71.

Schwandt, H. 1987b. Iterative methods for systems of equations with interval coefficients and linear form. *Computing* **38**, 143–61.

Schwetlick, H. 1979. *Numerische Lösung nichtlinearer Gleichungen.* Oldenbourg, München-Wien.

Shearer, J. M. & Wolfe, M. A. 1985. Some algorithms for the solution of a class of nonlinear algebraic equations. *Computing* **35**, 63–72.

Shen, Zuhe & Neumaier, A. 1988. A note on Moore's interval test for zeros of nonlinear systems. *Computing* **40**, 85–90.

Shi, Z. & Gao, W. 1988. Stability of interval parameter matrices, Kexue Tongbao. *Sci. Bull. Ed.* **33**, 1498–500. (1988); Zbl. 662.65035.

Siemens AG, 1987. *ARITHMOS Benutzerhandbuch.* Best.Nr. U 2900-J-Z 87-1, Siemens, München.

Šik, F. 1982. Solution of a system of linear equations with given error sets for coefficients. *Apl. Mat.* **27**, 319–25.

Smith, L. B. 1969. Interval arithmetic determinant evaluation and its use in testing for a Chebyshev system. *Comm. ACM* **12**, 89–93.

Spellucci, P. & Krier, N. 1985. Untersuchungen der Grenzgenauigkeit von Algorithmen zur Auflösung linearer Gleichungssysteme mit Fehlererfassung. In *Interval Mathematics 1985* (ed. K. Nickel), pp. 288–97. Springer Lecture Notes in Computer Science 29, Berlin.

Spreuer, H. 1981. A method for the computation of bounds with convergence of arbitrary order for ordinary linear boundary value problems. *J. Math. Anal. Appl.* **81**, 99–133.

Steckelberg, J. 1980. Über die Taylor-Methode für lineare algebraische Gleichungssysteme mit Daten- und Koeffizientenmengen. Manuscript, Inst. f. Angew. Math., Univ. Karlsruhe.

Stetter, H. J. 1988. Inclusion algorithms with functions as data. *Computing Suppl.* **6**, 213–24.

Stewart, N. F. 1973. Interval arithmetic for guaranteed bounds in linear programming. *J. Optim. Theory Appl.* **12**, 1–5.

Stoer, J. & Bulirsch, R. 1980. *Introduction to Numerical Analysis.* Springer, Berlin.

Stummel, F. 1984. Rounding error analysis of interval algorithms. *Z. Angew. Math. Mech.* **64**, 341–54.

Thiel, S. 1986. Intervalliterationsverfahren für diskretisierte elliptische Differentialgleichungen. Freiburger Intervall-Ber. 86/8.

Thiel, S. 1989. A generalization of the interval Newton single step method for nonlinear systems of equations. *Computing* **43**, 73–84.

Thieler, P. 1975. Verbesserung von Fehlerschranken bei iterativer Matrizeninversion. In *Interval Mathematics* (ed. K. Nickel), pp. 306–10. Springer Lecture Notes in Computer Science 29, Berlin.

Thieler, P. 1978. On componentwise error estimates for inverse matrices. *Computing* **19**, 303–12.

Tost, R. 1970. Lösung der 1. Randwertaufgabe der Laplace-Gleichung im Rechteck mit intervallanalytischen Methoden. Berichte der GMD Bonn 28.

Toth, D. L. 1985. On ray tracing parametric surfaces. *Computer Graphics* **19**, 171–9.

Varga, R. S. 1962. *Matrix Iterative Analysis.* Prentice-Hall, Englewood Cliffs, N.J.

Walter, W. 1988. FORTRAN-SC, a FORTRAN extension for engineering/scientific computation with access to ACRITH: language description with examples. In *Reliability in Computing* (ed. R. E. Moore), pp. 43–62. Academic Press, London.

Walster, G. W., Hansen, E. & Sengupta, S. 1985. Test results for a global optimization algorithm. In *Numerical Optimization 1984* (eds. P. T. Boggs, R. H. Byrd & R. B. Schnabel), pp. 272–87. SIAM Publications, Philadelphia.

Wilf, H. S. 1978. A global bisection algorithm for computing the zeros of polynomials in the complex plane. *J. Assoc. Comput. Mach.* **25**, 415–20.

Wilkinson, J. H. 1965. *The Algebraic Eigenvalue Problem.* Clarendon Press, Oxford.

Wolfe, M. A. 1988. Interval methods for algebraic equations. In *Reliability in Computing* (ed. R. E. Moore), pp. 229–48. Academic Press, London.

Wongwises, P. 1975. Experimentelle Untersuchungen zur numerischen Auflösung von linearen Gleichungssystemen mit Fehlererfassung. In *Interval Mathematics* (ed. K. Nickel), pp. 316–25. Springer Lecture Notes in Computer Science 29, Berlin.

Young, R. C. 1931. The albegra of many-valued quantities. *Math. Ann.* **104**, 260–90.

Zangwill, W. I. & Garcia, C. B. 1981. *Pathways to Solutions, Fixed Points, and Equilibria.* Prentice-Hall, Englewood Cliffs, N.J.

AUTHOR INDEX

SUBJECT INDEX